T0188631

Seismic Design and Analysis of Tanks

Seismic Design and Analysis of Tanks

Gian Michele Calvi
Roberto Nascimbene

IUSS - University School for Advanced Studies
Pavia
Italy

Published by John Wiley & Sons, Inc., Hoboken, New Jersey.
Published simultaneously in Canada.

Trademarks
Wiley and the Wiley logo are trademarks or registered trademarks of John Wiley & Sons, Inc. and/or its affiliates in the United States and other countries and may not be used without written permission. All other trademarks are the property of their respective owners. John Wiley & Sons, Inc. is not associated with any product or vendor mentioned in this book.

Limit of Liability/Disclaimer of Warranty
While the publisher and author have used their best efforts in preparing this book, they make no representations or warranties with respect to the accuracy or completeness of the contents of this book and specifically disclaim any implied warranties of merchantability or fitness for a particular purpose. No warranty may be created or extended by sales representatives or written sales materials. The advice and strategies contained herein may not be suitable for your situation. You should consult with a professional where appropriate. Neither the publisher nor author shall be liable for any loss of profit or any other commercial damages, including but not limited to special, incidental, consequential, or other damages. Further, readers should be aware that websites listed in this work may have changed or disappeared between when this work was written and when it is read. Neither the publisher nor authors shall be liable for any loss of profit or any other commercial damages, including but not limited to special, incidental, consequential, or other damages.

For general information on our other products and services or for technical support, please contact our Customer Care Department within the United States at (800) 762-2974, outside the United States at (317) 572-3993 or fax (317) 572-4002.

Wiley also publishes its books in a variety of electronic formats. Some content that appears in print may not be available in electronic formats. For more information about Wiley products, visit our web site at www.wiley.com.

Library of Congress Cataloging-in-Publication Data applied for:

Hardback ISBN: 9781119849810

Cover Design: Wiley
Cover Images: Courtesy of Gian Michele Calvi and Roberto Nascimbene

SKY10042871_021723

To our dads

Contents

Preface

Not many books dealing with designing shell structures are available in the international literature. This was the main motivation inducing us to write a book on this subject, published in 2011, in Italian. That book found its roots in classical texts and in well-established university courses. First of all, the fundamental text, *Stresses in Shells*, published by W. Flügge in 1960, possibly a compendium of the monumental *Statik und Dynamik der Schalen*, written when he was still living in Germany. Together with Vlasov, Reissner, Dischinger, and a few others, he had a fundamental role in developing the membrane and flexural solutions for most kinds of shells between the 1930s and the 1950s.

In the Preface to his first book in English, mentioned above and written at Stanford, where he moved before the Second World War, he wrote: "At first sight it may look to many people like a mathematics book, but it is hoped that the serious reader will soon see that it has been written by an engineer and for engineers … The author wishes to assure his readers that nowhere in this book has an advanced mathematical tool been used just for the sake of displaying it. No matter which mathematical tool has been used, it had to be used to solve the problem at hand."

This book contains all the fundamental equations to solve any static problem of simple and complex shell structures, being clearly and overtly not to be used as class notes, but rather to find specific solutions or as a basis for further research. The kind of book that cannot be ignored by designers of complex shells that hide the complicated mathematical nature of their structural responses behind their apparent intuitive simplicity and their aesthetic appeal.

Quite to the contrary, another fortunate book, *Thin Shell Concrete Structures*, by D. Billington, had been expressly written as a textbook for a graduate course, allowing an easier and faster reading. This obviously came at a price, sometimes leaving the reader with an impression of vague or missing information, or with the feeling of some missing link between theory and practice.

Obviously, it was easy for good teachers to bridge this gaps. A. Scordelis, at the University of California, Berkeley, integrated this text with his notes and papers, but he had participated in the design and analysis of spectacular shell structures, such as the elliptical paraboloid of the Oklahoma State Fair Arena (120 m), the circular paraboloid of the Arizona State Fairgrounds Coliseum at Phoenix (114 m), the reverse dome of the Garden State Art Center in New Jersey (78 m), the roof of the San Juan Coliseum in Puerto Rico (94 m), the roof of St. Mary Cathedral in San Francisco, conceived by Pier Luigi Nervi, and made of eight hyperbolic paraboloids with a height of 42 m. It is not often that a student has a teacher with such experience.

Our book in Italian was something in between, with an extensive presentation of the mathematical apparatus and a number of design examples presented in some detail. However, part of its success (it still sells well) was due to the total absence of any competing reference in Italian.

When we started considering the preparation of an English version, it immediately became clear that there was much less point in revisiting what was available in other books, while the section on seismic design and assessment could have been profitably expanded, since very little information is available on the subject.

The relevance of the subject had recently been confirmed by the occurrence of two shocks in northern Italy, on 20 and 29 May 2012 (with a magnitude M_W = 6.11 and 5.96). The affected region, in the Po Valley, is one of the most industrialized zones of Northern Italy. The majority of structures severely damaged were industrial facilities: one-storey pre-cast reinforced concrete structures and nearby storage steel tanks, causing the economic loss of approximately 5 billion Euros, mostly due to the interruption of industrial production. The large number of industrial facilities in the stricken area, in combination with their intrinsic deficiencies, induced damage and losses disproportionately high, compared to the relatively moderate seismic intensity of the events.

In the aftermath of the earthquakes, a large reconnaissance effort was undertaken and a clearinghouse (http://www.eqclearinghouse.org/2012-05-20-italy/), hosted by the Eucentre Foundation and the Earthquake Engineering Research Institute (EERI), was prepared. The most common types of failures observed in tanks were fracture of anchors and elephant's foot buckling near the base of the tanks. In general, elephant's foot buckling was experienced in squat tanks, while some of the slender tanks surveyed developed diamond-shaped buckling. Total and partial collapse of legged tanks was another common occurrence, induced by shear failure and/or buckling of their legs due to axial forces, resulting from the overturning moment. In some cases, flat-bottomed, steel cylindrical tanks, typically larger than legged tanks, failed in tension at the bottom of the tank wall, where they met the anchor rods or massive concrete pads.

It appears that we are still struggling to reach an acceptable quality in design, assessment, and strengthening of tanks and silos, and "competing against time", as G.W. Housner entitled the report on the Loma Prieta earthquake (17 Oct. 1989) to the Governor of California. The damage to infrastructures, freeways, industrial plants had been severe and the scope of the report had been extended from what happened to the measures to be taken to prevent such destruction in future earthquakes. After some thirty years, it is evident that the report title still applies: we can still state "earthquakes will occur, whether they are catastrophes or not depends on our actions", but our actions in the past three decades have not been as effective as they should have been.

This book is based on the evidence emerging from a number of structures surveyed following earthquake events, on some significant consulting activity developed in the field of industrial plants, on research developed and published by the authors and other colleagues.

The design and assessment of the expected performance of tanks and silos are presented, considering the following cases:

- above-ground cylindrical and rectangular anchored rigid tanks;
- above-ground cylindrical and rectangular unanchored tanks;
- underground rigid tanks;
- elevated tanks on shaft and frame-type towers;
- flexible tanks.

This possibly artificial categorization has been found to be convenient with reference to the main response parameters to be considered.

The effects of liquid viscosity, non-homogeneous liquids, soil–structure interaction, the introduction of damping devices and isolation systems are presented and discussed.

This book is intended primarily for teaching courses on seismic design and analysis of tanks to graduate students and for professional training courses. However, it is expected that this text can be effective and practical as a design and analysis reference for researchers and practising engineers.

Gian Michele Calvi
Roberto Nascimbene
IUSS - University School for Advanced Studies
Pavia
Italy

Acknowledgements

This book emerges from a long story that dates back to one of the first Italian university courses on design of shells, taught at the University of Pavia in the early 2000s. At that time, one of the authors of this book was the professor in charge, the other one his assistant. A friendly relationship between the two of them survived the challenges of life and made this book possible, favouring a continuous, enjoyable atmosphere over the long years of its gestation and completion.

Many students have been fundamental in providing criticism and suggestions to improve the class notes that were used as the origin of this endeavour and for several years. The list of their names cannot be made explicit here, but each one of them is felt to be part of this effort.

We mention here two names only, not of students or colleagues, but rather of the two people who have been essential to complete a decent product: Giulia Fagà and Gabriele Ferro, for their continuous support in drawing, refining, and commenting on the figures, which make this book more understandable and reading it more pleasurable.

1

Appealing Shell Structures

After reading this chapter you should be able to:

- List the main subsectors and components of a tank's design and analysis
- Explain the function of each element
- Identify the behaviour related to seismic and static performance

1.1   Beams and Arches

The structural design process has traditionally been, and still is, essentially carried out for elements subjected to bending actions (beams and slabs), generally controlled by a flexural behaviour, upon which the design is based. Once flexural resistance has been ensured, these members are verified to prevent excessive deformation or shear failure. In the case of members loaded by a combination of bending moments and axial compressive loads (columns or walls), the preliminary design is often based on the axial component only and the combination with flexural action is then verified (in the case of tanks, bending is sometimes considered just to check the possibility of buckling).

These simplified approaches assume the presence of structural components able to resist either tensile or compressive stresses, such as concrete and steel bars in reinforced concrete elements. The design is thus based on an estimate of the loads to be carried by each member, and subsequently on the design of sections where the maximum resulting bending moment is expected. As an example, consider a beam where the moment acting on a section is estimated as $M = \frac{ql^2}{\alpha}$, where q is the applied load per unit length, l is the length of the element and α is a coefficient that depends on the end constraints. This acting moment has to be balanced by a couple, estimated as the result of internal tensile and compressive actions, equal to each other and multiplied by the distance between their approximate points of application to compute a resisting moment.

It is thus quite understandable that in ancient times the material mainly used for roofing systems was timber, combined in boards, joists, beams, and girders with progressively increasing capacity.

The only viable, though more complex, alternative was to resort to an arch, which was able to cover long spans using materials able to resist only compressive actions, such as brick or stone masonry. Though

Seismic Design and Analysis of Tanks, First Edition. Gian Michele Calvi and Roberto Nascimbene.
© 2023 John Wiley & Sons, Inc. Published 2023 by John Wiley & Sons, Inc.

widely applied in ancient times, it was Robert Hooke[1] in 1675 who first clearly expressed the basic concept that allows the design of a fully compressed arch. His statement was simple, though very comprehensive: a perfectly compressed arch shall have a shape in reverse and identical to that which a suspended cable would assume under the same load combination. For example, under a uniformly distributed load, the cable would assume a parabolic shape, with an upward concavity and that should be the geometry of a compressed arch, with the concavity oriented downward.

As stated, this conceptual solution does not offer a relevant clue as to how to design an arch when several different load configurations are considered, nor takes into account the effects of abutment constraints, etc. The problem is thus far more complex and, as common in the past (and present) building engineering practice, crucial simplifications were adopted for sizing and preliminary design, e.g. assuming that the horizontal reaction at the abutment (P_H) can be approximated by the equation applicable to a three-hinged arch case:

$$P_H = \frac{ql^2}{8h} \tag{1.1}$$

where q and l are the load per unit length and the span length, and h represents the height of the arch.

It is interesting to compare an arch and a beam used to span a similar length supporting similar weights. Consider thus a three-hinged arch, assume the horizontal forces are eliminated at the abutments by means of some horizontal tie, and compare it to a beam of equal span, simply supported at both ends, assuming that it is made by an elastic material with a similar behaviour in tension and in compression.

Immediately one notes that the same external moment induced at midspan by the applied loads ($M_m = \frac{ql^2}{8}$, assuming a uniformly distributed load per unit length), must be equilibrated by internal action couples characterized by quite different arms. For the case of an arch, the internal couple results from forces located in the centre of the mass of the arch (compression) and of the tie (tension), while for the case of a elastic beam, they are applied at points located at a distance of two-thirds of the beam height. When a fully plastic response is assumed, and thus a constant value is assumed for both tensile and compressive stresses, the distance between the resultant forces is one half of the section depth (d, Figure 1.1(a)). In this case, the beam's internal action would be:

$$P_t = P_c = \frac{ql^2}{4d} \tag{1.2}$$

Consequently, assuming an identical strength in compression and in tension, f_w, and a given width of the beam section, b_b, the required section depth (d_b) could be derived as::

$$A = \frac{b_b d_b}{2} = \frac{ql^2}{4d_b f_w} \quad \text{and thus} \quad d_b = \sqrt{\frac{ql^2}{2b_b f_w}} \tag{1.3}$$

For the sake of simplicity, assume now that the arch and the tie are also made with materials with the same compression capacity (the arch) and tensile capacity (the tie). Assuming that both compression and tensile forces will act at the centre of the corresponding element, each force can be derived from Equation (1.1), and consequently the required depth of arch (d_a) and tie (d_t) would be:

$$A_a = A_t = b_{a,t}d_{a,t} = \frac{ql^2}{8hf_w} \quad \text{and thus} \quad d_{a,t} = \frac{ql^2}{8b_{a,t}f_w} \tag{1.4}$$

---

1 Ut pendet continuum flexile, sic stabit contiguum rigidum inversum.

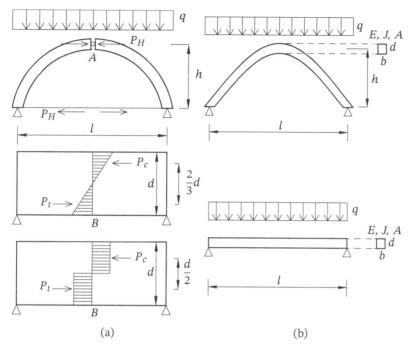

Figure 1.1 (a) Three-hinged arch with a uniform load on top; (b) two-hinged parabolic arch and simply supported beam.

Assuming that all considered elements have the same width ($b_a = b_t = b_b$), then the depth of the arch and the tie can be computed as a function of the depth of the beam, combining Equations (1.3) and (1.4):

$$d_{a,t} = \frac{d_b^2}{4h} \tag{1.5}$$

It can immediately be verified that for reasonable values of the rise of the arch compared to its span (l_a, e.g. $h = \frac{l_a}{4}$) and of the height of the beam, compared to its span (l_b, e.g. $d_b = \frac{l_b}{15}$), the depth required for the arch and the tie is at least 10 times less than the one required for the beam.

Applying the same uniform load on a two-hinged parabolic arch (as shown in Figure 1.1(b)) and on a simply supported beam with the same span (both with a rectangular section $A = bd$), the deflection of the arch at the keystone (Point A) and of the beam at midspan (Point B) can be calculated as follows:

$$w_A = 2.3 \cdot 10^{-5} \frac{ql^4}{EJ} \tag{1.6a}$$

$$w_B = \frac{5}{384} \frac{ql^4}{EJ} = 0.013 \frac{ql^4}{EJ} \approx 565 \cdot w_A \tag{1.6b}$$

The apparent overall stiffness differs by two or three orders of magnitude.

This rather trivial example is just a first case study in which the superiority of curved geometry structures is shown in terms of the required material to obtain similar strength or deformation capacities under gravity loads, when compared to similar structures based on straight geometry. The more complex case of cylindrical vs. rectangular tanks will offer more, possibly not as trivial, evidence.

1.2 Plates and Vaults

As already mentioned, a common technology to cover a rectangular area was based on the properties of timber, a material readily available, easy to work, and structurally attractive. This technology is made by a combination of linear elements, overlaying girders, beams and joists, until reasonable span dimensions are achieved to apply boards of reasonable thickness.

Covering the same area using a single plate would require some homogeneous material capable of carrying shear and bending moments in two directions. Clearly this is feasible, though impossible in practical terms, using a steel plate, but became a viable alternative only with the advent of reinforced concrete. Its potential for an isotropic (or rather orthotropic) behaviour, the possibility of shaping its geometry and tapering its thickness, the separation of the internal elements countering compression and tensile stresses, appear to be an ideal combination to build an efficient horizontal slab.

Consider first a simple comparison between a simply supported beam and a similar one-way slab of indefinite width. The bending moment will be expressed by the same equation, while the slab stiffness will increase because of the hindered transversal dilatation. This effect will be accounted for by a correction factor to be applied to the beam stiffness equal to $1 - v^2$, where v is the Poisson coefficient, in the range of 0.15 for concrete. The correction will thus be in the range of 2% not as relevant.

A much more relevant effect will become evident if comparing a one-way and a two-way response, particularly when the two sides of the slab will not differ much.

Take the example of a simply supported square plate, with a uniform load p, and assume an isotropic elastic response (i.e. in the case of concrete, neglecting any cracking phenomenon). In the case of a two-way response, the maximum bending moment and the deflection at the centre of the plate will be calculated as:

$$m_{max} = 0.04416\, pa^2 \tag{1.7a}$$

$$f = 0.00406\frac{pa^4}{B} = 0.04677\frac{pa^4}{Es^3} \tag{1.7b}$$

where a is the side of the plate, s its thickness and $B = \frac{Es^3}{12(1-v^2)}$ its flexural stiffness.

Considering the same geometry and the same load, but hinged supports on two opposite sides only, the bending moment and flexural deflection will be those of a simply supported beam (possibly with the minor stiffness correction mentioned above, not applied in the equations):

$$m_{max} = \frac{1}{8}pa^2 = 0.125\, pa^2 \tag{1.8a}$$

$$f = \frac{5}{384}\frac{pa^4}{EJ} = 0.15625\frac{pa^4}{Es^3} \tag{1.8b}$$

The values of bending moments and deflection calculated for the beam are thus approximately three times those obtained for the bidirectional plate.

It is easy to observe that a barrel vault sustained by continuous supports on two sides can be regarded as a tri-dimensional transformation of an arch with the corresponding transversal section. Its basic structural behaviour under gravity loads can thus be derived from that of an arch (Figure 1.2(a)). Things are quite different when a barrel vault is not sustained along the support lines perpendicular to the arch section, but rather along the other two sides (Figure 1.2(b)) or even by punctual vertical support located

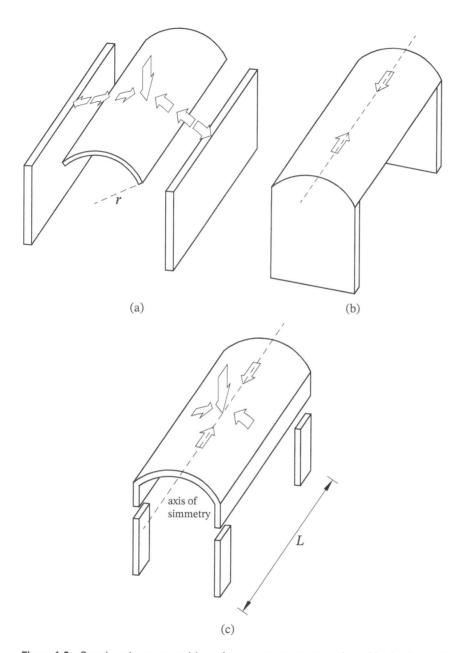

Figure 1.2 Barrel vault response: (a) continuous support on two edges; (b) edge beam without continuous lateral support, with columns and (c) edge beam with columns.

Figure 1.3   Flat roof plate on side beams.

at the corners (Figure 1.2(c)). The structural system becomes much more complex, with the barrel vaults forced to act somehow as a longitudinal beam, coupled with transversal arches. This response can be compared with that of a structure made of longitudinal beams and a transverse slab (Figure 1.3).

The structural response of the barrel vault can be decomposed into two interacting mechanisms: an arch action in the transversal direction and a beam action in the longitudinal direction.

While in the case of slab and beams the components' responses are essentially decoupled, the behaviour of the barrel vault is more complex, because the two systems interact between them, with a resulting combined response that depends mainly on the ratio between the longitudinal length of the vault (L) and its radius of curvature (r).

The traditional beam theory can be applied as a first approximation to calculate internal forces and deformation of the vault structure in the longitudinal direction. The applied bending moment can thus be readily computed and section equilibrium will allow the determination of stresses along the arch-shaped section of the beam. In general, if free rotation is assumed at the ends, the stress distribution will look like that shown in Figure 1.4(a). Unfortunately, one fundamental assumption of beam theory is that

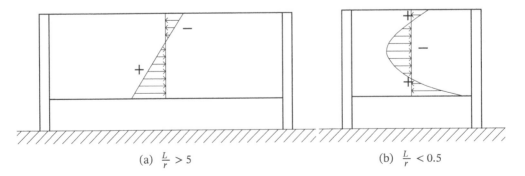

(a) $\frac{L}{r} > 5$ (b) $\frac{L}{r} < 0.5$

Figure 1.4   Qualitative distribution of internal stresses in a transversal arch section of a barrel vault, for (a) large or (b) small $\frac{L}{r}$ ratios.

the transversal plane section will remain plane in the deformed configuration and this is not always an acceptable approximation for barrel vaults. Actually, refined numeral analysis has shown that this assumption is valid only for relative values of the fundamental parameters mentioned above, i.e. essentially for $\frac{L}{r} > 5$. Obviously, this can be obtained also by taking measures to restrain section deformations, such as, for example, inserting transversal walls or ties [499]. It is clear that in these cases all the advantages of a deep beam will apply, with a large lever arm between compression and tension resultants and small displacements.

In the case of unrestrained relatively short vaults, the section deformation may differ significantly from a straight line and the related internal stress distribution will be affected (Figure 1.4(b)).

This problem is no different from that of any deep beam with a complex transversal section, which counts on shape more than on mass material to resist the applied forces. As pointed out, a main design problem in these cases is to take measures to control the section deformations. A second fundamental problem (in the case of reinforced concrete) is related to the high potential for significant cracking of the parts subjected to tension, which may not be adequately controlled by means of a proper reinforcement distribution. A possible viable solution may be found in a rational application of post-tensioning, which will not only be effective in controlling cracking, but also in reducing the deformation. Possible qualitative arrangements of cables are illustrated in Figure 1.5, as a function of the presence of beam-like elements at the edges of the vault. It is evident in Figure 1.5(a) that a rational longitudinal disposition of the cables in a vault will imply potentially significant effects on the transversal response since the sectional arch sections will be subjected to varying transverse forces and bending moments, not necessarily negligible nor necessarily favourable.

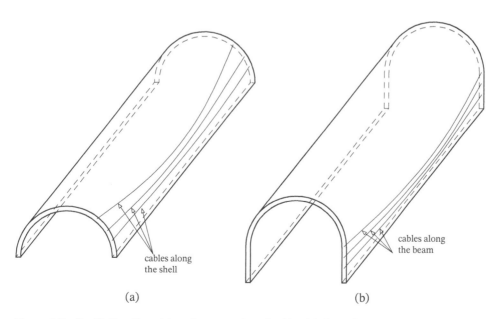

cables along
the shell

cables along
the beam

(a)   (b)

Figure 1.5 Qualitative disposition of post-tensioned cables (a) along the shell and (b) along the edge beams.

1.3   Rectangular and Cylindrical Tanks

The vertical walls of a rectangular tank can be regarded as a series of vertical plates, usually fixed on three sides and free, simply supported or fixed on the fourth one (the side shared with the roof). Each plate is generally subjected to its own weight (in plane) and to the forces generated by the contained material (mainly out of plane), which generate a variable pressure along the height. In the absence of internal friction, the maximum horizontal pressure at the base of each plate is $P_0 = \gamma H$, where γ is the weight per volume unit of the fluid and H is the height of the tank. This pressure varies linearly with the height, becoming zero at the top. As such, the pressure along the height can be evaluated as [184]:

$$p_z = -\gamma(H - y) \tag{1.9}$$

For squat and large tanks, the acting forces are mainly equilibrated by a vertical cantilever action while the contribution of the plate effect in the horizontal direction can be ignored. If this is the case, and it is further assumed that no constraint is provided at the top side, the maximum bending moment per unit length at the base is:

$$M_0 = -\frac{\gamma H^3}{6} = -\frac{1}{6}P_0 H^2 = -0.1667 P_0 H^2 \tag{1.10}$$

With a progressive reduction of the horizontal measure of each plate with respect to their height, the contribution of the transverse reaction becomes progressively more important. For example, when the height is equal to the horizontal span (i.e. the tank assumes a cubic shape), the same maximum moment around the base line of of each plate is reduced to less than one-fifth, becoming:

$$M_0 = -0.0299 P_0 H^2 \tag{1.11}$$

The estimation of this maximum bending moment is one of the crucial issues when the flexural response controls the design of the structure and is often used as a basic preliminary parameter to define the required wall thickness at the base of the tank.

As an example, with the aim of getting some feeling about figures, consider the case of a cubic tank with side length of 20 m, with a consequent total capacity of 8000 m^3 of liquid, assumed to be water. The resulting bending moment M_0 at the base of each wall is:

$$M_0 = -0.0299 P_0 H^2 = -0.0299 \gamma H^3 = -0.0299 \cdot 10 \cdot 20^3 = -2392 \text{ kNm/m} \tag{1.12}$$

Assuming that the design moment at the ultimate limit state is factorized to 1.3 times ($M_0^{slu} = -1.3 \cdot 2392 = -3110$ kNm/m) and that the neutral axis depth is approximately equal to $\xi = \frac{x}{d} = 0.35$, the required concrete thickness s and the corresponding amount of vertical reinforcement A_s can be estimated as:

$$d = \sqrt{\frac{|M_0|}{0.8\, \xi\, b(1 - 0.4\xi) f_{cd}}} = \sqrt{\frac{3110 \cdot 10^6}{0.8 \cdot 0.35 \cdot 1000(1 - 0.4 \cdot 0.35)17}} = 872 \text{ mm} \tag{1.13a}$$

$$s = d + 35 \approx 900 \text{ mm} \tag{1.13b}$$

$$A_s = \frac{|M_0|}{0.9 d f_{yd}} = 10276 \text{ mm}^2/\text{m} \tag{1.13c}$$

where f_{cd} is the concrete design strength, f_{yd} is the yield strength of steel, b is the unitary width of the wall, and d is the depth of the section excluding the concrete cover, assumed to be 35 mm. The resulting vertical reinforcement percentage is 1.14%. The acting bending moment reduces to about 1000 kNm/m at midheight, hence the section can be tapered, the reinforcement can be reduced, or both. A possible solution is to reduce the wall thickness to 550 mm and the vertical reinforcement ratio to about 1%. Considering a linear tapering, the wall depth at the top will be 200 mm. The bending moment acting at midheight around a vertical axis is about 1400 kNm/m and will require a horizontal reinforcement ratio approximately equal to 1.4%. Cracking width and distance at the base can be checked by applying one of the several formulations proposed in different codes of practice. Considering as an example the equations recommended by the Italian code, NTC 2018, the serviceability limit state bending moment ($\sim$2400 kN/m) will be used, estimating the maximum distance between cracks as $\Delta_{max} = 162$ and their width as approximately 0.2 mm.

Simple calculations can estimate the total amount of concrete required to cast the four walls of the tank, of about 880 m^3. A proper evaluation of the required reinforcement will imply some assumptions about appropriate minimum percentages for the compression sides of the sections, for corner detailing, for low stress regions. A reasonable total amount will be around 2200 kN, which corresponds to an average steel weight of 2.5 kN/m^3 or to an average reinforcement percentage in both directions of about 1.6%.

A tank with a similar volume capacity can be designed as a cylindrical structure, with a diameter of 20 m and a height of 25 m. In order to complete some simplified design as done above in the case of a square tank, and thus be able to compare the required material quantities, it is necessary to anticipate the basic features of the response of cylindrical tanks. This can be regarded as one of the simplest cases of a shell structure, which can be used as a case study for the development of a more general theory, to be applied to more complex geometrical shapes.

For this purpose, a number of key assumptions have to be accepted (similar to those applied to the simplified theory of plates):

1) the thickness is small compared to the smallest radius of curvature;
2) the resulting displacements will be small with respect to the shell thickness;
3) the cross-sections normal to the midsurface of the shell will remain straight and normal to the midsurface after deformation;
4) the component of internal stress normal to the midsurface is negligible if compared to the tangential components;
5) the stress distribution along the shell thickness is linear, and its value is zero at mid-surface.

Under these assumptions, the response of a cylindrical tank can be simplified in the interaction of two different reaction systems: a flexural response of vertical cantilevers, and a retaining action of a system of horizontal rings. The solution of this redundant problem must obviously respect both equilibrium and compatibility and was ingeniously developed between the end of the nineteenth and the beginning of the twentieth century.

To solve the problem it was first assumed one should ignore any flexural reaction at the base of the tank, i.e. to allow the base of the cylinder to rotate and to move freely along the horizontal plane. In this case, the horizontal acting forces will have to be equilibrated by the ring action only, while the vertical action will be simply equilibrated by a vertical reaction component. This situation is defined as a *membrane solution*, since bending is excluded, and shear in the plane tangent to the shell must also be zero because of the axi-symmetric condition of both loads and structure. Considering a tank containing a liquid with unit weight γ with no internal friction or cohesion, with height with no internal friction or cohesion, with height H, radius r, and a total self weight (or total applied vertical load) R, the solution described can be expressed by the following equations, as a function of internal vertical (N_y') and horizontal (N_θ') actions per unit length:

$$N_y' = \frac{R}{2\pi r} \tag{1.14a}$$

$$N_\theta' = \gamma(H - y)r \tag{1.14b}$$

where y is the vertical coordinate, being equal to zero at the base and to H at the top.

The horizontal displacement (w) and rotation around the tangent horizontal line (ϕ_y') required by compatibility in this specific case were also calculated, as:

$$w = -\frac{\gamma r^2}{Eh}(H - y) \tag{1.15a}$$

$$\phi_y' = \frac{\gamma r^2}{Eh} \tag{1.15b}$$

where h is the thickness of the shell and E is the Young modulus of its constituent material. Note that the rotation is constant along the height and the displacement, consistently, varies linearly, from a maximum value at the base to zero at the top. Assuming a concrete shell, with a Young modulus $E = 30000$ MPa, a thickness h of 200 mm and uncracked sections, the displacement at the base would be 4.2 mm and the constant rotation 16.7×10^{-5} rad, which will lead to zero displacement at the top. In the case of some flexural and shear constraint at the base that would not allow either relative displacement or rotation or both, some flexural action will occur, and the displacement and rotation resulting from this action will have to properly combine with the membrane results to assure compatibility.

The flexural solution of the problem is fully derived and explained in [66], here we will simply mention and use it, as expressed by the following fourth-order differential equation and its solution:

$$\frac{d^4w}{dy^4} + 4\beta^4 w = \frac{p_z}{B} \tag{1.16a}$$

$$w = e^{-\beta y}(C_3 \cos \beta y + C_4 \sin \beta y) + f(y) \tag{1.16b}$$

where p_z is the horizontal pressure, $B = \frac{Eh^3}{12(1-v^2)}$ is the flexural stiffness of a plate with thickness equal to that of the shell and $4\beta^4$ is the ratio between the horizontal stiffness of the rings (k_r) and the flexural stiffness of the plate, expressed by the following equations:

$$k_r = \frac{Eh}{r^2} \tag{1.17a}$$

$$4\beta^4 = \frac{k_r}{B} = \frac{12(1 - v^2)}{r^2 h^2} \tag{1.17b}$$

In a mathematical context, the function $f(y)$ is defined as "particular integral" and can here be represented by the membrane solution found earlier, which is "particular" in the sense that corresponds to the specific situation described above with reference to the base constraints. Again, in mathematical terms, the first term of the solution is the "general integral" of the associated homogeneous equation (i.e. without any applied pressure). This part of the solution depends on the boundary conditions (in this case, from the base restraints, which will determine the values of the constant C_3 and C_4).

It is intuitive that the parameter $4\beta^4$, i.e. the ratio between the cantilever and the ring stiffness, determines which one of the two systems will dominate the structural response. Equation (1.17b) is thus indicating that large radius, thicker tanks are more affected by the flexural response, while small, thin shells are dominated by the ring's reaction. It is interesting to observe that this is not the only relevant role played by the parameter β.

Actually, the wavelength of the damped sinusoid expressed by the general integral is $\lambda = \frac{2\pi}{\beta}$ and the damping factor expressed by the negative exponent of e is a linear function of β. Both effects have a clear physical meaning: if the stiffness of the rings system prevails on the flexural response, β is higher and consequently the period of the sinusoid is shorter and its effect is damped more rapidly. The damping factor can be expressed as a function of the wavelength, replacing β with $\frac{2\pi}{\lambda}$, and obtaining $e^{-\frac{2\pi}{\lambda}y}$. This expression indicates the reduction factor to be applied to the sinusoid is around 5, a distance equal to one quarter of the wavelength (i.e. at $y = \frac{\lambda}{4}$), around 23 at a distance of half wavelength ($y = \frac{\lambda}{2}$), and around 535 at $y = \lambda$.

In the case study we were considering, it is immediately obvious that to calculate that $4\beta^4 = 2.87 \text{ m}^{-4}$ and $\beta = 0.92 \text{ m}^{-1}$ (a Poisson ratio $v = 0.2$ has been adopted), the wavelength of the sinusoid is thus around 6 metres, and the flexural effect can certainly be ignored at a distance of about 3 metres from the base. While the values of bending moment (M) and shear force (Q) along the height can be calculated by applying the usual relations between displacement, curvature, bending moment and shear, i.e.:

$$M = -B\frac{d^2w}{dy^2} \tag{1.18a}$$

$$Q = -B\frac{d^3w}{dy^3} \tag{1.18b}$$

It appears that only the maximum values of bending moment (Figure 1.6) and shear force are relevant to design the required amount of vertical reinforcement at the base, a reinforcement that will be needed for a few metres only.

As already pointed out, the combination of the membrane and the flexural solutions aims to obtain total displacement and rotation at the base compatible with the restraint of each specific case. For a fixed-base

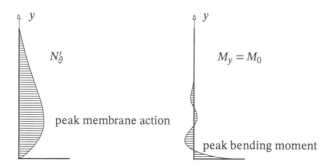

Figure 1.6 Qualitative variation of membrane action N'_θ (hoop force) and bending moment M_y or M_0 along the height of the container.

tank, the displacements and rotations generated by the flexural effect (M_0 and Q_0) must be equal and opposite to those generated by the membrane solution, i.e.:

$$-\frac{\gamma r^2}{Eh}H = \frac{1}{2\beta^3 B}(\beta M_0 + Q_0) \tag{1.19a}$$

$$\frac{\gamma r^2}{Eh} = -\frac{1}{2\beta^2 B}(2\beta M_0 + Q_0) \tag{1.19b}$$

In the case study, the base bending moment and shear, calculated by applying again a 1.3 protection factor, are $M_0 = 183$ kNm/m and $Q_0 = -345$ kN/m. These values are compatible with the assumed section depth (0.20 m), requiring a total amount of reinforcement of 3254 mm^2, or a geometrical steel percentage $\rho = 1.63\%$. At heights above 5 m from the base and on the compression side, the minimum allowed reinforcement percentage can be adopted (possibly 0.3%). The bending moment computed at the base of the cylindrical tank is about seventeen times lower than that obtained for the cubic tank, but unfortunately this is not the governing action to dimension the shell. Actually, when considering the circumferential membrane forces, it is immediately realized that a relevant crack pattern will be induced. For example, at a height of about 3 metres the membrane action is $N'_\theta = 1.3 \cdot 2300 = 3000$ kN/m (Figure 1.6), which would require about 7670 mm^2 of steel, corresponding to a very high geometrical percentage of 3.8%. The resulting theoretical crack opening width would be 0.4 mm with an average crack maximum distance of $\Delta_{max} = 315$ mm. This crack pattern would not be acceptable in most cases.

The most obvious correction of the preliminary design to redirect the result to performances similar to those obtained in the case of a rectangular tank with 900 mm thick walls will simply be to increase the shell thickness. In fact, increasing the shell depth from 200 to 400 mm will induce a modification of the coefficient $\beta = 0.65$ m^{-1}, thus increasing the flexural stiffness with respect to the membrane one. The previous calculations will be repeated, obtaining $M_0 = 358$ kNm/m, $Q_0 = -480$ kN/m and $N'_\theta = 1.3 \cdot 2000 = 2600$ kN/m. While bending moment and shear action will be easily absorbed by the increased resisting section, the theoretical mean crack width will now be 0.17 mm. The total required amounts of concrete and steel will still be around 30% lower than those estimated in the case of a rectangular tank, thus resulting in a competitive design.

A second, more sophisticated, design solution could be based on the insertion of post-tensioning cables to reduce the tensile membrane forces and displacement along the circumferential direction. A possible way of estimating an effective post-tensioning force could be to equilibrate the tensile circumferential resultant due to the internal fluid pressure $p = 1.3 \cdot 10 \cdot 25 \cdot 1 = 325$ kN/m acting on each ring of unitary height. In this case, a preliminary estimate of the required post-tensioning steel area (A_{sp}) can be obtained as follows:

$$A_{sp} = \frac{pr}{f'_{syd}} = \frac{325 \cdot 10}{1.1} = 2955 \text{ mm}^2$$

where f'_{syd} is the steel stress after prestress losses occurred, estimated around 65% of the steel cable yield strength. A reasonable corresponding concrete section is then obtained assuming that at yield conditions of the steel cables the concrete average stress is still acceptable, i.e., lower than the compression design strength:

$$A_c = \frac{A_s f_{syd}}{f_{cd}} = \frac{2955 \cdot 1452000}{17000} = 252352 \text{ mm}^2$$

For a strip of unitary height of the vertical wall, this corresponds to a thickness of about 25 cm. As a consequence, in this case the total amount of concrete required is 393 m^3, i.e. less than one-third of the amount estimated for the rectangular tank, with some 55 tons of steel reinforcement. These calculations are obviously preliminary estimates, since a more refined solution would require reasoning in terms of induced deformations rather than brutally equating forces.

An additional further alternative would be to modify the connection between the foundation and the cylinder, allowing relative displacement and rotation. In such conditions, the membrane boundary constraints would be respected and the flexural response would be eliminated. The consideration related to the circumferential stresses will still apply, with the consequent need of post-tensioning or adequate thickness. To allow rotation and displacement, a rubber pad should be designed and inserted, properly connecting it to both sides.

1.4   Seismic Behaviour of Tanks

A cylindrical tank is evidently a stiff structure, the simplest assumption to define its response to a ground motion is to consider it as a rigid body. As such, the tank will be possibly modelled as a mass, equal to a fraction of the total mass of structure and contents, rigidly connected to the ground. The mass relative displacement will be assumed as null, the maximum base shear (Q) equal to the mass multiplied by the peak ground acceleration (a_g) and the maximum overturning moment (M) equal to the base shear multiplied by some equivalent height.

In many cases, and particularly for empty tanks, this model will provide acceptable results, provided that reasonable values have been assumed for the fraction of the total mass and for the equivalent height at which the mass is lumped.

Considering first the case of an empty tank with a light roof, a possible choice is to consider the mass of the wall (m_w) only, to ignore the mass of the base plate (m_b) and to assume an equivalent height (H_{eq}), equal to one-half of the height of the tank:

$$Q = m_w a_g \qquad (1.20\text{a})$$
$$M = Q H_{eq} = m_w a_g H_{eq} \qquad (1.20\text{b})$$

Considering once more the examples of the rectangular and cylindrical tanks previously analyzed (Section 1.4) and a peak ground acceleration $a_g = 0.35g$, with reference to Figure 1.7(a), the base shear (Q) and the total bending moment (M) at the base of the wall will be:

Cylindrical tank	Rectangular tank	
$Q = 2749$ kN	$Q = 10762$ kN	(1.21a)
$M = 34361$ kNm	$M = 107625$ kNm	(1.21b)

These values give an immediate impression of the significant difference between the action on the cylindrical and rectangular tank, but will be of little use, since no mention has been made of the action induced by the contents. In any case, even limiting our consideration to the empty case, these values apply to the walls and to their connection to the foundation. In order to design the foundation and to evaluate the soil response, including the possibility of rocking or sliding phenomena, the shear (Q') and moment (M')

generated by the foundation mass need to be added. In this case, an assumption of a rigid response is even more justified, thus Equations (1.20a) and (1.20b) can be applied, using the foundation mass (m_b) and an equivalent height H_{eq} equal to one half of the thickness of the foundation plate. This thickness is assumed to be equal to 0.40 m for the cylindrical tank and to 1.0 m for the rectangular one. The different thicknesses are justified by the similar depth of the walls to be connected to the plate and by the different values of shear and bending moment to be transmitted. The resulting actions will be:

Cylindrical tank	Rectangular tank	
$Q' = 3849$ kN	$Q' = 14262$ kN	(1.22a)
$M' = 34581$ kNm	$M' = 109375$ kNm	(1.22b)

As expected, considering the small equivalent height where the mass is lumped, the additional actions induced by the foundation have a significant effect on the shear values and a negligible one on the overturning moment.

In the case of the rectangular tank, the consequence of this simple model on the design of each wall would be to consider a plate, restrained on three or four sides (depending on the roof type and connection) and to apply to the plate a uniform load equal to 35% of its own weight.

In the case of the cylindrical tank, a uniformly distributed load p_s (Figure 1.7(b)) will be applied along the height of the cylinder, equal to its total mass per unit length, again multiplied by the peak ground acceleration $a_g = 0.35g$, i.e.:

$$p_s = a_g h \gamma_c \tag{1.23}$$

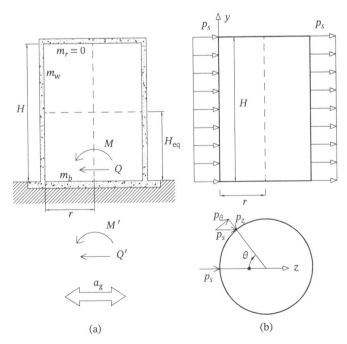

(a)   (b)

Figure 1.7 (a) Empty (cylindrical or rectangular) tank with the corresponding masses and heights and (b) an example of the seismic load p_s statically acting on the cylindrical tank's wall.

where h is the thickness of the tank and γ_c the unit weight of the structural material (concrete in our case).

To estimate the membrane actions, by analogy to what has been done in the case of the horizontal action induced by the pressure of the content, it is necessary to derive the tangent and radial components of p_s (the vertical component p_y is obviously null):

$$p_\theta = p_s \sin \theta \tag{1.24a}$$

$$p_z = p_s \cos \theta \tag{1.24b}$$

The solution of the problem is not as simple as in the case of an internal pressure, since the load is now not axisymmetric, and is discussed in detail in [39, 40]. From the integration of the equations which govern the membrane equilibrium, presented in [39, 40, 66], applying the boundary conditions at the unrestrained top of the tanks (i.e.: $N'_{y\theta} = N'_y = 0$ in $y = H$), results in:

$$N'_\theta = -p_s r \cos \theta \tag{1.25a}$$

$$N'_{y\theta} = 2p_s(H - y) \sin \theta \tag{1.25b}$$

$$N'_y = \frac{p_s}{r}(H - y)^2 \cos \theta \tag{1.25c}$$

In these equations, N'_θ and N'_y have the same meaning as in Equations (1.14a) and (1.14b), i.e. N'_θ is the horizontal reaction force per unit length tangent to the tank surface and N'_y is the vertical reaction force per unit length. $N'_{y\theta}$ is the in-plane shear force per unit length, which was null in the case of internal pressure because of the axisymmetric nature of the problem. The formulation of the equations is showing that:

- the membrane ring force (N'_θ) is constant over the height and varies along a ring from zero in the elements parallel to the acting force to a maximum compression on the side of the action to a maximum tension on the opposite side;
- the membrane vertical force (N'_y) varies quadratically along the height, from zero at the top to a maximum at the base, and varies along each ring similarly to N'_θ;
- the membrane shear ($N'_{y\theta}$) varies linearly along the height, from zero at the top to a maximum at the base, while along each ring it shows the opposite trend with respect to N'_θ and N'_y, being null in the elements perpendicular to the applied force and maximum in the parallel elements.

These trends are shown graphically in Figure 1.8.

The distributions of the axial and the shear force per unit length (N'_y and $N'_{y\theta}$) seem to indicate some similarities with the internal axial (σ) and shear (τ) stresses of a cantilever beam fixed at the base and subjected to the same horizontal applied load. In the case of the beam, the internal stresses at any height (y) will be expressed as functions of the bending moment (M_y) and shear (Q_y) acting at the corresponding section as:

$$\sigma = \frac{M_y}{J}x \quad \text{and} \quad \tau = \frac{Q_y S}{bJ} \tag{1.26}$$

where J is the moment of inertia of the section, x is the distance between the fibre considered within the section and the neutral axis, S is the first moment of the area above the considered fibre, b is the

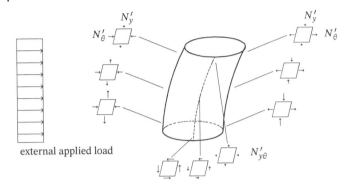

Figure 1.8 Qualitative distribution of N'_θ, N'_y and $N'_{y\theta}$.

width of the section at the fibre. Considering the case of the cylindrical tank, all the variables appearing in Equation (1.26) can be derived as follows (H is the height of the cylinder, h the shell thickness, r the radius of the generating ring):

$$Q_y = 4 \int_y^H \int_0^{\pi/2} p_s\, r\, d\theta\, dy = 2p_s \pi r(H - y) \qquad \text{Shear force at height } y \tag{1.27a}$$

$$M_y = Q\frac{H - y}{2} = p_s \pi r(H - y)^2 \qquad \text{Bending moment at height } y \tag{1.27b}$$

$$S = 2 \int_0^\theta hr^2 \cos\theta\, d\theta = 2hr^2 \sin\theta \qquad \begin{array}{l}\text{First moment of area of a ring} \\ \text{sector subtended by an angle } 2\theta\end{array} \tag{1.27c}$$

$$J = 4 \int_0^{\pi/2} hr^3 \cos^2\theta\, d\theta = \pi hr^3 \qquad \text{Moment of inertia of a ring} \tag{1.27d}$$

$$b = 2h \qquad x = r\cos\theta \qquad \text{Width and distance from the neutral axis} \tag{1.27e}$$

The internal stresses at any point of the cylindrical tank can be related to the membrane forces of Equations (1.25b) and (1.25c), assuming that these are constant over the thickness (h), in which case it results in:

$$N'_y = \sigma \cdot h \qquad \text{and} \qquad N'_{y\theta} = \tau \cdot h \tag{1.28}$$

Substituting Equations (1.27a)–(1.27e) into (1.26) and multiplying by h to obtain the membrane forces, the following equations are obtained:

$$N'_{y\theta} = \frac{2hp_s \pi r(H - y)2hr^2 \sin\theta}{2h\pi hr^3} = 2p_s(H - y)\sin\theta \tag{1.29a}$$

$$N'_y = \frac{hp_s \pi r(H - y)^2 r\cos\theta}{\pi hr^3} = \frac{p_s}{r}(H - y)^2 \cos\theta \tag{1.29b}$$

These results are identical to those expressed in Equations (1.25b) and (1.25c), indicating that the pure membrane theory applied to a cylindrical tank subjected to a uniform horizontal load and the beam theory applied to a vertical cantilever with the same section subjected to the same load will produce identical results. It is known that one of the fundamental assumptions for the application of the beam theory is that horizontal sections remain plane and undeformed. As a consequence, the application of this analogy is possible if the diameter is small compared to the height.

All the considerations and calculations developed up to this point have been related to empty tanks, and this appears rather limiting, and possibly useless, when it is observed that the mass of the contained liquid is normally several times larger than the mass of the containing structure. In the cylindrical tank considered as an example, the mass of the liquid is likely to be of the order of 7–8 times the mass of the concrete shell. On the other hand, it is immediately verified that considering the entire mass of the liquid as rigidly connected to the shell will result in unreasonably large actions.

Actually, a careful observation of the response of the liquid contained in a cylindrical tank to a horizontal acceleration (which could easily be simulated by moving a glass of water on a table), will lead one to note that the upper part of the liquid will respond with a kind of a vertical wave, as shown in Figure 1.9. This periodic variation of the height of the liquid at the opposite sides of the tank will induce a variation of the pressure on the walls and on the base plate. While the response of the shell was assumed to be characterized by a very short period of vibration (as a consequence of its high stiffness), this kind of motion, sometimes called "sloshing mode", is characterized by a long period of vibration, in the range of several seconds for normal size tanks.

In addition, it appears that only part of the liquid will participate in this mode of vibration, in the upper part of the tank, while the remaining lower part will essentially participate in the rigid response discussed and assumed for the shell structure. The problem is thus how to define the fraction of liquid participating in each mode and to estimate a reasonable period of vibration of the sloshing mode. The solution to this problem will easily allow an estimate of the induced actions, since the very significant difference of the periods of vibration of the two modes will facilitate a combination of their effects. The details of this problem are presented in Chapter 2, however, it is here anticipated that the fundamental parameter to determine the percentage of fluid participating in either mode is the ratio of the height of the fluid in the tank over its radius, $\gamma = \frac{H}{R}$. For the same ground motion intensity, a higher mass participating in the sloshing mode (i.e. for smaller values of γ) also implies a higher convective wave, and consequently higher values of the actions induced by the corresponding mode. This effect can be again intuitively verified by moving with similar input two different glasses of water over a table, one wide and low (like a highball), the other one taller and narrower (like a tumbler). The water will more easily overflow from the highball, because a greater fraction of the fluid will participate in the convective motion and the wave will be higher. The problem described above is graphically represented in Figure 1.10.

Figure 1.9 Qualitative trend of the convective wave and variation of the dynamic pressure on walls and base plate induced by a horizontal ground motion in a tank full of liquid. Source: Adapted from [412].

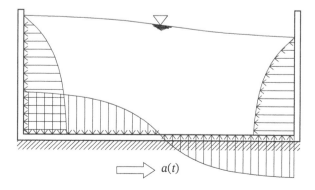

$a(t)$

for actions below the tank's base

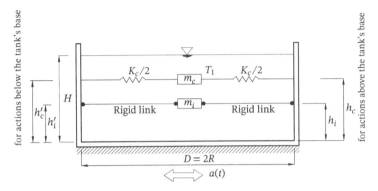

for actions above the tank's base

Figure 1.10 Equivalent mechanical model for a tank full of water under seismic excitation. Source: From [150]/American Society of Civil Engineers.

A lower mass is rigidly linked to the shell structure, while a higher mass is linked to the walls with elastic links, whose stiffness (together with the associated mass fraction) will define the period of vibration of the sloshing mode. In general, in tall and narrow tanks, characterized by a value of γ greater than 3, the fraction of liquid participating in the convective motion is less than 15%. As a consequence, in these cases, it is acceptable to compute the actions considering the whole liquid mass as rigidly linked to the walls. The solution will thus be identical to that discussed above, only the total mass located at mid-height of the tank and subjected to the ground acceleration will be larger. In the case of low and wide tanks, with values of γ smaller than 0.3, more than 80% of the mass of the liquid will likely participate in the convective motion.

As anticipated, the analytical solution to the problem of estimating the mass fractions participating in either mode and their height of application is provided in Chapter 2, however, the basic resulting equations are depicted in a graphical form in Figure 1.11. It is immediately noted that the mass participating in the two modes varies rapidly with γ, while the equivalent height of application of the two masses is much less sensitive to the geometry of the tank.

In Chapter 2 how the convective component of the response motion is actually composed of several modes, with distinct participating mass and different periods of vibration will be discussed, however, it will be clear that most of the mass will participate in the first mode, and on this we will now focus attention. The period of vibration of the first sloshing mode can be estimated using the following expression:

$$T_1 = \frac{2\pi}{\omega_{c1}} = \frac{2\pi\sqrt{\dfrac{R}{g}}}{\sqrt{\lambda_1 \tanh\left(\dfrac{\lambda_1 H}{R}\right)}} \tag{1.30}$$

where, as usual, r is the radius and H the height of the fluid in the tank, g is the constant of gravity and λ_1 is a numerical parameter equal to 1.8412. The resulting period of vibration T_1 is thus varying between 2.3 and 6.7 seconds for any tank with height and diameter between 5 and 25 m. In the example case considered above, with a diameter equal to 20 m and a height equal to 25 m, the period of vibration of the sloshing mode will be estimated as $T_1 = 4.7$ sec.

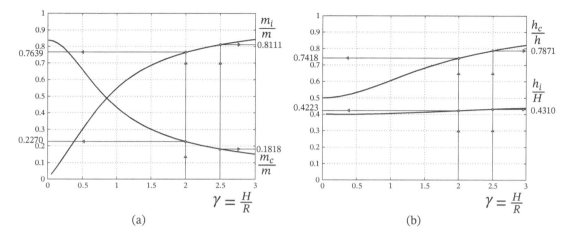

Figure 1.11 New Zealand Recommendation – NZSEE-09: (a) percentages of total mass for impulsive (m_i) and convective (m_c) components; (b) equivalent height of application of the impulsive (h_i) and convective (h_c) mass.

It is thus confirmed that the convective period of vibration is quite high, and certainly much greater than the period of the tank. The two motions will therefore be completely uncoupled and the actions resulting from each one can be computed separately and simply added together. The theory derived to treat the sloshing modes of cylindrical tanks can be adapted to rectangular tanks by introducing minor modifications (see Chapter 2), to estimate the following masses and equivalent heights of application for the two example used in this introduction, as follows:

Cylindrical tank $H = 25$ m, $r = 10$ m	Rectangular tank $H = 2B = 20$ m	
$m_i = 6361$ ton	$m_i = 6104$ ton	(1.31a)
$m_c = 1429$ ton	$m_c = 1816$ ton	(1.31b)
$h_i = 10.8$ m	$h_i = 8.4$ m	(1.31c)
$h_c = 19.7$ m	$h_c = 14.8$ m	(1.31d)

These values can be approximately derived from the plots in Figure 1.11. In both cases the impulsive mass is about 80% of the total liquid mass. For the cylindrical tank, this corresponds to almost eight times the mass of the shell, while for the cubic tank the liquid mass is only approximately two times the mass of the walls.

It is now possible to estimate the approximate value of the seismic actions, recurring to an acceleration spectrum, which will be here assumed to have the typical shape adopted for stiff soil properties, as depicted in Figure 1.12. To enter the spectrum and thus define the multipliers to be applied to the effective masses of each mode, it is necessary to estimate the period of vibration of each mode, as follows:

- as already discussed, the impulsive mode of the cylindrical tank will be assumed to be infinitely rigid, the corresponding period of vibration will be thus assumed to be zero ($T_{cyl,i} = 0$) and the corresponding spectral acceleration will be identical to the peak ground acceleration ($S_e(T_{cyl,i}) = a_g = 0.35g$);

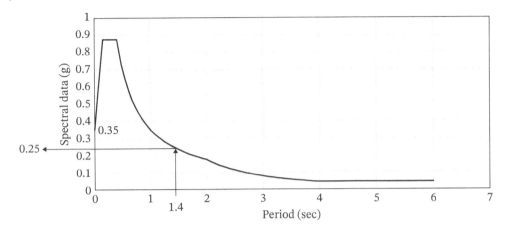

Figure 1.12 Acceleration spectrum having the typical shape adopted for stiff soil properties: soil A, spectrum Type 1, $S = 1.0$, $T_B = 0.15$ sec, $T_C = 0.4$ sec and $T_D = 2.0$ sec, $a_g = 0.35g$. Source: Point 3.2.2.2 in UNI EN 1998-1:2004.

- the impulsive action on one of the walls of the rectangular tanks should not be assumed to be acting at zero period, since it should account for the flexibility of a 20 by 20 m plate, 0.9 m thick, restrained along three sides. The fundamental period of vibration of such a plate can be estimated as around 1.4 sec and the corresponding spectral acceleration approximately equal to: $S_e(T_{rec,i}) = 0.25g$ (see Figure 1.12);
- the convective (sloshing) motion in the cylindrical and rectangular tanks will be characterized by slightly different periods of vibration, equal to $T_{cyl,c} = 4.7$ sec and $T_{rec,c} = 5.1$ sec and by a conventionally lower damping ratio, taken as 0.5%. The corresponding spectral acceleration, including an amplification factor $\eta = 1.34$, to account for the lower damping, will be low and similar: $S_e(T_{cyl,c}) = 0.04g$ and $S_e(T_{rec,c}) = 0.03g$.

The shear force and bending moment induced at the base of the tanks by the liquid response can be calculated by applying the spectral acceleration to the corresponding equivalent mass, obtaining the results reported below. It is immediately observed that the actions induced by the sloshing modes are negligible with respect to those generated by the impulsive modes. Also noticeable is the very favourable effect resulting from the flexibility of the walls of the rectangular tank:

Cylindrical tank	Rectangular tank	
$Q_i = 21818$ kN	$Q_i = 15260$ kN	(1.32a)
$Q_c = 572$ kN	$Q_c = 545$ kN	(1.32b)
$M_i = 235637$ kNm	$M_i = 128184$ kNm	(1.32c)
$M_c = 11261$ kNm	$M_c = 8063$ kNm	(1.32d)

The total impulsive actions at the base of the tanks can be calculated considering the contributions of the structure (Equation (1.21)) simply added to those of the impulsive liquid response:

Cylindrical tank	Rectangular tank	
$Q_b = 24567$ kN	$Q_b = 26022$ kN	(1.33a)
$M_b = 269998$ kNm	$M_b = 235809$ kNm	(1.33b)

Note that to obtain the maximum values of the action on the foundation plate (and on the soil, obviously), the contribution of the bending moment due to the height variation of the sloshing mode has to be combined with the total impulsive term (same for the shear contribution).

It is thus of interest to evaluate the height of the convective wave, also to estimate the possibility of liquid spilling or the potential pressure on the roof. To gain a feeling of the wave height, the following approximate expression can be used, applicable to the case of a cylindrical tank:

$$d_{max} = 0.84rS_e(T_1) \tag{1.34}$$

where, as usual, r is the radius of the tank, $S_e(T_1)$ is the ordinate of the appropriate acceleration response spectrum, evaluated for the period of vibration of the first convective mode, expressed as a fraction of g. In the case considered above, the wave height would be $d_{max} = 0.34$ m, while applying a similar equation to the case of the rectangular tank the wave height would result in $d_{max} = 0.31$ m. This wave will induce an additional bending moment on the foundation and possibly pressure on the wet area of the roof, with a consequent increment of the impulsive mass due to the roof weight acting on the liquid. These problems are discussed in Section 2.2.8.

1.5 Field Observation of Damage to Tanks Induced by Seismic Events

The damage to and possible failure of a storage tank during an earthquake may induce consequences, and namely losses, far exceeding those connected to the repair or re-construction of the structure and the proper value of its contents [67]. A tank could be the primary source of local public water and its loss could result in disruption of water use both for domestic consumption and for post-earthquake recovery. Fire services may be impaired when they will be most needed. Alternatively, a tank could contain products that are toxic and dangerous for the environment [573], possibly contaminating the atmosphere and the groundwater layer or resulting in conflagration determining an immediate flash of vapor and a toxic gas cloud (e.g. Figure 1.13).

Failure modes induced by ground motion can be quite diverse (e.g. Figure 1.14) and are usually grouped into several categories [328, 419, 469, 508]:

Figure 1.13 Burned tanks following the Izmit earthquake in the Kocaeli Province of Turkey (17 August 1999, Magnitude 7.4). Source: From [573]/Taylor & Francis.

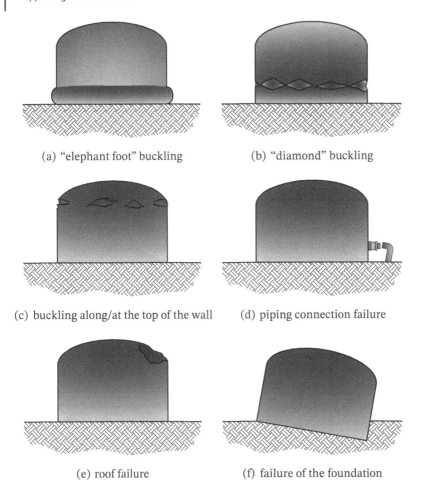

(a) "elephant foot" buckling

(b) "diamond" buckling

(c) buckling along/at the top of the wall

(d) piping connection failure

(e) roof failure

(f) failure of the foundation

Figure 1.14 Global overview of observed damage in tanks in past earthquakes.

- the most common collapse mechanism, observed in steel tanks, is a form of elastopastic buckling [481, 488, 489, 491, 493] commonly known as "elephant's foot" buckling (Figure 1.14(a)). The characteristic outward bulge above the base of the tank (Figure 1.15(a)), extending partly or completely around the circumference, results from the combination of the large circumferential tensile stresses induced by the internal hydrostatic and hydrodynamic pressures and by the axial compression due to the overturning moment caused by the horizontal seismic loads. At the attainment of the tensile yield strength in the annular ring strip, the applied compression vertical stress will easily induce buckling, even for small stress increments, due to the reduced elastic modulus. This kind of elastoplastic instability tends to assume an axisymmetrical shape due to the cyclic nature of the action and is generally associated with large tanks, with a relatively low height-to-radius ratio (say $\frac{H}{r} \leq 1$) [328].
- a second common failure mode, again related to instability problems, is known as "diamond-shaped" buckling. It is induced by an elastic form of instability, again in the proximity of the base, where the stresses are higher (Figure 1.14(b)). This damage mode (Figure 1.15(b)) takes place at relatively low

Figure 1.15 Observed damage to tanks in past earthquakes: (a), (c), (e) Emilia earthquake, Italy (20 and 29 May 2012, Magnitude 6.1 and 5.9 respectively); (b) Izmit earthquake, Turkey (17 August 1999, Magnitude 7.4) (from Izmit Collection, University of California, Berkeley); (d) San Fernando earthquake, California (9 February 1971, Magnitudo 6.6) (from Karl V. Steinbrugge Collection, University of California, Berkeley); (f) Edgecumbe earthquake, New Zealand (2 March 1987, Magnitude 6.3) (from Edgecumbe Collection, University of California, Berkeley).

values of tensile ring stresses and it is mainly caused by high vertical compression. It typically takes place in more slender tanks, with a height-to-radius ratio greater than one [328]. Elastic buckling considerations may govern the design of cylindrical shells with a relatively low internal pressure [481].

- local bucking has also been observed in the upper areas of the shell (Figures 1.15(c) and 1.15(e)), even close to the top (Figure 1.14(c)) [419, 420]. This latter failure (Figure 1.15(d)) is believed to have been caused by some rupture close to the base followed by a rapid loss of the tank contents and consequently resulting in a partial vacuum in the upper part, leading to a pressure decrease or even to a change of sign of the ring stress. Note that near the top the tank shell can be rather thin, since the theoretical acting forces and stress are low [469].

- an uncommon damage mechanism has been observed in the Emilia earthquake sequence (Italy, 2012, main shocks magnitudes 6.1 and 5.9). A diamond-shape elastic buckling was observed in the proximity of the circumferential welding, at an intermediate height (Figure 1.15(e)). The location of buckling is certainly related to the presence of imperfections induced by the welding and by the variation of shell thickness above and below the welding line.

- relative displacements between tank and connected piping and equipment have induced damage and failure to connections (Figures 1.15(f) and 1.14(d)). This cannot be regarded as proper tank damage, but it is undoubtedly inducing an operational break and the consequent economic loss. This kind of problem can easily be solved by providing some adequate deformation capacity at all connections.

- tanks are usually designed to allow sloshing waves to move within the freeboard between the still water surface and the roof, even if this solution has undesirable side effects on the system capacity and corrosion prevention and, ultimately, in the case of potable water, on water quality. As shown in Figure 1.16(a), an insufficient freeboard may cause an upward pressure on the tank roof that will influence the vertical force in the shell and increase the impulsive mass, modifying the impulsive and convective periods of vibration and the hydrodynamic pressure against the shell. Sloshing waves impacting on the roof (Figure 1.13(a)) have sometimes resulted in roof failure or in damage to the shell-roof joint or seam (Figure 1.14(e)). Similar problems have resulted from internal pressure induced by increased temperature related to burning of nearby tanks (Figure 1.16(b), [573]).

- in tanks containing inflammable material, an explosion can be caused by bouncing of the floating roof against the shell, producing sparks that may ignite the inflammable material [573].

- tanks are often constructed in areas close to the sea, with possibly loose sand soil and high ground water layer. These conditions may be prone to liquefaction (Figure 1.16(c)) and to consequent global or or partial settlement of the foundation (Figure 1.14(f)). Sliding (Figure 1.16(d)) or walk-off of foundations due to progressive rocking has also been observed.

- some examples of damage to the connection between the tank and the foundation are shown in Figures 1.17(a)–1.17(c). In some cases, this damage is associated with some shell damage, as shown Figure 1.17(a), where some diamond-shaped buckling of the tank wall is clearly visible. Often, excessive tensile demand causes the fracture of anchor bolts or their extraction from the concrete foundation. Spalling of concrete, induced by insufficient distance between the anchor bolt and the edge of the foundation and low resistance of the concrete, is shown in Figure 1.17(b). A flexural failure that occurred in the anchor plates is depicted in Figure 1.17(c). All these damage modes are clearly not pertaining to the shell structure, but are related to the actions induced by the tank response and to poor dimensioning and steel reinforcement, anchor bolts and plates, and foundation concrete.

- leg-supported tanks, usually smaller than flat-bottomed systems, have often evidenced problems in the legs (Figure 1.17(d), with little consequence on the functionality of the tank) or in their connections

(a)   (b)

(c)   (d)

Figure 1.16 Observed damage to tanks in past earthquakes: (a) Kern County, California earthquake (21 July 1952, Magnitude 7.7) (from Karl V. Steinbrugge Collection, University of California, Berkeley); (b) Tupras refinery, Izmit earthquake, Turkey (17 August 1999, Magnitude 7.4) (Izmit Collection: IZT-174, Courtesy of the National Information Service for Earthquake Engineering, EERC, University of California, Berkeley; photo by Andrew S. Whittaker); (c) and (d) Kobe earthquake, Japan (17 January 1995, Magnitude 6.7) (Peter W. Clark Kobe Collection, University of California, Berkeley).

(Figure 1.17(e), where a relatively stiff leg is directly welded to a thin shell, leading to an undesirable "strong leg-weak tank wall" mechanism [187]). In some cases they have also been stepping away (see in Figure 1.17(f) the footprint of the former position of the leg). It is noticeable how unanchored legged systems often performed considerably better (like the case in Figure 1.17(f)), with no damage to the tank or the foundation. In these cases, the possibility of a combination of rocking and sliding has limited the transmitted shear force and bending moment and the piping system was flexible enough to accommodate the relative displacement demand. This effect can be regarded as a sort of not-designed base-isolation response.

- in elevated tanks, damage and failure are often concentrated in the supporting steel reinforced concrete structure, as shown in Figures 1.18(a)–1.18(g).

(a) (b) (c)

(d) (e) (f)

Figure 1.17 Observed damage to tanks in past earthquakes: (a)–(d) and (f) Emilia earthquake, Italy (20 and 29 May 2012, Magnitude 6.1 and 5.9 respectively); (e) Maule Earthquake, Chile (2 February 2010, Magnitude 8.8). Source: From [187]/Elsevier.

Certainly, it is important to highlight that codes/standards and recommendations (and the design therein suggested as highlighted in Section 1.6) cannot prevent extreme events from occurring, but they can provide good practice in order to offer a measure of protection against the earthquake. The following main options can be followed to avoid or reduce the damage to an acceptable level [16, 581]:

- a fixed geometry of the tank (radius R, height H, slenderness) may be maintained by increasing the wall thickness and by appropriately dimensioning the anchors at the base; it is possible also to add dissipative steel anchors at the base as suggested by Malhotra in [369]; dampers mainly control the moment transmitted to the foundation and, in the case of unanchored tanks, they consistently reduce the hoop compressive stress in the wall;
- depending on the boundary conditions, it may be possible to upgrade the geometry of the tank by changing its slenderness $\frac{H}{R}$ or by replacing a lower shell course, causing the tank to meet the seismic requirements. The annular plate could also be added or thickened, which can increase the seismic resistance or the tank's wall stiffened be using vertical strings;
- in order to reduce the hydrodynamic loading on the shell, tanks in seismic regions can be mounted on multilayered elastomeric bearing isolators as suggested by many authors [73, 74, 93, 304, 344, 366, 367, 378, 545]. When fluid containers are located at the top of a supporting structure or in the upper stories of a building, base isolation is still very favourable [531, 544]. The elevated tank can be isolated by two techniques, namely, by placing the bearings between the base of supporting structure and the foundation [531], and by placing the bearings between the bottom of the liquid container and the top of the supporting structure [544]. As an alternative to elastomeric isolators, curved surface sliding bearings, which are usually called a friction pendulum system (FPS) are commonly used for base isolation of liquid storage tanks since the period of the isolation system is independent of the storage level [1, 65, 93, 344, 436, 540, 660]. It is important to highlight that sloshing in tanks commonly

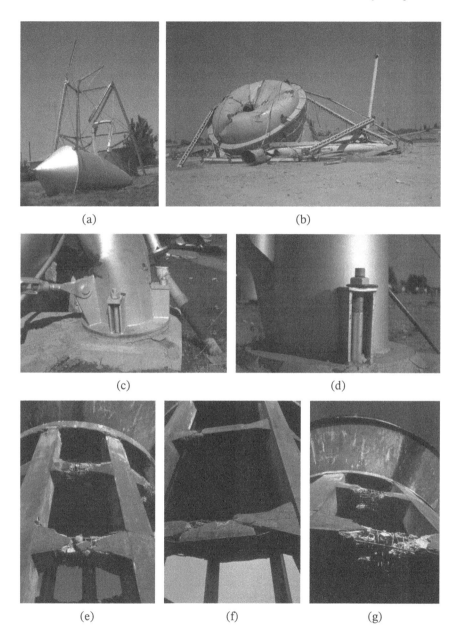

Figure 1.18 (a), (b) and (d) Kern County, California earthquake (21 July 1952, Magnitude 7.7); (c) Imperial Valley earthquake (21 October 1979, Magnitude 7.0); (e)–(g) Chile earthquake (22 May 1960, Magnitude 8.5). ((a)–(f) from Karl V. Steinbrugge Collection, University of California, Berkeley.)

used in engineering practice is a phenomenon of relatively low frequency and for this reason base isolation might increase it, causing a slightly higher convective pressure component and water elevation response. However, this problem can easily be solved by appropriate selection of the base isolation system or by adding dampers at the location of the maximum water particle velocities, as suggested in Section 6.3. So, generally, base isolation has more effect on the impulsive component, thus leading to the reduction of the impulsive pressure component of the hydrodynamic motion.

One more suggestion should be added to the above options, that the tank should be situated within a fluid tight bund [414] and, in order that it can be inspected externally for corrosion or leaks, a minimum distance between the tank's wall and the bund wall is recommended where possible [387].

1.6 Design Considerations

Storage containments are facilities used as a source of supply for essential lifelines such as potable water, fuel, and sewage disposal. Hence, the seismic performance of storage tanks is a matter of special importance in earthquake-prone countries beyond the economic value of the tanks themselves. As clearly highlighted in past earthquakes (Sections 1.5 and 5.5), severe damage or collapse, causing uncontrolled fires, spillage of toxic chemicals or liquefied gasses or loss of drinking water supplies, may cause substantially more damage than the effects on structures caused by the earthquake itself [431, 459]. In order to reach a level of structural safety of containments, according to Point 2.1.6(2), Eurocode 8, Part 4, the tank should be conceptually designed in such a way that it is able to sustain all the actions calculated using structural elements detailed for local ductility and constructed from ductile materials. Once the hydrodynamic pressures, exerted against the tank wall and bottom, are known (Chapters 2 and 3 for above-ground anchored and unanchored rigid tanks respectively; Chapter 4 for elevated tanks and Chapter 5 for flexible tanks; Section 6.5 for underground rigid tanks), they can be integrated so the tanks' designer can find the critical variables [414]:

- shear and bending moments, hoop tension, axial compression and tension, in the tank's wall, upper shell courses included;
- sloshing wave effects and eventual impact on the roof;
- shear and bending moments (overturning moments) transmitted to the foundation and subsequently to the soil.

The design according to the above quantities should also take into consideration, in order to mitigate earthquake effects, four main requirements (Point 2.1.6(2), Eurocode 8, Part 4): (1) redundancy of the system; (2) absence of interaction between mechanical/electrical and structural components; (3) easy access for inspection, maintenance and repair; and (4) quality control of all the components. A further consideration in the design of tanks is to establish the appropriate codes and standards to be used (mainly dependent on the tank supplier/designer in agreement with the buyer of the tank itself and the country of installation) and the material which is more effective and acceptable: steel or concrete (usually the supplier should take into account factors such as allowable stresses, welded or bolted connection cost compared to ordinary reinforced concrete or precast, availability as well as purchasing terms and mill delivery) [414]. Depending on the nature and amount of the contents and associated potential danger (as well as the combination of the actions) (Point 2.1.1(1), Eurocode 8, Part 4), the functional requirements during and after an earthquake, the environmental conditions, the materials used to build the tank and its corresponding typology, two main limit states are defined:

- the ultimate limit state (ULS) corresponds to structural failure or in the case of complete collapse would entail severe consequences, the ultimate limit state is defined as the one corresponding to a state prior to structural collapse. The design seismic action, a_{Ed}, for which the ultimate limit state may not be exceeded shall be expressed in terms of the reference seismic action, a_{Ek}, associated with a reference probability of exceedance,[2] P_{NCR}, in 50 years or a reference return period, T_{NCR} (Points 2.1(1) and 3.2.1, Eurocode 8, Part 1). Furthermore, the design seismic action, a_{Ed}, shall be expressed in terms of the importance factor γ_I which takes into account the reliability differentiation (Points 2.1(2), (3) and (4), Eurocode 8, Part 1):

$$a_{Ed} = \gamma_I a_{Ek} \tag{1.35}$$

from which can also be derived the following relation:

$$a_g = \gamma_I a_{gR} \tag{1.36}$$

where the value of the reference peak ground acceleration on type A soil,[3] a_{gR} (Point 3.2.1(2), Eurocode 8, Part 1), corresponds to the reference return period T_{NCR} of the seismic action for the no-collapse requirements; and a_g is the design ground acceleration on type A ground. For tanks it is appropriate to consider four main different importance classes, depending on the potential loss of life due to the failure of the containment structure and on the economic and social consequences of failure (Table 1.1 and Points 2.1.4(4)–(8), Eurocode 8, Part 4):

1) Class I refers to situations where the risk to life is low and the economic and social consequences of failure are small or negligible.
2) Class II refers to situations with medium risk to life and to local economic and social life.

Table 1.1 Importance Classes I to IV depending on the use and contents of the facility and the implications for public safety. Here are suggested values of γ_I, but it may be different for the various seismic zones of the country considered, depending on the seismic hazard conditions and on the public safety considerations.

Class	γ_I	Risk to life	Economic/social consequences
I	0.8	Low	Small/Negligible
II	1.0	Medium	Medium
III	1.2	High	Great
IV	1.6	Exceptionl	Extreme

---

2 The recommended values are $P_{NCR} = 10\%$ and $T_{NCR} = 475$ years.
3 Rock or other rock-like geological formation, including at most 5 m of weaker material at the surface, with a value of the average shear wave velocity $v_{s,30} > 800$ m/s (Point 3.1 and Table 3.1 in Eurocode 8, Part 1).

3) Class III refers to situations with a high risk to life and great economic and social consequences of failure.
4) Class IV refers to situations with exceptional risk to life and extreme economic and social consequences of failure.

- the damage limit state (DLS) (or serviceability limit state (SLS) as called by Greiner in [193]) corresponds to requirements such as integrity and minimum operating level (Point 2.1.3, Eurocode 8, Part 4). The integrity requirements of the tank are satisfied when the system itself, including a specified set of accessory elements integrated with it (floating roofs, discharge devices, attached pipings, etc.), remains fully serviceable and leakproof under the relevant seismic action. Furthermore, the minimum operating level requirements of the containment are satisfied when the extent and amount of damage of the system itself, including some of its main components, are limited, so that, after the operations for damage checking and control have been carried out, the capacity of the tank can be restored up to a predefined level of operation. The relevant seismic action for which the damage limit state may not be exceeded shall have a probability of exceedance,[4] P_{DLR}, in 10 years and a return period, T_{DLR} (Point 2.1.3(5), Eurocode 8, Part 4).

It should be noted that the above preliminary considerations have to be specified and further explained in case of steel and concrete tanks.

In accordance with Point 2.1, Eurocode 3, Part 1.6, a steel tank should be designed in such a way that it will sustain all actions and satisfy the following requirements:

- overall equilibrium such as sliding, uplifting, and overturning. These three verifications may be found at Point 13.5.4, AWWA D100 (2011) and at Point 14.3.4, AWWA D103 (2009) in case of welded and factory-coated bolted steel tanks respectively. Hydrodynamic actions of the tank's liquid contents (subdivided into a portion of the liquid that acts as a solid in the bottom part and a portion of the fluid near the surface creating waves) produce forces and moments that want to tip the tank over, called an overturning behaviour. The response of the tank to this overturning action is to develop a pressure acting on the wall and on the bottom slab. With a sufficient acceleration due to the earthquake motion, a portion of the tank starts to uplift from the foundation or to slide. Usually, as well documented in [414], overturning, uplift, or sliding is not damaging the tank so much but the attached piping or equipment.
- equilibrium between actions and internal forces and moments, that should be checked with reference to the plastic (LS1) and buckling (LS3) limit state. As clearly defined by Rotter in [490] the plastic limit state (LS1) represents the state of extensive rupture or unacceptably large plastic deformation in tension or compression, when stability phenomena do not intervene. The simplest way to evaluate this limit state is to derive the plastic collapse load obtained from a mechanism based on small displacement theory (Point 4.1.1, Eurocode 3, Part 1.6). It is important to emphasize that the plastic limit state covers two main conditions in the membrane and bending shell design: (1) tensile rupture or compressive yield through the thickness (membrane); and (2) plastic collapse mechanism (involving bending). The limit state of buckling should be taken as the condition in which all or part of the structure suddenly develops large displacements normal to the shell surface, caused by loss of stability under compressive

---

4 The recommended values are $P_{DLR} = 10\%$ and $T_{DLR} = 95$ years.

membrane or shear membrane stresses in the shell wall, leading to inability to sustain any increase in the stress resultants, possibly causing total collapse of the tank (Point 4.1.3(1), Eurocode 3, Part 1.6). The tank's behaviour exhibited in the wall of a containment structure before, during, and after buckling appears depends mainly on the geometry, imperfections, boundary conditions, quality of construction, and loading pattern.

- limitation of cracks due to cyclic plastification, or better called the cyclic plasticity (LS2) limit state. The cyclic plasticity limit state, following the definition given by Rotter in [485], is concerned with a few cycles of loading that induce large strains (reversing in sign) and then producing plastic damage in both directions. Furthermore, according to Point 4.1.2(1), Eurocode 3, Part 1.6, this limit state should be taken as the condition in which repeated cycles of loading and unloading produce yielding in tension and in compression at the same point, thus causing plastic work to be repeatedly done on the tank's wall, eventually leading to local cracking by exhaustion of the energy absorption capacity of the material.
- limitation of cracks due to fatigue (LS4 or fatigue limit state). The limit state of fatigue should be taken as the condition in which repeated cycles of increasing and decreasing stress lead to the development of a fatigue crack (Point 4.1.4(1), Eurocode 3, Part 1.6).

Furthermore, API 620, API 650, API 12B, and UNI EN 14015 cover, in more detail, the design requirements for steel storage tanks in various standard sizes and capacities for different internal pressures, shop/field erected, product stored, and operating temperatures. Particularly, in order to assure structural stability and integrity, they establish minimum requirements for the design of tank sidewalls, roofs (fixed or floating) and bottoms, bolted or welded joints, foundation anchorage, openings (manhole or nozzle), and stiffening rings.

The two limit states for concrete tanks are the ultimate limit state (ULS) and the serviceability (or damage) limit states (SLS) as highlighted at Points 3.5.2.2(b) and 3.5.1 in EC8-4, respectively. Reinforced concrete or precast containments, including water and waste water treatment structures, belong to the category of reinforced concrete systems where minimal cracking is a prime requisite [307]. This means that the serviceability limit states consistently govern the design and usually leakage (Point 7, Eurocode 2, Part 3, where four tightness classes are introduced to classify liquid-retaining structures) and durability (Point 4 in Eurocode 2, Part 3) and govern deflection limitations [471]. In the seismic design situation relevant to the damage limitation state, tanks should be checked to satisfy the relevant serviceability limit state verifications required by Eurocode 2, Part 3, which covers additional rules to those in Part 1. According to Point 1.4.9, ACI 318-19 does not govern design and construction of concrete tanks and reservoirs, but indicates ACI 350-06 "Code Requirements for Environmental Engineering Concrete Structures" as the reference to be used in the design (and construction) of reinforced concrete reservoirs and other structures commonly used in water and waste treatment works, where dense, impermeable concrete with high resistance to chemical attack is required. In the case of concrete pedestal, elevated water storage tanks, ACI 371.R-08 describes how to apply ACI 318-19 and ACI 350-06 in designing concrete elements, including foundations, geotechnical requirements, appurtenances, and accessories. Over the years also the Portland Cement Association (PCA) has published many reports [128, 411, 412] on the design of rectangular and cylindrical reinforced concrete tanks, providing tables that assist practitioners in structural analysis of plates and shells. The prevention of the formation of through cracks due to tensile stresses can be done by following the prestressing philosophy [580]. The existing design approaches can be found in:

- ACI 372R-13 and AWWA D110-04 where three types of core walls (cast-in-place, shotcrete and precast) are considered with circumferential prestressing, using wrapped wire or strand systems, and vertical prestressing (or a vertically fluted steel diaphragm), using single or multiple high strength strands, bars, or wires;
- ACI 373R.97 and AWWA D115-06 for tanks prestressed with circumferential post-tensioned tendons placed within or on the external surface of the wall and vertical prestressed reinforcement.

As for steel tanks, the maximum action effects induced in the seismic design situation should be less or equal to the resistance of the shell evaluated as in the persistent or transient design situations. For concrete shells, the calculation of resistances and the verifications should be carried out in accordance with the ultimate limit state in bending with axial force and the ultimate limit state in shear for in-plane or radial shear (Point 3.5.2.2, EC8-4).

1.7 A Simplified Description of the Seismic Response of Tanks

The dynamic fluid pressures developed during an earthquake are important in the design of containment structures, such as tanks and silos. The analysis of a tank–fluid system requires preliminary considerations of the motions of both the structure and the contained liquid. As shown in Figure 1.19 when the base of a tank containing liquid is subjected to a lateral acceleration $a(t)$, the hydrodynamic pressures exerted by the fluid on the wall and base change in intensity and distribution from those corresponding to a state of hydrostatic equilibrium [214]. For a horizontally excited upright containment structure, a portion of the fluid close to the upper free surface tends not to displace laterally with the tank walls, experiencing vertical sloshing or rocking oscillations about a horizontal axis normal to the direction of the excitation (essentially antisymmetric modes) [66]. Nearer the floor base and along the walls at the bottom, the fluid is unable to move out of the way as the tank displaces and moves synchronously with the wall acting as an added mass rigidly attached to the wall [626]. The motion of the fluid in a storage container may be considered to be composed of two main parts: the first that moves in unison with the tank and is known as the *rigid impulsive* mass component, and the second undergoing sloshing motion that is known as

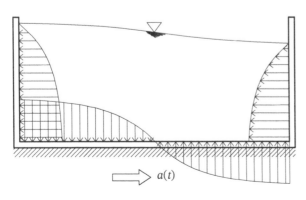

Figure 1.19 Hydrodynamic pressure distribution on the wall and base of a upright tank induced by ground acceleration $a(t)$. Source: Adapted from [412].

the *convective* mass component. The proportions of the liquid mass that participate in the two types of motion depend on the geometry of the tank: height of the free surface of the fluid H, diameter $D = 2R$ and radius R. The portion of the fluid that acts in a convective fashion decreases as the tank aspect ratio $\gamma = \frac{H}{R}$ increases, with the rigid mode becoming more predominant. In the case of tall/slender tanks, the pressure induced by the impulsive mass component is maximum near the tank base and is associated with high–frequency oscillations. Conversely, for tanks of very low aspect ratio, such as shallow/squat/broad tanks, only about 30% of the total fluid acts with the walls, with the remainder responding in sloshing modes [250]. In this case, the pressure induced by the convective component is maximum at the liquid surface and is associated with low–frequency oscillations.

It is often advantageous to replace the liquid, conceptually, by an equivalent mechanical model [251] based on a system of rigid bodies, spring masses, and stiffness [119, 646]. Work by Graham and Rodriguez [189], Westergaard [664], Housner [240–242], Werner and Sundquist [663], Jacobsen, Hoskins and Ayre [238, 263, 265] has resulted in the simple mechanical analogy illustrated in Figure 1.20 in the case of a rigid tank wall. As discussed in Section 5.1, with flexible walls, the deformability of the shell is such that the natural cantilever period of the tank, acting with the fluid inertia mass, is significant and furthermore the flexibility of the tank walls also interacts with the sloshing behaviour, modifying the response. Usually the mechanical model parameters, between the equivalent and the actual/real system, are determined and based on the following conditions [251]:

- the equivalent masses and moments of inertia must be preserved between the equivalent and the actual/real system;
- the center of gravity must be the same for small oscillations;
- the equivalent system must possess the same modes of vibration and produce the same damping forces if compared with the real system;
- the forces (shear) and moments under a certain excitation at the base of the equivalent system must be equivalent to that produced by the actual system.

When the walls of the tank in Figure 1.20 accelerate back and forth, a certain fraction of the liquid, usually called the impulsive mass m_0 or m_i, is forced to participate in this motion and exerts a reactive force on the tank. The mass m_0 is rigidly linked to the wall at a height h_0 or h_i so that the horizontal force exerted by it is collinear with the resultant force exerted by the equivalent impulsive fluid [242]. A complete mechanical analogy for transverse sloshing must include an infinite number of oscillating

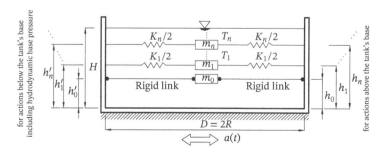

Figure 1.20 Equivalent mechanical model for response of contents of a rigid tank: impulsive and convective motions. Source: From [150]/American Society of Civil Engineers.

masses $m_1, m_2, \ldots, m_n, \ldots$, one for each of the infinity of normal sloshing modes, connected to the tank walls by springs of stiffness [88, 100, 646]:

$$K_n = 4\pi^2 \frac{m_n}{T_n^2} \tag{1.37}$$

The flexibility of the attached masses represents the various antisymmetric slosh frequencies of the fluid in the tank associated with vertical displacements. The sum of the impulsive and convective masses $(m_0, m_1, m_2, \ldots, m_n, \ldots)$ is equal to the total mass of the liquid m. As shown by Malhotra in [370], for most tanks in the range $0.3 < \gamma = \frac{H}{R} < 3$ (from shallow to very slender containers), the impulsive mode m_0 and the first convective mode m_1 together account for 85–98% of the total liquid mass in the tank. The results obtained using only the first impulsive and first convective modes are considered satisfactory in most cases. However, due to the fact that the sloshing liquid determines the freeboard requirements, to estimate fluid displacements it has been shown that higher modes should be considered as highlighted in Section 2.2.8 [377]. The behaviour of tanks containing fine granular material, such as cement or grains, may be expected to have very limited convective response. Furthermore many silos containing fine granular materials have high aspect ratios (usually $\frac{H}{R} > 2$) and for these containers the behaviour can be well approximated to that of similar tanks containing fluid of the same density as the granular material, since the fluid response of such tanks is dominated by impulsive motions (Point 1.4, NZSEE, 2009 Edition).

1.8 Discussion of the Existing Codes

Table 1.2 lists various international codes [659], standards, guidelines and recommendations, in order to highlight the main details on the types of tanks considered in each of the following documents [252, 270]:

- UNI EN 1998-4:2006 *Design of structures for earthquake resistance. Part 4: silos, tanks and pipelines* in the following abbreviated as a EC8-4. This standard specifies principles, analyses, and design rules for the seismic evaluation of the structural aspects of industrial facilities composed of above-ground and buried pipelines and of storage tanks and silos of different types and uses. The standard includes criteria and rules without restriction on sizes, structural materials (reinforced concrete, prestressed concrete, steel), types (above-ground, underground, elevated, rigid, flexible, cylindrical and rectangular, anchored and unanchored/uplifting), stored materials (either fluid or granular) and soil–structure interaction. Furthermore, UNI EN 14015:2006 is a specification for the design and manufacture of site-built, vertical, cylindrical, flat-bottomed, above-ground, welded, steel tanks for the storage of liquids at ambient temperature and above (mainly for static loads). Annex G of the UNI EN 14015:2006 gives recommendations for the seismic design of storage tanks and is mainly based on the requirements of Appendix E of API 650.
- *Seismic Design of Storage Tanks, Recommendations of a Study Group of the New Zealand National Society for Earthquake Engineering*, first edition, December 1986 by Prof. M.J.N. Priestley and then revised in November 2009 by Prof. D. Whittaker. The recommendation, for the sake of brevity denoted as NZSEE-09, has been revised to bring it in line with the New Zealand loading code NZS 1170.5 (2004) "Structural design actions" [667]. The outline of the procedure proposed by the study group is given by Whittaker and Jury in [666]. Herein the same tanks described in EC8-4 are considered and deeply investigated.

Table 1.2 Main categories of tanks considered in earthquake codes and standards.

Document	Year	Nation	Support[a]	Material[b]	Tank's type[c]	Shape[d]
EC8-4	2006	Europe	A, U	S, RC, P	1, 2, 3, 4, 5	R, C
UNI EN 14015	2006	Europe	A, U	S	1	C
NZSEE	2009	New Zealand	A, U	S, RC, P	1, 2, 3, 4, 5	R, C
ACI 350.3	2006	USA	A, U	RC, P	1, 2, 3	R, C
ACI 371.R	1998	USA	A	S	3	C
API 650	2012	USA	A, U	S	1, 5	C
API 650	2012	USA	A, U	Al	1, 5	C
API 620	2012	USA	A, U	S	1, 5	C
ASCE 7	2010	USA	A, U	S, RC, P	1, 2, 3, 4	R, C
AWWA D100	2011	USA	A, U	S	1, 3, 4	C
AWWA D103	2009	USA	A, U	S	1	C
AWWA D110	2004	USA	A, U	P	1, 2	C
AWWA D115	2006	USA	A, U	P	1, 2	C
IITK-GSDMA	2007	India	A, U	S, RC, P	1, 2, 3, 4	R, C
COVENIN 3623	2000	Venezuela	A, U	S	1	C
AIJ	2010	Japan	A, U	S, RC, P	1, 2, 3, 4	C

a) A = anchored or mechanically-anchored, U = unanchored or self-anchored or uplifting.
b) S = steel, RC = reinforced concrete, P = prestressed concrete, Al = Aluminium or aluminium alloys.
c) 1 = above-ground rigid tanks, 2 = underground rigid tanks, 3 = elevated tanks on shaft-type tower, 4 = elevated tanks on frame-type tower, 5 = flexible tanks.
d) C = cylindrical, R = rectangular.

- ACI (American Concrete Institute), *Seismic Design of Liquid-Containing Concrete Structures and Commentary*, (ACI 350.3-06) prescribes procedures for the seismic analysis and design of liquid-containing concrete structures: rectangular and circular, on-grade and below grade. Concrete pedestal-mounted structures and composite-style elevated water tanks, that consist of a steel water storage tank supported by a cylindrical reinforced concrete pedestal, are covered by ACI 371R-98, *Guide for the Analysis, Design and Construction of Concrete-Pedestal Water Towers*.
- API 650 2012 Edition: "Welded tanks for oil storage", American Petroleum Institute Standards, Washington, DC. This standard establishes minimum requirements for material, design, fabrication, erection, and testing for vertical, cylindrical, above-ground, closed- and open-top, welded storage tanks, used in the storage of petroleum and petroleum products, in various sizes and capacities for internal pressures approximating atmospheric pressure (internal pressure not exceeding the weight of the roof plates). Appendix E provides requirements for the design of welded on-grade steel flat-bottom tanks that may be subject to seismic ground motion. The Annex G of the European Standard UNI EN 14015:2006 gives recommendations for the seismic design of storage tanks and is mainly based on the requirements of Appendix E of API 650; API 620 2012 Edition covers a higher

pressure (not more than $15\,lbf/in^2$) and temperature (not greater than 250°F) rating than API 650 (atmospheric storage tanks). Appendix L provides minimum requirements for the design of welded storage tanks that may be subject to seismic action according to Appendix E of API 650. API 12B 2009 covers material, design, fabrication, and testing requirements for vertical, cylindrical, above-ground, closed- and open-top, bolted steel storage tanks in various sizes and capacities for internal pressures approximately atmospheric.

- AWWA D100 2011 Edition: "Welded carbon steel tanks for water storage" and AWWA D103 Edition 2009 "Factory-coated bolted carbon steel tanks for water storage", American Water Works Association. They provide minimum requirements for the design, construction, inspection, and testing of new cylindrical, welded and factory-coated bolted steel tanks for the storage of water at atmospheric pressure. The AWWA D115 2006 Edition "Tendon-prestressed concrete water tanks" describes recommended practice for the design, construction, and field observation of concrete tanks using tendons for prestressing. Instead AWWA D110 Edition 2004 "Wire and strand-wound, circular, prestressed concrete water tanks" applies to circular, prestressed concrete water-containing structures with the following four types of core walls: (1) cast-in-place concrete with vertical prestressed reinforcement (2) shotcrete with a steel diaphragm acting as a water barrier and assuring water tightness (3) precast concrete with a steel diaphragm; and (4) cast-in-place concrete with a steel diaphragm.
- ASCE 7 2010 Edition includes nonbuilding structures that may be required to resist the effects of an earthquake. Chapter 15 focuses on tanks and vessels: reinforced concrete, prestressed concrete, steel, aluminium, and fibre-reinforced plastic materials.
- IITK-GSDMA (2007) "Guidelines for Seismic Design of Liquid Storage Tanks, Previsions with Commentary and Explanatory Examples" freely downloaded from NICEE, National Information Centre of Earthquake Engineering (Indian Institute of Technology, Kanpur and Gujarat State Disaster Management Authority).[5] Two main documents can also be obtained entitled "Review of Code Provisions on Design Seismic Forces for Liquid Storage Tanks" and "Review of Code Provisions on Seismic Analysis of Liquid Storage Tanks" where IBC, ACI (371R and 350.3), AWWA (D100, D103, D110, D115), API 650, Eurocode 8, Part 4, and NZSEE are reviewed and compared in order to highlight similarities, discrepancies, and limitations in design, analysis, and modelling of tanks.
- COVENIN 3623 2000 Edition, "Diseño sismorresistente de tanques metálicos", is a Natinal Code from Venezuela. This standard establishes minimum requirements for design and analysis of vertical and cylindrical steel storage tanks with internal pressures approximating atmospheric pressure.
- NCh2369 2003 Edition, "Diseño sismico de estructuras e instalaciones (industriales)" (Seismic design for industrial structures and facilities), is a National Code from Chile. This standard at Chapter 11.7 and 11.8 focuses on above-ground and elevated tanks by referring to the American standard introduced above.
- AIJ (Architectural Institute of Japan), *Design Recommendation for Storage Tanks and their Supports with Emphasis on Seismic Design*, 2010 Edition. This Design Recommendation is applied to the structural design, mainly the seismic design, of water storage tanks, silos, spherical storage tanks (pressure vessels), flat-bottomed, cylindrical, above-ground storage tanks, and underground storage tanks.

---

5 All the documents may be obtained from http://www.nicee.org/IITK-GSDMA_Codes.php from the National Information Center of Earthquake Engineering in India.

Though the seismic analyses of liquid-containing tanks are retained by various codes of practices, their implementation strategy is rather varied, leading to significantly different design forces in some cases. Nine documents (IBC 2006 Edition, Eurocode 8, Part 4 - 1998 Edition, NZSEE, ACI 350.3 2001 Edition, ACI 371 1998 Edition, AWWA D-100-05, AWWA D-110-95, AWWA D-115-95, and API 650 - 2005 Edition) have been reviewed and compared by Jaiswal et al. [270].

1.9 Structure of the Book

The response quantities related to the equivalent model depicted in Figure 1.20 [250, 451] and the seismic design codes compared in Table 1.2 are investigated corresponding to the following main typologies [200] (Point 4.2, HSE Research Report RR527):

1) above-ground anchored rigid tanks: cylindrical (Section 2.2) and rectangular (Section 2.3);
2) above-ground unanchored tanks: cylindrical (Section 3.2) and rectangular (Section 3.3);
3) underground rigid tanks (Section 6.5);
4) elevated tanks on a shaft and frame-type tower (Section 4.1);
5) flexible tanks (Section 5.1).

Particular sections will be introduced in order to highlight some important aspects regarding:

1) effects of liquid viscosity (Section 2.2.6);
2) effects of inhomogeneous liquids (Section 2.2.7);
3) effects of soil-structure interaction (Section 6.2);
4) flow-dampening device systems (Section 6.3);
5) isolation systems (Section 6.4).

2

Above-Ground Anchored Rigid Tanks

After reading this chapter you should be able to:

___

- Understand the equivalent mechanical model to analyse anchored tanks
- Evaluate impulsive and convective modes and masses for tanks with rigid walls
- Identify the critical components in a tank's design

___

2.1 Introduction

This section provides information on seismic analysis procedures for rigid tanks subjected to horizontal and vertical action, mainly using EC8-4, NZSEE, and USA Standards, having the following characteristics:

- upright cylindrical shape (Section 2.2) and rectangular cross-section (Section 2.3);
- rigid foundation (Sections 2.2 and 2.3) and flexible foundation or soil–structure interaction (Section 6.2);
- fully anchored at the base (Sections 2.2 and 2.3) and unanchored (Section 3.1).

The procedures presented in the following consider a theoretical and analytical approach to treating the seismic design of liquid storage rigid containments, based on the seminal papers by Jacobsen [263], Housner [240], Epstein [150] and Wozniak and Mitchell [672]. Although such provisions represent a significant advantage in the design of tanks, EC8-4 in Annex A, Point A.1 explicitly highlights that the dynamic interaction between the fluid, the tank walls, and the underlying foundation soil, is a problem of considerable complexity usually requiring high computational resources and software such as LS-DYNA [351] using the Arbitrary Lagrangian Eulerian (ALE) method [132, 650].

2.2 Vertical Cylindrical Tanks Fully Anchored at the Base

The basic system investigated is shown in Figure 2.1. It is a ground-supported, vertical, circular tank of radius R fixed to a rigid base and to be excited horizontally. The tank is filled with a fluid of mass density

Seismic Design and Analysis of Tanks, First Edition. Gian Michele Calvi and Roberto Nascimbene.
© 2023 John Wiley & Sons, Inc. Published 2023 by John Wiley & Sons, Inc.

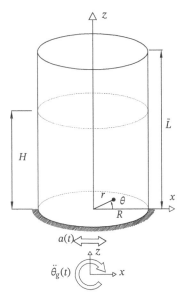

Figure 2.1 Tank-liquid geometry investigated: coordinate system (r, θ, z), ground acceleration $a(t)$ and rigid base rotation $\ddot{\theta}_g(t)$ [204, 330, 337].

ρ_l (kg/m^3) to a level H. The upper surface of the liquid is considered to be free. The acceleration at any time t is denoted by $a(t)$ and the rigid base rotation by $\ddot{\theta}_g(t)$ [204, 215, 221, 330, 337, 498, 593]. The locations of points in the liquid and tank are univocally defined by the cylindrical coordinate system (r, θ, z), with the origin taken at the centre of the tank base and $\theta = 0$ taken in the direction of the horizontal excitation. Two main nondimensional coordinates are then introduced as $\xi = \frac{r}{R}$ and $\zeta = \frac{z}{H}$. This system is first considered under the assumption that the motion experienced by the base tank is the same as the free-field ground motion. Furthermore, it is assumed that the liquid is initially at rest, incompressible, and inviscid, that the flow $\Phi(r, \theta, z, t)$ is irrotational, and that all structural and liquid motions remain within the linearly elastic range of response (no separation or cavitation between the liquid and the tank). Under these assumptions the fluid flow must satisfy the Laplace's equation subject to the appropriate boundary conditions along the tank wall $(r = R)$, base $(z = 0)$, and the free liquid surface $(z = H)$ [247, 251, 263, 626, 638]:

$$\frac{\partial^2 \Phi}{\partial r^2} + \frac{1}{r}\frac{\partial \Phi}{\partial r} + \frac{1}{r^2}\frac{\partial^2 \Phi}{\partial \theta^2} + \frac{\partial^2 \Phi}{\partial z^2} = \nabla^2 \Phi = 0 \tag{2.1}$$

Then, the component of the liquid velocity v_n in the direction of a generalized n-coordinate for an arbitrary point at a certain time t, and the hydrodynamic pressure p at the same point and time, are given respectively by [263]:

$$v_n = -\frac{\partial \Phi}{\partial n} \tag{2.2a}$$

$$p = \rho_l \frac{\partial \Phi}{\partial t} \tag{2.2b}$$

The solution of Equation (2.1) may be expressed as the sum of two separate components, previously defined in Section 2.1 as rigid impulsive and convective. The first component exactly satisfies the actual boundary conditions along the tank walls and tank bottom, and the condition of zero hydrodynamic pressure at the original position of the free surface of the fluid in the static situation:

$$\left.\frac{\partial \Phi_i}{\partial z}\right|_{z=0} = 0 \qquad\qquad \text{tank base} \qquad (2.3a)$$

$$\left.\frac{\partial \Phi_i}{\partial r}\right|_{r=R} = -v(t)\cos\theta \qquad\qquad \text{tank wall} \qquad (2.3b)$$

$$\left.\frac{\partial \Phi_i}{\partial t}\right|_{z=H} = 0 \qquad\qquad \text{liquid surface} \qquad (2.3c)$$

where the radial velocity component of the fluid in the case of a rigid tank must be equal to the corresponding component $v(t)$ of the ground motion $a(t)$; the subscript $_i$ stands for the impulsive component; Equation (2.3a) requires that the vertical component of the liquid velocity must be zero for a horizontally excited tanks, Equation (2.3b) specifies that the radial velocity component of the fluid and the tank wall must be the same motion as the ground, and Equation (2.3c) implies that the pressure at the free surface is zero. The second component (convective denoted by the subscript $_c$) accounts for the effects of the surface waves associated with the sloshing action of the liquid. The convective term does not alter relations (2.3) that are already satisfied, while fulfilling the correct equilibrium condition at the free surface [263, 680]:

$$\left.\frac{\partial \Phi_c}{\partial z}\right|_{z=0} = 0 \qquad\qquad \text{tank base} \qquad (2.4a)$$

$$\left.\frac{\partial \Phi_c}{\partial r}\right|_{r=R} = 0 \qquad\qquad \text{tank wall} \qquad (2.4b)$$

$$\left.\left(\frac{\partial^2 \Phi_c}{\partial t^2} + g\frac{\partial \Phi_c}{\partial z}\right)\right|_{z=H} = -g\left.\frac{\partial \Phi_i}{\partial z}\right|_{z=H} \qquad \text{liquid surface} \qquad (2.4c)$$

in which g is the gravitational acceleration.

2.2.1 Impulsive Pressure Component

The impulsive solution $\Phi_i(\xi,\zeta,\theta,t) = C_i(\xi,\zeta)Hv(t)\cos\theta$, obtained by the method of separation of variable [680], can be substituted in Equation (2.2b) to obtain the spatial-temporal variation (ξ,ζ,θ,t) of the rigid impulsive pressure (Point A.2.1.2, EC8-4 and Point 4.3.2 in HSE Research Report RR527):

$$p_i(\xi,\zeta,\theta,t) = C_i(\xi,\zeta)\rho_l Ha(t)\cos\theta \qquad (2.5)$$

where $a(t)$ is the free-field ground acceleration, with (design ground; Point 3.2.1(2), Eurocode 8, Part 1 and Section 1.6) peak value denoted by a_g (Equation (1.36)) and C_i is given by [408, 645][1]:

$$C_i(\xi,\zeta) = 2\sum_{n=0}^{\infty} \frac{(-1)^n}{v_n^2} \frac{I_1\left(\frac{v_n}{\gamma}\xi\right)}{I_1'\left(\frac{v_n}{\gamma}\right)}\cos(v_n\zeta) \qquad (2.6a)$$

$$v_n = \frac{2n+1}{2}\pi \qquad (2.6b)$$

$$\gamma = \frac{H}{R} \qquad (2.6c)$$

---

1 According to Veletsos in [626], the minus sign in front of the right side of expression (2.6a) has been omitted for simplicity.

The maximum value of the impulsive pressure p_i may then be determined by replacing $a(t)$ by its spectral value S_e using the natural frequency and damping value according to Equations (2.18) and (2.19) (Point C3.3.1, NZSEE-09) [635]. In Equation (2.6a), the symbols $I_1(\cdot)$ and $I_1'(\cdot)$ denote respectively the modified Bessel function of the first kind of order 1 and its first derivative. Usually Bessel functions of a higher order can be expressed by Bessel functions of lower orders using the recurrences formulas such as $I_1'(x) = \frac{dI_1(x)}{dx} = I_0(x) - \frac{I_1(x)}{x}$, where I_0 is the modified Bessel function of the first kind of order 0 [561]:

$$I_0(x) = 1 + \frac{x^2}{2^2} + \frac{x^4}{2^2 \cdot 4^2} + \frac{x^6}{2^2 \cdot 4^2 \cdot 6^2} + \cdots \tag{2.7a}$$

$$I_1(x) = \frac{x}{2} + \frac{x^3}{2^2 \cdot 4} + \frac{x^5}{2^2 \cdot 4^2 \cdot 6} + \frac{x^7}{2^2 \cdot 4^2 \cdot 6^2 \cdot 8} + \cdots \tag{2.7b}$$

Relation (2.5) highlights that the temporal variation of the impulsive pressure is the same as that of the ground acceleration $a(t)$. According to the mechanical analogy, introduced in Section 1.7, this pressure component may be visualized as being due to part of the liquid vibrating synchronously with the tank wall as a rigidly attached mass (see Figure 1.20). Function $C_i(\xi, \zeta)$ in Equation (2.6a) is displayed in Figure 2.2 for four different values of $\gamma = \frac{H}{R}$. Figure 2.2(a) for $\xi = 1$ at the wall of the tank and $\cos\theta = 1$ (horizontal seismic direction) shows the variation of p_i normalized to $\rho_l Ra(t)$ [646]. Note that, irrespective of the value of γ involved, the maximum value of C_i occurs at the base, and that the latter value increases rapidly with $\gamma = \frac{H}{R}$ approaching values greater then unity in the case of slender and tall tanks. Furthermore, the more slender the tank ($\gamma = 5$), the more nearly uniform is the distribution of C_i; this means, according to Point 14.4.7.5.1, FEMA 450, that in the case of $\gamma > 4$, the overturning begins to approach "rigid mass" behaviour (the sloshing mass is really small). It follows that the entire volume of the liquid in the lower part of a slender tank moves as a rigidly attached mass, whereas a small fraction of the liquid in a large tank ($\gamma = 0.5$) behaves in this manner. Figure 2.2(b) shows the radial variation of p_i on the tank bottom. As $\frac{H}{R} \to \infty$ the ratio $\frac{I_1(x)}{I_1'(x)} = x$ [680] and expression (2.5) reduces to:

$$p_i(\xi, \zeta, \theta, t) = \rho_l ra(t) \cos\theta \tag{2.8}$$

which states that the hydrodynamic pressure becomes linear along the radial direction (Figure 2.2(b)) and there is no variation along the z direction. As $\frac{H}{R} \to 0$ the ratio $\frac{I_1}{I_1'} \to 1$ and relations (2.5) and (2.6a)

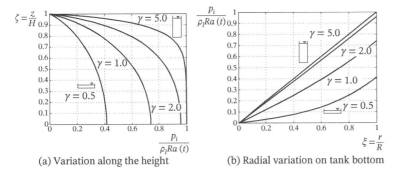

(a) Variation along the height   (b) Radial variation on tank bottom

Figure 2.2 Coefficient C_i for impulsive pressure variation, normalized to $\rho_l Ra(t)$, in rigid tanks: four values as a function of the tank aspect ratio $\gamma = \frac{H}{R}$. Source: Adapted from [206].

reduces to the expression given by Chopra in [87] for the hydrodynamic pressure exerted by the fluid on a rigid dam.

The resultant base horizontal shear $Q_i(t)$ is obtained by integrating the impulsive component of the hydrodynamic wall pressure, p_i given by Equation (2.5), over the vertical wetted surface of the tank [222]:

$$Q_i(t) = \int_0^{2\pi} \int_0^H p_i(R, z, \theta, t) R \cos\theta \, dz \, d\theta = m_i a(t) \tag{2.9}$$

in which m_i is the impulsive mass that is the portion of the total contained mass of the fluid, $m = \rho_l \pi R^2 H$, that may be considered to move synchronously with the tank [645]:

$$m_i = 2m\gamma \sum_{n=0}^{\infty} \frac{I_1\left(\dfrac{\nu_n}{\gamma}\right)}{\nu_n^3 I_1'\left(\dfrac{\nu_n}{\gamma}\right)} \tag{2.10}$$

According to Figure 2.3, a clear distinction must be made between the hydrodynamic moment $M_i'(t)$ immediately below the tank bottom, induced on the foundation, and the hydrodynamic moment $M_i(t)$ induced on a section of the tank wall immediately above the base (Point 14.4.7.5.2, FEMA 450). The moment $M_i(t)$ is due only to the pressures exerted on the tank wall (Figure 2.3(a)), whereas the moment $M_i'(t)$ also incorporates the contribution of the pressure on the tank base (Figure 2.3(b)). Let $\Delta M_i(t)$ denote the moment increment induced by the pressure on the tank base, the total foundation moment $M_i'(t)$ can then be written as [222]:

$$M_i'(t) = M_i(t) + \Delta M_i(t) = \underbrace{\int_0^{2\pi} \int_0^H p_i(R, z, \theta, t) R z \cos\theta \, dz \, d\theta}_{\text{wall contribution}}$$

$$+ \underbrace{\int_0^{2\pi} \int_0^R p_i(r, 0, \theta, t) r^2 \cos\theta \, dr \, d\theta}_{\text{base contribution}} = m_i h_i a(t) + m_i \Delta h_i a(t) = m_i h_i' a(t) \tag{2.11}$$

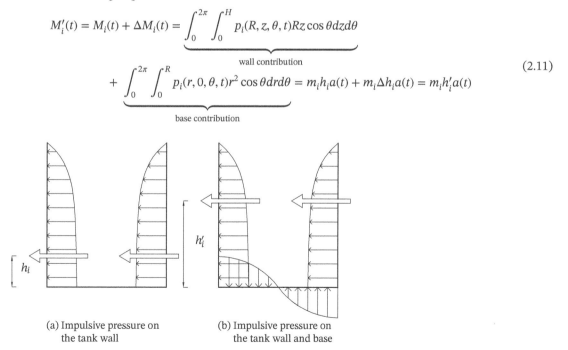

(a) Impulsive pressure on the tank wall

(b) Impulsive pressure on the tank wall and base

Figure 2.3 Impulsive hydrodynamic pressures exerted on the wall and the base of the tank.

The total foundation moment may conveniently be expressed in the form:

$$M_i'(t) = m_i h_i' a(t) \tag{2.12}$$

in which h_i' represents the height at which the impulsive component of the base shear $Q_i(t)$ must be applied to yield the moment immediately below the tank bottom [626, 680]:

$$h_i' = H \frac{\dfrac{1}{2} + 2\gamma \displaystyle\sum_{n=0}^{\infty} \frac{I_1(\nu_n/\gamma)}{\nu_n^4 I_1'(\nu_n/\gamma)}(\nu_n + 2(-1)^{n+1})}{2\gamma \displaystyle\sum_{n=0}^{\infty} \frac{I_1(\nu_n/\gamma)}{\nu_n^3 I_1'(\nu_n/\gamma)}} = h_i + \Delta h_i \tag{2.13}$$

The impulsive moment M_i immediately above the tank bottom is given by:

$$M_i(t) = m_i h_i a(t) \tag{2.14}$$

where [626, 680]:

$$h_i = H \frac{\displaystyle\sum_{n=0}^{\infty} \frac{(-1)^n I_1(\nu_n/\gamma)}{\nu_n^4 I_1'(\nu_n/\gamma)}(\nu_n(-1)^n - 1)}{\displaystyle\sum_{n=0}^{\infty} \frac{I_1(\nu_n/\gamma)}{\nu_n^3 I_1'(\nu_n/\gamma)}} \tag{2.15}$$

The increment of the base moment $M_i(t)$ due to the hydrodynamic pressure exerted on the tank base can be expressed as [626]:

$$\Delta M_i(t) = m_i \Delta h_i a(t) \tag{2.16}$$

where:

$$\Delta h_i = H \frac{\displaystyle\sum_{n=0}^{\infty} \frac{(-1)^n I_2(\nu_n/\gamma)}{\nu_n^3 I_1'(\nu_n/\gamma)}}{\gamma \displaystyle\sum_{n=0}^{\infty} \frac{I_1(\nu_n/\gamma)}{\nu_n^3 I_1'(\nu_n/\gamma)}} \tag{2.17}$$

in which I_2 denotes the modified Bessel function of the first kind of order 2.

The quantity m_i, h_i', h_i and Δh_i are displayed graphically in Figure 2.4 as a function of the nondimensional ratio $\gamma = \frac{H}{R}$ and listed for several different values of $\frac{H}{R}$ by Veletsos in [626]. The variation of the rigid impulsive mass is shown in Figure 2.4(a): the proportion of the liquid mass that participates in the impulsive action increases with increasing γ tending asymptotically to the total mass m. The impulsive height h_i (Figure 2.4(b)) varies within a relatively narrow range, whereas h_i' varies widely. Note that for squat tanks h_i' exceeds significantly the total liquid depth H, indicating that the base pressure is a major contributor to the foundation moment M_i'.

According to the NZSEE-09 Recommendations, the maximum impulsive pressure component on the walls of a tank can be written as a spatial function of (z, θ) (Eq. (C3.3)–(C3.4) Point C3.3.1, NZSEE-09) [5]:

$$p_i(z, \theta) = q_0(z) \frac{S_e(T_0)}{g} \gamma_l R \cos\theta \tag{2.18}$$

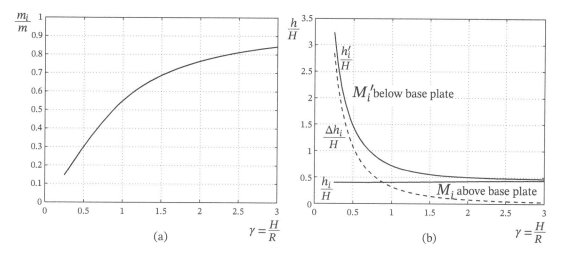

Figure 2.4 Nondimensional impulsive mass m_i/m (a) and heights h_i/H, h_i'/H and Δh_i as a function of γ (b).

or analogously:

$$p_i(z, \theta) = q_0'(z)\frac{S_e(T_0)}{g}\gamma_l H \cos\theta \tag{2.19}$$

which corresponds to Equation (2.5) from EC8-4 by using:

$$q_0'(z)\big|_{\sum_{n=1}^{\infty}} = C_i(1, \zeta)\big|_{\sum_{n=0}^{\infty}} \tag{2.20}$$

In Equations (2.18) and (2.19), γ_l is the unit weight of the liquid (in kN/m³); $S_e(T_0)$ is the spectral acceleration obtained from a horizontal elastic response spectrum corresponding to the impulsive period with or without soil–structure interaction (Section 2.2.5) and to the equivalent viscous damping level for horizontal mode consistent with the limit state (Point 2.3.3.1, EC8-4 and Point 4.1.3(1), EN 1998-2:2005; Point 3.4.3, ASCE/SEI 43-05) [25], tank type (bolted and welded steel or reinforced and prestressed concrete [25]; Point 4.1.3(1), EN 1998-2:2005) and interaction with the soil (Section 6.2); $q_0(z)$ or $q_0'(z) = q_0(z)\frac{R}{H}$ are the dimensionless impulsive pressure functions of the height variable z and usually for containers of very small tank aspect ratios $\gamma = \frac{H}{R}$, it is more instructive to use expression (2.19) with $q_0'(z)$. The function $q_0(z)$, with r taken equal to R, has been defined by Yang in [680]:

$$q_0(z) = \frac{H}{R}\sum_{n=1}^{\infty}\frac{8(-1)^{n+1}}{[(2n-1)\pi]^2}\frac{I_1\left[(2n-1)\frac{\pi}{2}\frac{R}{H}\right]}{I_1'\left[(2n-1)\frac{\pi}{2}\frac{R}{H}\right]}\cos\left[(2n-1)\frac{\pi}{2}\frac{z}{H}\right] = \frac{H}{R}q_0'(z) \tag{2.21}$$

and it is displayed in Figure 2.5(a) normalized with respect to the value of $q_0(z)$ at $z = 0$. Note that, irrespective of the value of the tank aspect ratio $\gamma = \frac{H}{R}$ involved, the maximum value of $q_0(z)$ occurs at the base and, in line with Figure 2.2(a), the more slender the tank, the more nearly uniform is the impulsive pressure function distribution along z. The values of $q_0(z)$ or $q_0'(z)$ for $z = 0$ are shown in Figure 2.5(b)

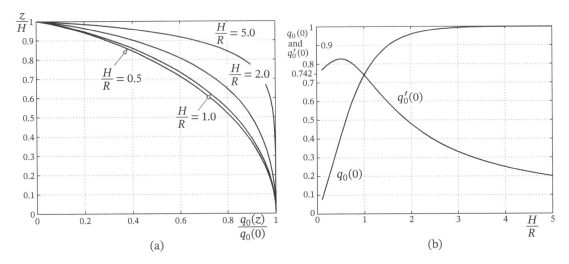

Figure 2.5 (a) Vertical distribution of the dimensionless impulsive pressure function $q_0(z)$ normalized with respect to the value of $q_0(z)$ at $z = 0$; (b) peak value of the dimensionless impulsive pressure functions at the base of the tank. Source: From [60]/American Society of Civil Engineers.

where we can observe that as $\frac{H}{R} \to 0$ the coefficient $q_0'(z) \to 0.742$ and relation (2.19) coincide with the expression for the hydrodynamic pressure exerted by the impounded liquid at the base of a horizontally excited, straight, and rigid wall proposed by Bustamante and Flores in [60]:

$$p_i(0,0) = 0.742 \frac{S_e(T_0)}{g} \gamma_l H \tag{2.22}$$

The rigid impulsive base shear Q_0 (Equation (2.9)) and the impulsive moment just above the base plate M_0 (Equation (2.14)) are given by (NZSEE-09, Eq. (C3.9)–(C3.10), Point C3.3.1):

$$Q_0 = S_e(T_0)m_0 \tag{2.23a}$$

$$M_0 = Q_0 h_0 \tag{2.23b}$$

where the mass m_0 can be obtained from relation (2.10) or Figure 2.4(a) and the height h_0 from Equation (2.15) or Figure 2.4(b).

The rigid impulsive moment M_0' below the base plate (Equation (2.12)) and including the base pressure component is given by (NZSEE-09, Eq. (C3.11), Point C3.3.1)):

$$M_0' = Q_0 h_0' \tag{2.24}$$

where the height h_0' (Figure 2.3(b)) at which the impulsive mass is placed to give the overturning moment, arising from the pressure on both the wall and the base, is given by expression (2.13) or Figure 2.4(b).

According to the relations originally derived by Housner in [240–242], the dynamic equivalent model (m_0, h_0, h_0'), defined in Chapter 9, ACI 350.3-06, Point 13.5, AWWA D100-11 and Point 14.3, AWWA

D103-09, can be computed in accordance with the following expressions to be used in Equations (2.23) and (2.24) [150, 344]:

- welded and bolted steel tanks (AWWA D100-11, Points 13.5.2.2.1/2 and 13.5.3.2.2; AWWA D103-09, Points 14.3.2.2.1/2 and 14.3.3.2.2; Point 8, IAEA-TECDOC-1347):

$$\frac{m_0}{m} = \frac{\tanh\left(\sqrt{3}\frac{R}{H}\right)}{\sqrt{3}\frac{R}{H}} \qquad \text{for} \quad \frac{D}{H} \geq 1.333 \qquad (2.25a)$$

$$\frac{m_0}{m} = 1 - 0.218\frac{D}{H} \qquad \text{for} \quad \frac{D}{H} < 1.333 \qquad (2.25b)$$

$$\frac{h_0}{H} = \frac{3}{8} \qquad \text{for} \quad \frac{D}{H} \geq 1.333 \qquad (2.25c)$$

$$\frac{h_0}{H} = 0.5 - 0.094\frac{D}{H} \qquad \text{for} \quad \frac{D}{H} < 1.333 \qquad (2.25d)$$

$$\frac{h_0'}{H} = \frac{3}{8}\left[1 + \frac{4}{3}\left(\frac{\sqrt{3}\frac{R}{H}}{\tanh\left(\sqrt{3}\frac{R}{H}\right)} - 1\right)\right] \qquad \text{for} \quad \frac{D}{H} \geq 1.333 \qquad (2.25e)$$

$$\frac{h_0'}{H} = 0.5 + 0.06\frac{D}{H} \qquad \text{for} \quad \frac{D}{H} < 1.333 \qquad (2.25f)$$

- liquid-containing concrete tanks (ordinary reinforced and prestressed structures) use Equation (2.25a) applied $\forall\frac{D}{H}$ and expression (2.25c) whereas relations (2.25d)–(2.25f) are modified in the following way (ACI 350.3-06, Point 9.3):

$$\frac{h_0}{H} = 0.5 - 0.09375\frac{D}{H} \qquad \text{for} \quad \frac{D}{H} < 1.333 \qquad (2.26a)$$

$$\frac{h_0'}{H} = \frac{\sqrt{3}\frac{R}{H}}{2\tanh\left(\sqrt{3}\frac{R}{H}\right)} - \frac{1}{8} \qquad \text{for} \quad \frac{D}{H} \geq 0.75 \qquad (2.26b)$$

$$\frac{h_0'}{H} = 0.45 \qquad \text{for} \quad \frac{D}{H} < 0.75 \qquad (2.26c)$$

In Equations (2.25), and (2.26), m is the total mass of the fluid contained in the tank and $D = 2R$ is the diameter. Relations (2.25) can be found also at Points E.6.1.1 and E.6.1.2.1/2 API 650 2012 Edition.

2.2.2 Convective Pressure Component

The spatial-temporal variation of the convective solution $\Phi_c(\xi, \zeta, \theta, t)$, which satisfies Equations (2.1) and (2.4), using the method of the separation of the variable [680], can be derived by applying to the response for a harmonic ground acceleration, the inverse Fourier Transform and the convolution integral [408, 626, 680] (Point A.2.1.3, EC8-4 and Point 4.3.1, HSE Research Report RR527):

$$p_c(\xi, \zeta, \theta, t) = \rho_l \cos\theta \sum_{n=1}^{\infty} \psi_n \cosh(\lambda_n\gamma\zeta) J_1(\lambda_n\xi) a_{cn}(t) \qquad (2.27)$$

where the infinite series represents the infinite modes of sloshing of the fluid; ρ_l (in kg/m^3) is the mass density of the liquid, γ can be derived from Equation (2.6c), J_1 is the Bessel function of the first kind of order 1 [561] and numbers λ_n are the $n-$roots of the first derivative of J_1. The first ten of these roots are $\lambda_1 = 1.8412$, $\lambda_2 = 5.3314$, $\lambda_3 = 8.5363$, $\lambda_4 = 11.7060$, $\lambda_5 = 14.8636$, $\lambda_6 = 18.0155$, $\lambda_7 = 21.1644$, $\lambda_8 = 24.3113$, $\lambda_9 = 27.4571$ and $\lambda_{10} = 30.6019$ [23, 221, 596, 646]. The function ψ_n reads:

$$\psi_n = \frac{2R}{(\lambda_n^2 - 1)J_1(\lambda_n)\cosh(\lambda_n\gamma)} \tag{2.28}$$

The antisymmetrical modes are depicted in Figure 2.6 by using the slosh wave shape [121, 287, 626]:

$$\frac{2RJ_1(\lambda_n\xi)}{(\lambda_n^2 - 1)J_1(\lambda_n)} \tag{2.29}$$

obtained from Equations (2.27) and (2.28) (and explained later in expression (2.85)). In the case where $n = 1$ the wave has zero amplitude at $r = 0$, a positive peak at one wall, and a negative peak at the other wall; this is usually the fundamental antisymmetric wave. The function $a_{cn}(t)$ in Equation (2.27) represents the instantaneous value of the pseudoacceleration induced by the prescribed free-field ground motion $a(t)$ in a single degree of freedom linear oscillator having the following circular frequency ω_{cn} or natural frequency f_{cn} (Section 2.2.5):

$$\omega_{cn} = \sqrt{g\frac{\lambda_n}{R}\tanh(\lambda_n\gamma)} \quad \text{or} \quad f_{cn} = \frac{1}{2\pi}\sqrt{g\frac{\lambda_n}{R}\tanh(\lambda_n\gamma)} \tag{2.30}$$

and a viscous damping ratio (Section 2.2.6), both equal to those of the $n^{\text{th}}-$sloshing mode of vibration of the liquid in the tank. The maximum value of the convective pressure p_c may then be determined

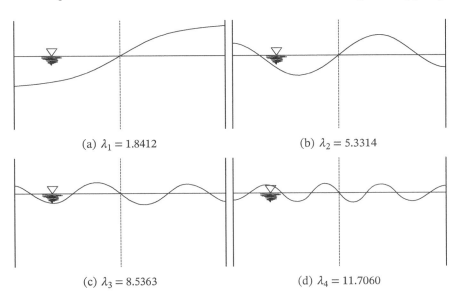

(a) $\lambda_1 = 1.8412$ (b) $\lambda_2 = 5.3314$

(c) $\lambda_3 = 8.5363$ (d) $\lambda_4 = 11.7060$

Figure 2.6 Slosh wave shapes for first four antisymmetrical modes of a cylindrical (Point 4.3.1, HSE Research Report RR527).

by replacing $a_{cn}(t)$ by its spectral value S_e using the natural frequency and damping value according to Equation (2.40) (Point C3.3.1, NZSEE-09) [635].

A complete mechanical analogy for transverse sloshing (see Figure 1.20) must include an n–number of oscillating masses, connecting to n–spring elements of stiffnesses $K_n = \omega_{cn}^2 m_{cn}$ (Equation (1.37)) [119] such that the natural frequency of the nth–mass is the same as the nth–sloshing frequency. The nth–term of the series in Equation (2.27) represents the hydrodynamic wall pressure contributed by the nth–mode of vibration of the fluid. For design purposes, only the first oscillating mode ($n = 1$) needs to be considered as clearly stated in Figure 2.7(a) where the magnitude of the pressures for the latter modes are only a small fraction of those for the fundamental mode. It can be seen from Figure 2.7(a) (using $\theta = 0$ and $\xi = 1$) that the vertical distribution of the convective pressure p_c decays much more rapidly with depth for slender tanks than for shallow tanks. It follows that in squat tanks the convective pressure maintains a relatively high value down to the base of the tank, while in tall tanks, the sloshing effect is confined to the vicinity of the free surface of the fluid. Figure 2.7(b) shows the nondimensional values of the first two lowest frequencies (Equation (2.30)) as a function of the tank aspect ratio $\frac{H}{R}$.

The instantaneous value of the convective base shear is given by (Point A.2.1.3 in EC8-4):

$$Q_c(t) = \sum_{n=1}^{\infty} m_{cn} a_{cn}(t) \tag{2.31}$$

in which m_{cn} is the convective component of the liquid mass associated with the nth–sloshing mode of vibration [419]:

$$m_{cn} = m \frac{2 \tanh(\lambda_n \gamma)}{\lambda_n \gamma (\lambda_n^2 - 1)} \tag{2.32}$$

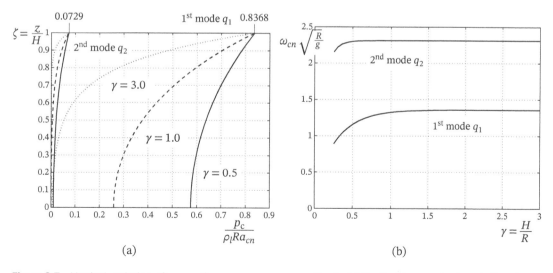

Figure 2.7 Vertical variation of convective pressure corresponding to (a) the first two modes ($n = 1, 2$) and (b) frequency distribution of the first two lowest circular frequencies as a function of γ.

The instantaneous value of the hydrodynamic overturning moment $M'_c(t)$, incorporating the contribution of the pressures on the tank base, at a section immediately below the base plate may conveniently be expressed in the form:

$$M'_c(t) = \sum_{n=1}^{\infty}(m_{cn}a_{cn}(t))h'_{cn} = \sum_{n=1}^{\infty}Q_{cn}(t)h'_{cn} = M_c(t) + \Delta M_c(t) \tag{2.33}$$

where the quantity h'_{cn} can be written as (Figure 2.8(b)) [638]:

$$h'_{cn} = H\left(1 + \frac{2 - \cosh(\lambda_n \gamma)}{\lambda_n \gamma \sinh(\lambda_n \gamma)}\right) = H\left[1 - \frac{1}{\lambda_n \gamma}\left(\tanh\frac{\lambda_n \gamma}{2} - \operatorname{csch}(\lambda_n \gamma)\right)\right] = h_{cn} + \Delta h_{cn} \tag{2.34}$$

In addition to the moment $M_c(t)$ induced by the wall pressure, the foundation moment $M'_c(t)$ includes the effect of the hydrodynamic pressures exerted on the tank base, usually denoted by $\Delta M_c(t)$ and corresponding height Δh_{cn}. The convective moment M_c, arising from pressures on the wall (Figure 2.8(a)), immediately above the tank base is given by:

$$M_c(t) = \sum_{n=1}^{\infty}(m_{cn}a_{cn}(t))h_{cn} = \sum_{n=1}^{\infty}Q_{cn}(t)h_{cn} \tag{2.35}$$

where h_{cn} represents the convective height at which the nth–term of the sloshing component of the base shear must be applied to yield the corresponding moment (Figure 2.8(a)):

$$h_{cn} = H\left(1 + \frac{1 - \cosh(\lambda_n \gamma)}{\lambda_n \gamma \sinh(\lambda_n \gamma)}\right) \tag{2.36}$$

The increment of the base moment $M_c(t)$ due to the hydrodynamic pressure exerted on the tank base can be expressed as [635]:

$$\Delta M_c(t) = \sum_{n=1}^{\infty}(m_{cn}a_{cn}(t))\Delta h_{cn} \tag{2.37}$$

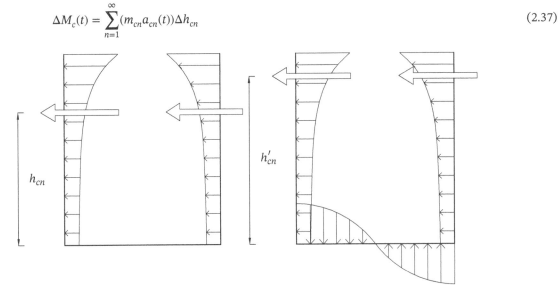

(a) Convective pressure on the tank wall

(b) Convective pressure on the tank wall and base

Figure 2.8 Convective hydrodynamic pressures exerted on the wall and the base of the tank.

where:

$$\Delta h_{cn} = \frac{R}{\lambda_n \sinh(\lambda_n \gamma)} \tag{2.38}$$

The convective masses m_{c1} and m_{c2} [626], normalized with respect to the total liquid mass m, and the heights h'_{c1}, h'_{c2}, h_{c1} and h_{c2} for the first two convective modes are shown in Figure 2.9 as a function of the slenderness ratio γ. By comparing Figures 2.4(a) and 2.9(a), it is observed that whereas the proportion of the impulsive mass increases with increasing γ (slender tank), the convective masses display exactly the opposite trend such that [419, 635]:

$$m_i + \sum_{n=1}^{\infty} m_{cn} = m \tag{2.39}$$

It is further observed that the masses corresponding to the higher sloshing modes of vibration are generally quite small in comparison with the first mode. For the larger values of $\frac{H}{R}$, heights h'_{c1} and h'_{c2} are close to the heights h_{c1} and h_{c2} corresponding to the moment immediately above the tank base. It follows that the contribution to the foundation moment $M'_c(t)$ of the base pressures is not important in this case [635]. By contrast for the shallower tank, small values of $\frac{H}{R}$, the heights h'_{c1} and h'_{c2} exceed the total liquid depth H indicating that the base pressure is a major contributor to the foundation moment in this case.

According to the NZSEE-09 Recommendations, the maximum convective pressure component on the walls of a tank can be written as a spatial function of (z, θ) and corresponding to the first $(n = 1)$ sloshing mode (Eq. (C3.6), Point C3.3.1, NZSEE-09) [5]:

$$p_{c1}(z, \theta) = q_1(z) \frac{S_e(T_1)}{g} \gamma_l R \cos \theta \tag{2.40}$$

which corresponds to Equation (2.27) from EC8-4 by putting $\xi = 1$ and using [626, 645, 680]:

$$q_n(z) = \frac{2 \cosh\left(\lambda_n \frac{z}{R}\right)}{(\lambda_n^2 - 1) \cosh\left(\lambda_n \frac{H}{R}\right)} \tag{2.41}$$

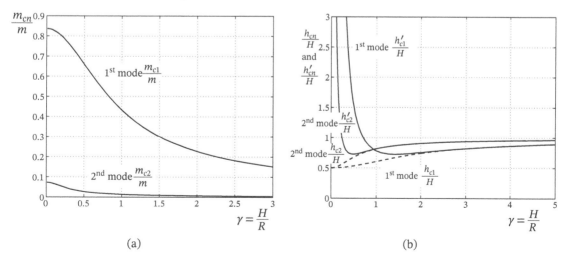

(a) (b)

Figure 2.9 Convective components of liquid mass (m_{c1}, m_{c2}) (a) and heights of convective masses (h_{c1}, h_{c2}, h'_{c1}, h'_{c2}) for the first two modes of sloshing (b).

In Equations (2.40) and (2.41), γ_l is the unit weight of the liquid (in kN/m^3); $S_e(T_1)$ is the spectral acceleration obtained from a horizontal elastic response spectrum corresponding to the convective period of the first mode (Equation (2.54)) and to the viscous damping ratio, as given in Section 2.2.6; for convective periods longer than 4.0 sec, a more complete definition of the elastic displacement spectrum should be used, according to Points 3.2.2.2 (5)P/(6) and Annex A, EC8-4; $q_j(z)$ are dimensionless pressure functions that define the heightwise variations of the nth−convective components of the wall pressures; λ_n are the n−roots of the first derivative of the Bessel function of the first kind of order 1 (as in Equation (2.27)). The variations of $q_1(z)$ ($n = 1$ and $\lambda_1 = 1.8412$) and $q_2(z)$ ($n = 2$ and $\lambda_2 = 5.3314$) for systems with different value of $\frac{H}{R}$ are shown in Figure 2.7(a).

The maximum convective base shear Q_1 (Equation (2.31)) and the convective moment just above the base plate M_1 (Equation (2.35)) arising from the first sloshing mode are given by (NZSEE-09, Eq. (C3.12)–(C3.13), Point C3.3.1)):

$$Q_1 = S_e(T_1)m_1 \tag{2.42a}$$

$$M_1 = Q_1 h_1 \tag{2.42b}$$

where m_1 and h_1 (Figure 2.9) can be obtained from Equations (2.32) and (2.36) respectively. The total convective moment at the base ($M'_1 = M_1 + \Delta M_1$), including the base pressure component (Equation 2.33), is given by:

$$M'_1 = Q_1 h'_1 \tag{2.43}$$

in which h'_1 is given by expression (2.34) or Figure 2.9(b).

According to the relations originally derived by Housner in [240–242], the dynamic equivalent model (m_1, h_1, h'_1), defined an Chapter 9, ACI 350.3-06, Point 13.5, AWWA D100-11 and Point 14.3, AWWA D103-09, can be computed in accordance with the following expressions to be used in Equations (2.42) and (2.43) [67, 150, 344, 414]:

- welded and bolted steel tanks (AWWA D100-11, Points 13.5.2.2.1/2 and 13.5.3.2.2; AWWA D103-09, Points 14.3.2.2.1/2 and 14.3.3.2.2):

$$\frac{m_1}{m} = \frac{1}{4}\sqrt{\frac{27}{8}\frac{R}{H}}\tanh\left(\sqrt{\frac{27}{8}\frac{H}{R}}\right) \tag{2.44a}$$

$$\frac{h_1}{H} = 1 - \frac{\cosh\left(\sqrt{\frac{27}{8}\frac{H}{R}}\right) - 1}{\sqrt{\frac{27}{8}\frac{H}{R}}\sinh\left(\sqrt{\frac{27}{8}\frac{H}{R}}\right)} \tag{2.44b}$$

$$\frac{h'_1}{H} = 1 - \frac{\cosh\left(\sqrt{\frac{27}{8}\frac{H}{R}}\right) - \frac{31}{16}}{\sqrt{\frac{27}{8}\frac{H}{R}}\sinh\left(\sqrt{\frac{27}{8}\frac{H}{R}}\right)} \tag{2.44c}$$

- liquid-containing concrete tanks (ordinary reinforced and prestressed structures) use Equations (2.44a) and (2.44b) applied $\forall \frac{D}{H}$ whereas expression (2.44c) is modified in the following way (ACI 350.3-06, Point 9.3):

$$\frac{h'_1}{H} = 1 - \frac{\cosh\left(\sqrt{\frac{27}{8}\frac{H}{R}}\right) - 2.01}{\sqrt{\frac{27}{8}\frac{H}{R}} \sinh\left(\sqrt{\frac{27}{8}\frac{H}{R}}\right)} \tag{2.45}$$

In Equations (2.44) and (2.45), m is the total mass of the fluid contained in the tank. Relations (2.44) can be found also at Points E.6.1.1 and E.6.1.2.1/2, API 650 2012 Edition.

2.2.3 Effects of Vertical Component of the Seismic Action

In 1979, a simplified study by Marchaj [384] focused attention on the importance of vertical acceleration in the analysis and design of tanks. The vertical component of the free-field ground acceleration $a_{vg}(t)$, induces an axisymmetric hydrodynamic wall pressure on a rigid tank p_{vr}, in addition to that induced by the horizontal component $a(t)$. For a rigid tank, the pressure is uniformly distributed in the circumferential direction and increases linearly from top to bottom. The spatial-temporal variation (ζ, t) is given by (Point A.2.2, EC8-4):

$$p_{vr}(\zeta, t) = \rho_l H(1 - \zeta)a_v(t) \tag{2.46}$$

where ρ_l (in kg/m³) is the mass density of the liquid and H is the height of the free surface of the fluid; in the case of a rigid tank on a rigid foundation, $a_v(t) = a_{vg}(t)$, whereas in the case of a rigid tank on a flexible foundation or soil–structure interaction, $a_v(t)$ represents the instantaneous value of the pseudo-acceleration induced by the prescribed free-field ground motion $a_{vg}(t)$ in a single degree of freedom linear oscillator having a circular frequency corresponding to the impulsive period with or without a soil–structure interaction (Section 2.2.5) and a damping factor consistent with the limit state (Point 2.3.3.1, EC8-4 and Point 4.1.3(1), EN 1998-2:2005; Point 3.4.3, ASCE/SEI 43-05) [25], tank type (bolted and welded steel; Point 4.1.3(1), EN 1998-2:2005) and interaction with the soil described in Section 6.2.

According to the NZSEE-09 Recommendations, the distribution of the pressure on the wall can be written as a spatial function of z (Eq. (C3.7), Point C3.3.1):

$$p_{vr}(z) = \gamma_l H\left(1 - \frac{z}{H}\right)\frac{S_{ve}(T_V)}{g} \tag{2.47}$$

where γ_l is the unit weight of the liquid (in kN/m³); $S_{ve}(T_V)$ is the spectral acceleration obtained from a vertical elastic response spectrum corresponding to the period T_V (Section 2.2.5) of the vertical mode of vibration with or without soil–structure interaction and to a damping ratio already described with reference to Equation (2.46). The hydrodynamic pressure p_{vr} (Equations (2.46) and (2.47)) induced by the vertical earthquake has been determined by the single mass analogy shown in Figure 2.10. The total vertical seismic force acting on the tank shall be calculated following Point 3.2, NZSEE-09 Recommendations:

$$V = S_{ve}(T_V)m_V \tag{2.48}$$

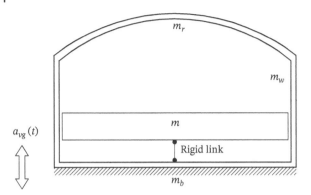

Figure 2.10 Equivalent mechanical model for vertical response of a rigid tank.

where $m_V = m + m_b + m_w + m_r$, with m equal to the total mass of the liquid, m_b the base mass, m_w the wall mass, and m_r the roof mass.

The maximum value of the circumferential hoop force in the tank wall is given by [25, 626]:

$$N_{\theta V} = \gamma_l HR \frac{S_{ve}(T_V)}{g} \tag{2.49}$$

where $\gamma_l H \frac{S_{ve}(T_V)}{g}$ is the maximum value of the pressure induced by the vertical component of the base shaking.

2.2.4 Effects of Tank Inertia

According to NZSEE-09 Recommendations (Point 3.4), the earthquake actions from the horizontal inertia of the tank walls and roof should be added to the effects of the impulsive hydrodynamic pressures. For steel tanks, usually the inertia forces on the wall and roof due to its own masses are small compared with the hydrodynamic forces and may be neglected; whereas for reinforced and prestressed concrete tanks, they should be included (Point 4.1, ACI 350.3-06). For both tanks, steel and concrete, unless there is an unusually heavy roof loading, the latter effects are generally small and may be neglected [626]. Horizontal inertia forces induce a pressure normal to the surface of the wall whose spatial-temporal variation (ζ, θ, t) is given by (Point A.2.1.5, EC8-4):

$$p_w(\zeta, t) = \rho_s s(\zeta) \cos \theta a(t) \tag{2.50}$$

where ρ_s is the mass density of the material's wall and $s(\zeta)$ the variable thickness of the wall itself. The inertia pressure p_w, which follows the variation of wall thickness, should be added to the impulsive pressure component given by Equations (2.5) or (2.18)/(2.19).

The instantaneous value of the base shear $Q_w(t)$ and corresponding bending moment $M_w(t)$ due to the tank inertia are given by:

$$Q_w(t) = (m_w + m_r)a(t) \tag{2.51a}$$

$$M_w(t) = (m_w h_w + m_r \bar{h}_r)a(t) \tag{2.51b}$$

The maximum values of these forces are obtained by replacing $a(t)$ with $S_e(T_0)$ (Eqs. (3.11) and (3.14), Point 3.5, NZSEE-09):

$$Q_w = (m_w + m_r)S_e(T_0) \tag{2.52a}$$
$$M_w = (m_w h_w + m_r \bar{h}_r)S_e(T_0) \tag{2.52b}$$

where m_w and m_r are the total masses of the tank wall and roof respectively (Figure 2.10), and h_w and $\bar{h}_r$ are the distances from the base to the respective mass centres. The effects of these masses on the impulsive mode period (T_0) should be considered (Point 3.4.1.1, NZSEE-09 Recommendations) as explicitly declared in Section 2.2.5 in Equation (2.55).

The maximum value of the circumferential hoop force in the tank wall is given by [626]:

$$N_{\theta w} = \rho_s R s S_e(T_0) \tag{2.53}$$

where $\rho_s s S_e(T_0)$ is the maximum value of the pressure induced by the horizontal inertia forces.

2.2.5 Periods of Vibration

The convective period of vibration of the nth–sloshing mode of vibration of the liquid in the tank can be obtained by the circular frequency ω_{cn} (or equivalently by the natural frequency f_{cn}) from Equation (2.30) (Eq. (C3.24), Point C3.6, NZSEE-09 and Eq. (A.9), Point A.2.1.3, EC8-4) [67]:

$$T_n = \frac{2\pi}{\omega_{cn}} = \frac{2\pi\sqrt{\dfrac{R}{g}}}{\sqrt{\lambda_n \tanh\left(\dfrac{\lambda_n H}{R}\right)}} \tag{2.54}$$

where $\lambda_1 = 1.8412$, $\lambda_2 = 5.3314$, $\lambda_3 = 8.5363$, $\lambda_4 = 11.7060$, $\lambda_5 = 14.8636$, $\lambda_6 = 18.0155$, $\lambda_7 = 21.1644$, $\lambda_8 = 24.3113$, $\lambda_9 = 27.4571$ and $\lambda_{10} = 30.6019$ are the first ten roots of the Bessel function of the first kind of order 1, as already defined in Section 2.2.2 [561, 645, 646].

In case of a rigid tank ignoring the soil–structure interaction, the impulsive period T_0 is theoretically zero,[2] and $S_e(T_0) = Sa_g$ where a_g is the peak value of the ground acceleration $a(t)$ and S the soil factor (Point 3.2.2.2, UNI EN 1998-1:2004). Unless calculated on the basis of a more rigorous analysis, Point CA2.2.1, NZS 3106 (2009), suggests for a ground-supported circular (or rectangular) tank's horizontal mode to assume $T_0 = 0.1$ sec. This puts the response within or near to the peak spectral response, consequently, the peak response spectrum ordinates are required to be used, unless it can be shown that other ordinates are appropriate.

The flexibility of the supporting media, like that of the tank wall itself, as clearly explained in Section 5.1, may be considered to affect only the impulsive components [635]. Accordingly, the convective components of the response may be evaluated from the expression for rigidly supported rigid tanks given by Equation (2.54). The impulsive period of vibration of the tank–foundation system including the soil–structure interaction has been derived by Jennings and Bielak in [36, 37, 274] and then simplified

---

2 AWWA D100-11, Point 13.5.1 and AWWA D103 Edition 2009, Point 14.3.1 explicitly declare that: "The natural period of the structure is very small and is assumed to be zero for the general procedure" based on the mapped maximum considered earthquake spectral response accelerations (AWWA D100-11, Point 13.2.7.1 and AWWA D103 Edition 2009, Point 14.2.7.1).

by Veletsos and Nair [626, 630, 631] (Eq. (C3.30), Point C3.6, NZSEE-09 and Eq. (A.52), Point A.7.2.2, EC8-4) [82]:

$$T_0 = 2\pi \sqrt{\frac{m_0 + (m_b + m_w + m_r)}{K_x} + \frac{m_0 h_0^2}{K_\theta}} \tag{2.55}$$

where m_0 is the impulsive mass (Equation (2.10) or (2.25a) and (2.25b)) and m_b, m_w and m_r are the base, wall, and roof mass, respectively (Figure 2.10); the last two masses should be considered in the case of inertia effects included in the mechanical model (Section 2.2.4); h_0 is the distance from the tank base to the point of concentration of the impulsive liquid mass (Equation (2.15) or (2.25c) and (2.25d) and (2.26a)). In Equation (2.55), K_x and K_θ are the horizontal and rocking stiffness of the foundation, respectively [271]. In the case of a circular rigid foundation supported at the surface of a homogeneous soil deposit [626], the horizontal force (in N/m) at the level of the tank's base needed to deflect the base by a unit amount in the direction of the force may be determined from [179, 342, 352, 644] (Point 5.6.2.1.1, FEMA 450):

$$K_x = \frac{8}{2 - \nu_s} G_s R_b \alpha_x \tag{2.56}$$

and accordingly the moment (in Nm) necessary to rotate the foundation basement by a unit amount is (Point 5.6.2.1.1, FEMA 450):

$$K_\theta = \frac{8}{3(1 - \nu_s)} G_s R_b^3 \alpha_\theta \tag{2.57}$$

in which R_b is the radius of the foundation, G_s and ν_s are the shear modulus and Poisson's ratio for the soil, respectively. The stiffnesses defined by Equations (2.56) and (2.57) are for dynamic conditions of loading due to the dimensionless coefficients α_x and α_θ which can be evaluated from Figures 2.11(a) and 2.11(b) or from expressions (2.61a) and (2.61b) as a function of the Poisson's ratio for the soil ν_s and a dimensionless frequency parameter:

$$\alpha = \frac{2\pi R_b}{T_0 \nu_s} \tag{2.58}$$

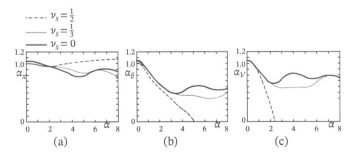

Figure 2.11 Dimensionless coefficients (or impedance functions so called by Gazetas in [179]) α_x, α_θ, and α_V of rigid circular footings on a homogeneous half-space [179, 352, 644, 662] used for the evaluation of impulsive and vertical periods of vibration in the case of soil–structure interaction. Source: Adapted from [179].

where $v_s = \sqrt{G_s / \rho_{soil}}$ is the shear wave velocity of the soil (in m/sec using G_s in N/m^2 and ρ_{soil} in kg/m^3) and ρ_{soil} the soil mass density. Since the period T_0 is not known at the outset, because it depends on the stiffnesses K_x and K_θ which depend on their own from α_x, α_θ and α, it must be computed by iteration. Specifically, values for α_x and α_θ are first assumed and then the associated values of K_x, K_θ, T_0 and α are computed. Finally, using α, values for α_x and α_θ are calculated from Figures 2.11(a) and 2.11(b) or from expressions (2.61a) and (2.61b). If the computed values differ from the assumed, the process is repeated with the derived values used as the next assumed value. Usually convergence values may be achieved in one or two cycles.

In the case of a rigid tank ignoring the soil–structure interaction, the vertical period T_V is theoretically zero, and $S_{ve}(T_V) = a_{vg}$ (Section 2.2.3), where a_{vg} is the vertical component of the free-field ground acceleration (Point 3.2.2.3, UNI EN 1998-1:2004). It is interesting to note that Point 4.10.1 in IITK-GSDMA 2007 suggests that, in the absence of a more refined analysis, the time period of a vertical mode of vibration for all types of tank may be taken as 0.3 sec. Furthermore, unless calculated on the basis of a more rigorous analysis, Point CA2.2.1, NZS 3106 (2009) suggests for a ground-supported circular (or rectangular) tank's vertical mode to assume $T_V = 0.1$ sec.

The vertical period of vibration of the tank–foundation system including the soil–structure interaction is given by (Eq. (C3.33), Point C3.6, NZSEE and Eq. (A.54), Point A.7.2.2, EC8-4):

$$T_V = 2\pi \sqrt{\frac{m_V}{K_V}} \quad \text{or} \quad \omega_V = \frac{2\pi}{T_V} \tag{2.59}$$

where $m_V = m + m_b + m_w + m_r$ (Figure 2.10), with m equal to the total mass of the liquid, m_b, m_w, and m_r are the base, wall, and roof mass, respectively and:

$$K_V = \frac{4}{1 - v_s} G_s R_b \alpha_V \tag{2.60}$$

The dimensionless factor α_V in Equation (2.60), required to convert the static stiffness value to a dynamic value, may be obtained from Figure 2.11(c) or expression (2.61c) as a function of α given by relation (2.58), in which the period T_0 must be substituted by the vertical period T_V.

The dynamic impedance functions α_x, α_θ and α_V, graphically depicted in Figure 2.11 for a rigid circular foundation, have been evaluated by Luco and Westmann in [352] and by Wei and Veletsos in [644, 662] as Fredholm integral non-linear equations of the second kind [454] and then numerically solved by transforming them to a system of simultaneous, linear, algebraic equations using the method of finite difference [288]. Approximate closed–form expressions were presented by Veletsos and Verbic in [643]:

$$\alpha_x = 1 \quad \forall v_s \tag{2.61a}$$

$$\alpha_\theta = 1 - b_1 \frac{(b_2 \alpha)^2}{1 + (b_2 \alpha)^2} - b_3 \alpha^2 \tag{2.61b}$$

$$\alpha_V = 1 - b_1 \frac{(b_2 \alpha)^2}{1 + (b_2 \alpha)^2} - b_3 \alpha^2 \tag{2.61c}$$

where b_1, b_2, and b_3 are dimensionless functions of α (Equation (2.58)) and v_s as shown in Table 2.1.

Table 2.1 Dimensionless coefficients b_1, b_2, b_3 and b_4 in order to evaluate α_θ, α_V, β_θ and β_V.

Coeff.	α_θ and β_θ				α_V and β_V		
	$v_s = 0$	$v_s = 1/3$	$v_s = 0.45$	$v_s = 0.5$	$v_s = 0$	$v_s = 1/3$	$v_s = 0.5$
b_1	0.525	0.5	0.45	0.4	0.25	0.35	0
b_2	0.8	0.8	0.8	0.8	1.0	0.8	0
b_3	0	0	0.023	0.027	0	0	0.17
b_4	–	–	–	–	0.85	0.75	0.85

Source: From [643]/American Society of Civil Engineers.

2.2.6 Effects of Liquid Viscosity

The viscosity of the liquid in tanks increases the liquid's viscous damping capacity during sloshing, and this increase, in turn, tends to decrease the corresponding sloshing wave amplitude to a bounded level [250]. According to Point 2.3.3.2, EC8-4, Points 3.2 and 3.7, NZSEE-09, Point E.1, API 650 2012 Edition, and Point 15.7.2, ASCE 7, the overall viscous damping ratio[3] ξ (in %), for convective modes in all types of tanks, should be taken as 0.5% for water and other fluids. For granular contents in the tank, the sloshing effects may be neglected (Point 3.3.5, NZSEE-09) and the total mass of the content should be assumed to act in the impulsive mode and taken as rigidly connected to the wall.

A more refined evaluation of the viscous damping ratio, confirmed theoretically and experimentally, can be obtained by the research conducted by Case and Parkinson in [71] as a function of the significant variables given by the liquid height H (in metres), liquid kinematic viscosity v_c (in m^2/sec) and tank size R (in metres):

$$\xi = \xi_1 + \xi_2 + \xi_3 \tag{2.62}$$

where [2, 250, 550]:

$$\xi_1 = \frac{1}{2\pi}\left[\frac{4\pi v_c}{\omega_{cn}}\left(\frac{\lambda_n}{R}\right)^2\right]\cdot 100 \tag{2.63a}$$

$$\xi_2 = \frac{1}{2\pi}\sqrt{\frac{v_c}{2\omega_{cn}}}\left(\frac{\pi}{R}\right)\left[\frac{1+\left(\frac{n}{\lambda_n}\right)^2}{1-\left(\frac{n}{\lambda_n}\right)^2}-\frac{2\lambda_n\frac{H}{R}}{\sinh\left(2\lambda_n\frac{H}{R}\right)}\right]\cdot 100 \tag{2.63b}$$

$$\xi_3 = \frac{1}{2\pi}\sqrt{\frac{v_c}{2\omega_{cn}}}\left(\frac{\pi}{R}\right)\frac{2\lambda_n}{\sinh\left(2\lambda_n\frac{H}{R}\right)}\cdot 100 \tag{2.63c}$$

The first term ξ_1 is the viscous dissipation at the free surface, ξ_2 is due to the sidewalls and ξ_3 to the tank bottom [250, 394]. One may observe that the contribution of the free surface given by expression (2.63a) is small compared to the wall damping ξ_2, particularly for large containers. Thus, approximate and

---

3 The ratio of the actual viscous damping to its critical value [2, 100, 550].

empirical expressions for the fundamental antisymmetric mode ($n = 1$ in Figure 2.6(a) with $\lambda_1 = 1.8412$ and ω_{c1} (in rad/sec) derived from Equation (2.30)) have been carried out at least in the following three main extensive experimental investigations [2, 3, 251, 550]:

- Mikishev and Dorozhkin in [396] proposed the following correlation from their tests:

$$\xi = \frac{4.98}{2\pi} \sqrt{\frac{v_c}{\sqrt{R^3 g}}} \left[1 + \frac{0.318}{\sinh\left(\lambda_1 \frac{H}{R}\right)} \left(1 + \frac{1 - \frac{H}{R}}{\cosh\left(\lambda_1 \frac{H}{R}\right)} \right) \right] \cdot 100 \tag{2.64}$$

For large depths, $\frac{H}{R} > 1$, relation (2.64) may be approximated by:

$$\frac{4.98}{2\pi} \sqrt{\frac{v_c}{\sqrt{R^3 g}}} \cdot 100 = \frac{4.98}{2\pi} G_A \cdot 100 \tag{2.65}$$

where the nondimensional term $G_A = \sqrt{\frac{v_c}{\sqrt{R^3 g}}}$ in expression (2.65) is called the Galileo number [3];

- a similarly extensive but independent experimental study by Stephens et al. in [567] found a slightly different correlation from their tests:

$$\xi = \frac{5.23}{2\pi} \sqrt{\frac{v_c}{\sqrt{R^3 g \tanh\left(\lambda_1 \frac{H}{R}\right)}}} \left[1 + 2 \frac{1 - \frac{H}{R}}{\sinh\left(3.68 \frac{H}{R}\right)} \right] \cdot 100 \tag{2.66}$$

For $\frac{H}{R} > 1$ expression (2.66) takes the form:

$$\frac{5.23}{2\pi} \sqrt{\frac{v_c}{\sqrt{R^3 g}}} \cdot 100 = \frac{5.23}{2\pi} G_A \cdot 100 \tag{2.67}$$

- the theoretical value given by Miles in [397] reads:

$$\xi = \frac{3.51}{2\pi} \sqrt{\frac{v_c}{\sqrt{R^3 g \tanh\left(\lambda_1 \frac{H}{R}\right)}}} \left[1 + 2 \frac{1 - \frac{H}{R}}{\sinh\left(3.68 \frac{H}{R}\right)} \right] \cdot 100 \tag{2.68}$$

where $\frac{H}{R} > 1$ can be simplified to obtain:

$$\frac{3.51}{2\pi} \sqrt{\frac{v_c}{\sqrt{R^3 g}}} \cdot 100 = \frac{3.51}{2\pi} G_A \cdot 100 \tag{2.69}$$

The square bracket's term in Equations (2.64), (2.66) and (2.68) is a dimensionless shape coefficient which decreases with increasing $\frac{H}{R}$ and reaches its limiting value of unity for values of $\frac{H}{R}$ of about one. The larger factors for the lower value of $\frac{H}{R}$ are attributed to the viscosity effects of the liquid moving along the tank base [25].

Kinematic viscosity (1 cst $= 1 \cdot 10^{-6}$ m^2/sec) for some commons fluids is indicated in Table 2.2.

Table 2.2 Values of the kinematic viscosity v_c in centistokes (1 cst $= 1\,mm^2/sec$) for different fluids and a range of temperatures.

Fluid	Temperature in °C					
	0	10	20	50	100	200
Mercury	0.13	0.12	0.12	0.11	0.09	0.08
Ammonia	0.31	0.29	0.27	0.2	–	–
Benzene (benzol)	0.8	0.7	0.6	0.51	0.31	–
Sea water	1.79	1.31	1.04	0.8	0.45	0.15
Water	1.78	1.31	1.00	0.55	0.29	–
Kerosene	4.2	2.8	2.71	1.3	0.9	–
SAE 30 oil	2000	600	393	55	12	2

2.2.7 Effects of Inhomogeneous Liquids

As highlighted in the previous sections, the response to earthquakes of tanks containing homogeneous liquids has been deeply investigated by many researchers [214, 240–242, 263, 370, 626] in the past, whereas there is a paucity of information concerning the corresponding response of tanks containing layers of liquid of different densities. Interest in the response of tanks with layered liquids could be related to the presence of many waste storage tanks in nuclear facilities, as reported by Bandyopadhyay in [25, 26], or to the process for the recovery and decontamination of discharge fuel material typically carried out in tanks containing two layered liquids [57]. The earliest contributions on this topic are due to Tang [585–590] and are mainly focused on tanks containing two liquids; here we will follow the analytical solution proposed by Veletsos and Shivakumar [538, 632–635] applicable to multi-layered systems with an arbitrary number of uniform layers of different thicknesses and densities.

The basic system investigated is shown in Figure 2.12. It is a ground-supported, vertical, rigid, circular tank of radius R anchored to a rigid base and to be excited horizontally, filled to a height H. The liquid layers are numbered sequentially starting with 1 at the bottom layer and terminating with N at the top layer; the jth–interface is the common boundary to the jth and $(j+1)^{th}$–layers; H_j and ρ_j are, respectively, the thickness and mass density of the jth–layer; furthermore the values of ρ_j are considered to decrease with increasing values of j from bottom to top. Points within the jth–layer are defined by a local coordinate system (r, θ, z_j) with the origin z_j taken at the $(j-1)^{th}$–interface. Three main nondimensional coordinates are then introduced as $\xi = \frac{r}{R}$, $\eta_j = \frac{z_j}{R}$, and the slenderness $\gamma_j = \frac{H_j}{R}$. This system is considered under the assumption that the motion experienced by the base tank is the same as the free-field ground motion. Furthermore, it is assumed that the liquid is initially at rest, incompressible, and inviscid, that the flow field in the jth–layer, $\Phi_j(r, \theta, z_j, t)$, is irrotational, and that all structural and liquid motions remain within the linearly elastic range of response (no separation or cavitation between the liquid and the tank). Under these assumptions the fluid flow in the jth–layer must satisfy the Laplace's Equation (2.1):

$$\nabla^2 \Phi_j = 0 \tag{2.70}$$

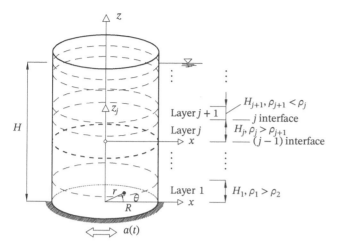

Figure 2.12 Tank-liquid geometry investigated: coordinate system (r, θ, z), ground acceleration $a(t)$ and multi-layered liquid main characteristics [538, 632–634].

Then, the component of the liquid velocity v_{jn} in the direction of a generalized n-coordinate for an arbitrary particle in the jth–layer at a certain time t, and the hydrodynamic pressure p_j at the same point and time, are given respectively by [632] (similar to expressions (2.2)):

$$v_{jn} = -\frac{\partial \Phi_j}{\partial n} \tag{2.71a}$$

$$p_j = \rho_j \frac{\partial \Phi_j}{\partial t} \tag{2.71b}$$

The solution of Equation (2.70) must satisfy the same boundary conditions (2.3) along the tank wall $(r = R)$, base $(z_1 = 0)$, and the free liquid surface $(z_N = H_N)$ [632]:

$$\left.\frac{\partial \Phi_1}{\partial z_1}\right|_{z_1=0} = 0 \qquad\qquad \text{tank base} \tag{2.72a}$$

$$\left.\frac{\partial \Phi_j}{\partial r}\right|_{r=R} = -v(t)\cos\theta \qquad\qquad \text{tank wall} \tag{2.72b}$$

$$\left(\frac{\partial \Phi_N}{\partial t} - g d_N\right)_{z_N=H_N} = 0 \qquad\qquad \text{liquid surface} \tag{2.72c}$$

where d_N represents the vertical surface displacement of the free liquid surface; the radial velocity component of the fluid in the case of a rigid tank must be equal to the corresponding component $v(t)$ of the ground motion $a(t)$; g is the gravitational acceleration. One more condition must be added to the three previously defined, stating that the vertical velocity of the liquid must be continuous at the interface of a pair of layers:

$$\left.\frac{\partial \Phi_j}{\partial z_j}\right|_{z_j=H_j} = \left.\frac{\partial \Phi_{j+1}}{\partial z_{j+1}}\right|_{z_{j+1}=0} \qquad\qquad \text{interface} \tag{2.72d}$$

Using the interpretation proposed by Lamb [322], expression (2.72d) states that the total hydrodynamic pressure (hydrodynamic plus the increment due to the vertical displacement at the interface between layers) must be continuous. The solution of Equation (2.70) under the appropriate boundary conditions (2.72) has been written by Veletsos and Shivakumar [634] in a classical fashion as the sum of an impulsive component in the jth–layer:

$$p_i^j(\xi, \eta_j, \theta, t) = - \left\{ \xi - \sum_{m=1}^{\infty} \left[e_{m,j} \frac{\cosh(\lambda_m \eta_j)}{\sinh(\lambda_m \gamma_j)} \right. \right.$$
$$\left. \left. - e_{m,j-1} \frac{\cosh \lambda_m (\gamma_j - \eta_j)}{\sinh(\lambda_m \gamma_j)} \right] \frac{J_1(\lambda_m \xi)}{J_1(\lambda_m)} \right\} \rho_j R a(t) \cos \theta \qquad (2.73)$$

and a convective component:

$$p_c^j(\xi, \eta_j, \theta, t) = - \left[\sum_{m=1}^{\infty} \sum_{n=1}^{N} c_{mn,j}(\eta_j) \frac{J_1(\lambda_m \xi)}{J_1(\lambda_m)} a_{mn}(t) \right] \rho_j R \cos \theta \qquad (2.74)$$

in which:

$$c_{mn,j}(\eta_j) = C_{mn}^2 \left[\frac{d_{mn,j} \cosh(\lambda_m \eta_j) - d_{mn,j-1} \cosh \lambda_m (\gamma_j - \eta_j)}{\sinh(\lambda_m \gamma_j)} \right] \qquad (2.75)$$

The impulsive component, which represents the effect of the part of the liquid that may be considered to move rigidly with the tank's wall, experiences the same acceleration as the ground. In Equation (2.73) J_1 is the Bessel function of the first kind of order 1 [561] and numbers λ_m are the m–roots of the first derivative of J_1 equal to $\lambda_1 = 1.8412$, $\lambda_2 = 5.3314$, $\lambda_3 = 8.5363$, $\lambda_4 = 11.7060$, $\lambda_5 = 14.8636$, $\lambda_6 = 18.0155$, $\lambda_7 = 21.1644$, $\lambda_8 = 24.3113$, $\lambda_9 = 27.4571$ and $\lambda_{10} = 30.6019$ [23, 221, 596, 646]; for a two-layered system with $j = 1$ and 2, $e_{m,j}$ is equal to:

$$e_{m,0} = 0 \qquad (2.76a)$$

$$e_{m,1} = \varepsilon_m \frac{\left[\left(1 - \frac{\rho_2}{\rho_1} \right) \cosh(\lambda_m \gamma_2) + \frac{\rho_2}{\rho_1} \right] \sinh(\lambda_m \gamma_1)}{\cosh(\lambda_m \gamma_1) \cosh(\lambda_m \gamma_2) + \frac{\rho_2}{\rho_1} \sinh(\lambda_m \gamma_1) \sinh(\lambda_m \gamma_2)} \qquad (2.76b)$$

$$e_{m,2} = \varepsilon_m \frac{\sinh(\lambda_m \gamma_2) + e_{m,1}}{\cosh(\lambda_m \gamma_2)} \qquad (2.76c)$$

where $\varepsilon_m = \frac{2}{\lambda_m^2 - 1}$. The maximum value of the impulsive pressure p_i^j may then be determined by replacing $a(t)$ by its spectral value S_e using an impulsive period with or without a soil–structure interaction given in Section 2.2.5 and an equivalent viscous damping level for horizontal mode consistent with the limit state (Point 2.3.3.1, EC8-4 and Point 4.1.3(1), EN 1998-2:2005; Point 3.4.3, ASCE/SEI 43-05) [25], tank type (bolted and welded steel or reinforced and prestressed concrete [25]; Point 4.1.3(1), EN 1998-2:2005) and interaction with the soil (Section 6.2).

The convective component (2.74) represents the effects of the sloshing action of the liquid and it is clear that there is an infinite m^{th}–number of horizontal sloshing modes of vibration ($\sum_{m=1}^{\infty}$) and that for each such mode, there exist N vertical modes ($\sum_{n=1}^{N}$). Furthermore, associated with each horizontal and vertical mode, there is a distinct acceleration function [634]:

$$a_{mn}(t) = \omega_{mn} \int_0^t a(t) \sin\left[\omega_{mn}(t-\tau)\right] d\tau \tag{2.77}$$

that is the instantaneous value of the pseudo-acceleration induced by the prescribed free-field ground motion $a(t)$ in the system with an appropriate viscous damping (Section 2.2.6) [634] and having the following circular frequency ω_{mn} for the mode of vibration considered:

$$\left(1 + \frac{\rho_2}{\rho_1} \tanh(\lambda_m \gamma_1) \tanh(\lambda_m \gamma_2)\right) \omega_{mn}^4 - \left(\tanh(\lambda_m \gamma_1) + \tanh(\lambda_m \gamma_2)\right) \frac{\lambda_m g}{R} \omega_{mn}^2$$
$$+ \left(1 - \frac{\rho_2}{\rho_1}\right)\left(\frac{\lambda_m g}{R}\right)^2 \tanh(\lambda_m \gamma_1) \tanh(\lambda_m \gamma_2) = 0 \tag{2.78}$$

valid for the case of a two-layered system without a soil–structure interaction. With the root of the first derivative of J_1 equal to $\lambda_1 = 1.8412$, we can derive from expression (2.78) the circular frequencies ω_{11} and ω_{12}, while with the root $\lambda_2 = 5.3314$ we obtain ω_{21} and ω_{22} corresponding to the mn^{th}–mode of vibration of the system. The dimensionless coefficient C_{mn} in Equation (2.75) depends on the tank shape, the relative thicknesses, densities, and number of liquid layers, and on the mode under consideration [634]:

$$C_{mn}(t) = \omega_{mn} \sqrt{\frac{R}{\lambda_m g}} \tag{2.79}$$

For a two-layered system with $j = 1$ and 2, $d_{mn,j}$ ($d_{mn,0} = 0$) may be represented as a vector's $\mathbf{d}_{mn}$ of dimensionless coefficients [634]:

$$\mathbf{d}_{mn} = \varepsilon_m \frac{\mathbf{D}_{mn}^T \mathbf{c}}{\mathbf{D}_{mn}^T \mathbf{B} \mathbf{D}_{mn}} \mathbf{D}_{mn} \tag{2.80}$$

where $\mathbf{D}_{mn}$ is the vector of the maximum vertical (or sloshing) displacement amplitudes of the liquid at the layer interfaces when the system is vibrating in the mth–horizontal and n^{th}–vertical mode. It can be derived by solving the characteristic value problem defined by Veletsos and Shivakumar in [632]:

$$\mathbf{B} \mathbf{D}_{mn} = \frac{\omega_{mn}^2 R}{\lambda_m g} \mathbf{A} \mathbf{D}_{mn} \tag{2.81}$$

where it is clear that for each horizontal m^{th}–mode of vibration ($\lambda_1, \lambda_2, \ldots$), there exist N vertical modes, each associated with a distinct frequency (ω_{mn}); using the circular frequency ω_{11} (determined previously by substituting λ_1 in expression (2.78)) in Equation (2.81), we can derive $D_{11,1}$ and $D_{11,2}$ while using the circular frequency ω_{12} (again determined by substituting λ_1 in expression (2.78)) to obtain $D_{12,1}$ and $D_{12,2}$; $D_{mn,j}$ denotes the jth–element (j^{th}–interfacial displacement) of the vector $\mathbf{D}_{mn}$ corresponding to the nth–frequency of the system for the mth–horizontal mode of vibration. In relations (2.80) and (2.81), $\mathbf{c}$ is a vector of size N the jth–element of which is c_j, $\mathbf{A}$ is a tridiagonal and symmetric matrix of size $N \times N$ for which the element of the jth–row are given by expressions (2.82a)–(2.82c), and $\mathbf{B}$ is a diagonal matrix of size $N \times N$ with the jth–element given by Equation (2.82d) [632]:

$$A_{j,j} = \frac{\rho_j}{\rho_1} \coth(\lambda_m \gamma_j) + \frac{\rho_{j+1}}{\rho_1} \coth(\lambda_m \gamma_{j+1}) \tag{2.82a}$$

$$A_{j,j-1} = -\frac{\rho_j}{\rho_1} \frac{1}{\sinh(\lambda_m \gamma_j)} \tag{2.82b}$$

$$A_{j,j+1} = -\frac{\rho_{j+1}}{\rho_1}\frac{1}{\sinh(\lambda_m\gamma_{j+1})} \tag{2.83c}$$

$$B_{jj} = c_j = \frac{\rho_j}{\rho_1} - \frac{\rho_{j+1}}{\rho_1} \tag{2.83d}$$

Figure 2.13(a) shows the dimensionless function for the impulsive component of the pressure, obtained by putting $\xi = 1$ and $\cos\theta = 1$ in Equation (2.73):

$$-\frac{p_i^j(1,\eta_j,0,t)}{\rho_1 Ra(t)} = \left\{1 - \sum_{m=1}^{\infty}\left[e_{m,j}\frac{\cosh(\lambda_m\eta_j)}{\sinh(\lambda_m\gamma_j)} - e_{m,j-1}\frac{\cosh\lambda_m(\gamma_j-\eta_j)}{\sinh(\lambda_m\gamma_j)}\right]\right\}\frac{\rho_j}{\rho_1} \tag{2.83}$$

It is interesting to note that for a single-layered homogeneous system the term in curly brackets in Equation (2.83) reduces to the following expression [634]:

$$1 - \sum_{m=1}^{\infty}\varepsilon_m\frac{\cosh(\lambda_m\frac{z}{R})}{\cosh(\lambda_m\frac{H}{R})} \quad \rightarrow \quad 1 - \sum_{m=1}^{\infty}(\text{Eq. (2.41)}) \tag{2.84}$$

which is a further verification of expression (2.39). The impulsive pressure as expected increases from zero at the top to a maximum at the base, exhibiting a slope discontinuity at the layer interface; furthermore, the impulsive pressure decreases with decreasing $\frac{\rho_2}{\rho_1}$. Figures 2.13(b) and 2.13(c) show the function $c_{mn,j}(\eta_j)\frac{\rho_j}{\rho_1} = -\frac{p_c^j(1,\eta_j,0,t)}{\rho_1 Ra(t)}$ (Equation (2.75)) for the convective components of the pressure associated with the first two horizontal modes of vibration. Usually the $c_{mn,j}(\eta_j)$ functions for the higher horizontal modes are negligibly small compared to the first and second components and, for practical reasons, are not included in the formulation. The convective pressure in the lower layer increases with decreasing $\frac{\rho_2}{\rho_1}$ and is discontinuous at the layer interface; the convective pressure associated with a density ratio $\frac{\rho_2}{\rho_1} = 1.0$ presents its maximum value, for the first sloshing mode, at the top $c_{11} = 0.8368$ which, for coherence, is equal to the value obtained with Equations (2.27) or (2.41) and shown in Figure 2.7(a) in the case of a homogeneous liquid.

2.2.8 Convective Wave Displacement and Pressure

In the case of tanks without roofs or with floating roofs, the maximum vertical displacement of the liquid surface is needed to determine the freeboard[4] that must be provided to prevent over-topping or spillage of the tank's contents [12]. The spatial-temporal variation of the vertical surface displacement $d(\xi,\theta,t)$ of the liquid may be written as [626]:

$$d(\xi,\theta,t) = R\cos\theta\sum_{n=1}^{\infty}\frac{2}{\lambda_n^2-1}\frac{J_1(\lambda_n\xi)}{J_1(\lambda_n)}\frac{a_{cn}(t)}{g} \tag{2.85}$$

where the symbols used have already been explained in Equation (2.27). The absolute maximum value $d_{\max}$, from the at-rest level of the liquid, of the displacement given in relation (2.85), may then be obtained

---

4 The freeboard represents the difference between the roof elevation and the elevation of the upper liquid surface at rest.

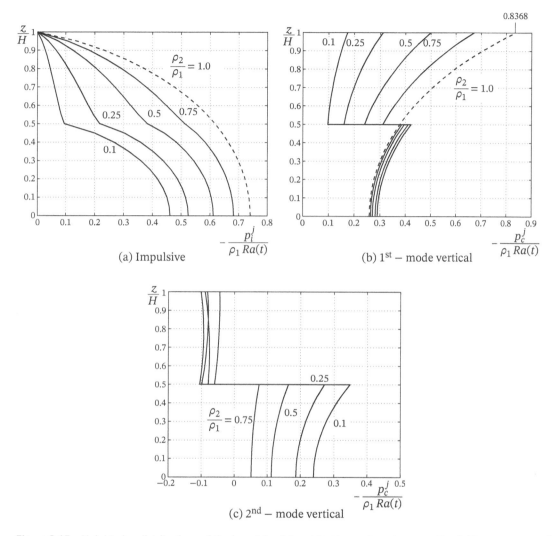

Figure 2.13 Heightwise distributions of the impulsive (a) and first/second mode convective (b/c) components of wall pressure for a tank with $\frac{H}{R} = 1$ and $\frac{H_2}{H_1} = 1$, containing two layers of liquids with different values of the density ratio $\frac{\rho_2}{\rho_1} = 0.1, 0.25, 0.5, 0.75$ and 1 [538, 632–635]; (b) first mode horizontal and first mode vertical (such as the first mode in Figure 2.7(a) in case of a homogeneous liquid); (c) first mode horizontal and second mode vertical.

approximately by application of the square root of the sum of the square rule, as [626] (Eq. (C3.36), Point C3.9, NZSEE-09):

$$d_{\max} = R\sqrt{\left(0.837\frac{S_e(T_1)}{g}\right)^2 + \left(0.073\frac{S_e(T_2)}{g}\right)^2 + \left(0.028\frac{S_e(T_3)}{g}\right)^2 + \cdots} \qquad (2.86)$$

where $S_e(T_n)$ is the spectral acceleration obtained from a horizontal elastic response spectrum corresponding to the convective period of the nth–sloshing mode (Equation (2.54)) and to the viscous damping ratio as given in Section 2.2.6. EC8-4 at Point A.2.1.4 suggests that the sloshing wave height is mainly provided by the first mode and this may be translated into the following simplification of Equation (2.86):

$$d_{\max} = 0.84R\frac{S_e(T_1)}{g} \tag{2.87}$$

If the available freeboard [25]:

$$d_{\text{free}} \geq 1.6 \cdot \text{SF} \cdot d_{\max} \tag{2.88}$$

is not adequate for the computed convective wave displacement $d_{\max}$, either the liquid level must be lowered or the tank roof and wall must be designed to withstand the impact of the sloshing liquid. In Equation (2.88), coefficient 1.6 accounts for increased slosh height due to non-linear effects [25] and SF is the appropriate seismic scale factor. If the hazard curve for a site is not available or highly uncertain for annual probabilities less than the seismic hazard exceedance probability P_H, a constant (site-independent) scale factor could be used as a function of the risk or probability reduction factor R_R (Table 2.3). For structural design considerations, the risk reduction factors may be used to increase the annual probability of hazard exceedance P_H above the performance goals P_F (Point 8.4.3, DOE-STD-1020-2012). Hence the risk reduction factors are the ratio of the design basis annual probability of hazard exceedance to the performance goal annual probability of exceedance $P_H = R_R \cdot P_F$.

Point C3.5 in NZS 3106 (2009) suggests considering the possibility of using a lower seismic return period factor when checking over-topping or roof pressure than that used for the design of the tank walls. Where loss of liquid must be prevented (for example, tanks for the storage of toxic liquids), or where overtopping may result in scouring of the foundation materials or cause damage to pipes or roof, then provisions should be made by freeboard allowance or designing the roof structure to resist the resulting uplift pressures.

Insufficient freeboard (Figures 2.15(b) and 2.16(a)), as clearly evidenced in Figure 2.14, causes an upward liquid pressure $p_{\max}$ acting against the tank roof that will change the longitudinal force $N_y^{\max}$ in the cylindrical walls of the tanks; furthermore, it causes an increase in impulsive mass due to the constraining action of the roof, and consequently an influence in impulsive and convective periods and hydrodynamic pressures against the tank wall. The maximum upward pressure acting on the tank roof

Table 2.3 Recommended value for constant (site-independent) scale factors SF to achieve various risk reduction factor.

Risk reduction factor R_R	Constant scale factors SF
20	1.60
10	1.25
5	1.00

Source: Adapted from [25].

Figure 2.14 Calexico earthquake (California, 14 June 2010): a 7.2 magnitude earthquake struck the water treatment tanks and damaged the massive reservoirs. Source: from Adam DuBrowa/FEMA.

and the associated impulsive $\bar{m}_i$ and convective masses $\bar{m}_c$, may be evaluated using the procedure proposed by Malhotra in [371, 372] in the case of flat and conical roofs (Appendix F in FM Approvals Class Number 4020):

Step 1 by assuming a flat, free surface sloshing profile (Figure 2.15), the angle θ with respect to the horizontal plane, is given by:

$$\theta = \tan^{-1}\left(\frac{S_e(T_1)}{g}\right) \tag{2.89}$$

where $S_e(T_1)$ is the spectral acceleration obtained from a horizontal elastic response spectrum corresponding to the convective period of the first mode (Equation (2.54)) and to the viscous damping ratio, as given in Section 2.2.6, and g is the gravitational acceleration. The vertical displacement of liquid surface due to sloshing is consequently equal to (Figure 2.15(a)):

$$d = R\tan\theta = R\frac{S_e(T_1)}{g} \tag{2.90}$$

According to Malhotra [371, 372], in Equation (2.90) it has been assumed that the entire convective liquid mass moves in unison, thus giving a somewhat conservative estimate of the sloshing wave height if compared with Equation (2.87).

Step 2 in the case of insufficient freeboard, when the actual value d_f in Figure 2.15(b) is less than the vertical displacement of the liquid surface d, obtained from Equation (2.90), a portion of the tank roof is wetted. The wetted width x_f may be evaluated by equating the empty volume above the inclined water surface after the earthquake to the empty volume in the tank before the earthquake

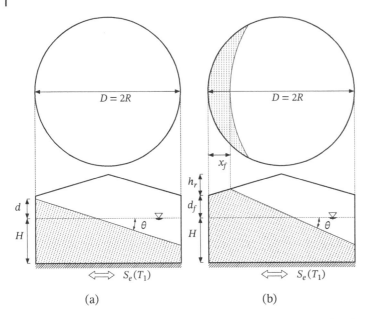

Figure 2.15 Horizontally excited cylindrical tank with conical roof under an acceleration $S_e(T_1)$: (a) sufficient freeboard and (b) insufficient freeboard.

(that is $\pi R^2 d_f + \frac{\pi R^2 h_r}{3}$). In the case of a flat roof (Figure 2.16(a)) by assuming $\frac{h_r}{d} = 0$, we obtain [371]

$$\frac{d_f}{d} = \frac{1}{\pi}\left(1 - \frac{x_f}{R}\right)\left(\psi_0 - \frac{\sin 2\psi_0}{2}\right) + \frac{2}{3\pi}\sin^3\psi_0 \qquad (2.91)$$

where $\psi_0 = \cos^{-1}\left(\frac{x_f}{R} - 1\right)$ can be obtained from Figure 2.16(a). Figure 2.16(b) shows the relationship between the nondimensional ratio $\frac{x_f}{R}$ as a function of $\frac{d_f}{d}$ for tanks with different $\frac{h_r}{d}$.

Step 3 for fixed values of $\frac{d_f}{d}$ and $\frac{h_r}{d}$, we can directly obtain from Figure 2.16(b) the wetted width x_f and then the maximum upward pressure on the tank roof:

$$p_{\max} = \rho_l g x_f \left(\tan\theta + \frac{h_r}{R}\right) \qquad (2.92)$$

where ρ_l is the mass density of the liquid. The pressure (2.92) on the roof is resisted by the vertical/longitudinal tensile force $N_y^{\max}$ per unit circumference of the cylindrical walls of the tanks, under the hypotheses that $x_f \ll R$ (however, in any case $x_f < 0.5R$) and that the upward force is resisted by the wet side of the tank shell only:

$$N_y^{\max} = \frac{1}{2}p_{\max}x_f = \frac{1}{2}\rho_l g x_f^2\left(\tan\theta + \frac{h_r}{R}\right) \qquad (2.93)$$

or by substituting Equations (2.89) and (2.90):

$$N_y^{\max} = \frac{1}{2}\rho_l g x_f^2\left(\frac{d + h_r}{R}\right) \qquad (2.94)$$

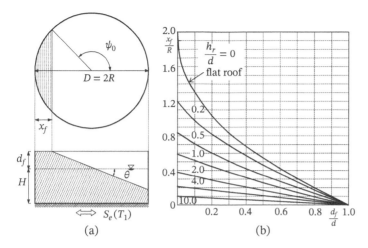

Figure 2.16 (a) Horizontally excited cylindrical tank with flat roof under an acceleration $S_e(T_1)$; (b) normalized wetted width $\frac{x_f}{R}$ as a function of nondimensional actual freeboard $\frac{d_f}{d}$ for flat ($\frac{h_r}{d} = 0$) and conical roof. Source: Adapted from [372].

Step 4 the modified values of the impulsive and convective masses are (Point E-4.3, FM Approvals Class Number 4020):

$$\bar{m}_i = \begin{bmatrix} m_i + m_c \left(1 - \frac{d_f + \frac{h_r}{3}}{d}\right) & d_f + \frac{h_r}{3} < d \\ m_i & d_f + \frac{h_r}{3} \geq d \end{bmatrix} \qquad (2.95a)$$

$$\bar{m}_c = m - \bar{m}_i \qquad (2.95b)$$

$$\bar{T}_0 = T_0 \sqrt{\frac{\bar{m}_i}{m_i}} \qquad (2.95c)$$

$$\bar{T}_1 = T_1 \sqrt{\frac{\bar{m}_c}{m_c}} \qquad (2.95d)$$

The constraint on the sloshing motion due to the impact of the liquid on the roof increases the mass participation in the impulsive mode and conversely decreases the convective mode. In the limiting case of a fully filled tank in which the surface sloshing section cannot develop, the entire liquid in the tank acts impulsively and the resulting wall pressure is uniformly distributed along the tank height. In the case of tanks with insufficient freeboard, modified masses $\bar{m}_i$ and $\bar{m}_c$ (Equations (2.95a) and (2.95b)) should be used instead of m_i (Equation (2.10)) and m_{c1} (Equation (2.32)) to compute the base shears and moments from Sections 2.2.1 and 2.2.2; in place of T_0 and T_1 at Section 2.2.5, the adjusted impulsive $\bar{T}_0$ (Equation (2.95c)) and convective $\bar{T}_1$ (Equation (2.95d)) periods must be used to evaluate the spectral acceleration S_e obtained from a horizontal elastic response spectrum.

2.2.9 Combination of Pressures and Behaviour Factor

The hydrodynamic fluid pressure exerted on the wall of a rigid tank due to a ground motion is given by the following contributions [224] (Point A.2.1.6, EC8-4):

1) the impulsive fluid pressure component p_i (Equation (2.5) or Equations (2.18) and (2.19)) which varies synchronously with the horizontal ground acceleration $a(t)$; according to Section 2.2.4 to the impulsive value must be added the inertia force of the tank itself, given by expression (2.50) [626];
2) the long period component contributed by the convective fluid pressure p_c driven by $a_{cn}(t)$ (Equation (2.27) or (2.40));
3) the hydrodynamic wall pressure p_{vr} (Equations (2.46) and (2.47)) induced by the vertical component of the ground acceleration $a_v(t)$.

In addition, the total hydrodynamic pressure referred to by the above components is in excess of the hydrostatic pressure p_{hs}, and must be added to the latter to obtain the total pressure exerted on the wall.

A conservative estimate of the maximum value of the horizontal hydrodynamic wall pressure p_{tot}^h may be obtained by taking the sum of the numerical values of the maximum impulsive pressure p_i and of the maximum modal components of the convective pressure p_c (Point 4.8, HSE Research Report RR527):

$$|p_{tot}^h| = |p_{hs}| + (|p_i| + |p_w|) + |p_c| = |p_{hs}| + (|p_i| + |p_w|) + \sum_{n=1}^{\infty} |p_{cn}| \tag{2.96}$$

This approach, in agreement with Point A.2.1.6, EC8-4 and Point 15.7.6.1, ASCE 7 2010 Edition,[5] effectively uses the upper bound rule of adding the absolute values of the maxima, assuming that all of the $a_{cn}(t)$ occur simultaneously with the maximum ground acceleration and that the maximum positive and the maximum negative values of $a_{cn}(t)$ are equally likely [626]. In most cases, it is adequate to consider only the contribution of the fundamental sloshing mode $a_{c1}(t)$:

$$|p_{tot}^h| = |p_{hs}| + (|p_i| + |p_w|) + |p_{c1}| \tag{2.97}$$

A preferable approach would be to add to the maximum value of the impulsive pressure the square root of the sum of the squares of the maximum modal contributions of the convective pressure:

$$|p_{tot}^h| = |p_{hs}| + (|p_i| + |p_w|) + \sqrt{\sum_{n=1}^{\infty} p_{cn}^2} \tag{2.98}$$

Alternatively, according to Point 4.2, NZSEE-09, Chapter 4, ACI 350.3-06, Point E-5, FM Approvals Class Number 4020, Point E.6, API 650 2012 Edition, Point 13.5, AWWA D100-11, Point 4.3.1, AWWA D110-04 and D115-06, the total response resulting from the combined effects of impulsive and convective actions should be found by combining the peak spectral responses of each action according to the square root of the sum of the squares method (SRSS):

$$|p_{tot}^h| = |p_{hs}| + \sqrt{(|p_i| + |p_w|)^2 + \sum_{n=1}^{\infty} p_{cn}^2} \tag{2.99}$$

---

5 To be clear, Note e. at Point 15.7.6.1 clarifies that impulsive and convective seismic components are permitted to be combined also using the square root of the sum of the squares (SRSS) method in lieu of the direct sum method.

or considering only the contribution of the fundamental sloshing mode:

$$|p_{tot}^h| = |p_{hs}| + \sqrt{(|p_i| + |p_w|)^2 + p_{c1}^2} \qquad (2.100)$$

The peak value of the hydrodynamic pressure on the tank wall due to horizontal (impulsive and convective) and vertical seismic action, separately applied, may be obtained by applying the rule at Point 4.3.3.5.2 (4), UNI EN 1998-1:2004 (that derives from Points 3.2 (2) and 4.2, EC8-4) [5, 267, 344]:

$$E_h(+)0.30 \cdot E_v \qquad \text{or} \qquad |p_{tot}^h|(+)0.30 \cdot |p_{vr}| \qquad (2.101a)$$

$$0.30 \cdot E_h(+)0.30 \cdot E_v \qquad \text{or} \qquad 0.30 \cdot |p_{tot}^h|(+)0.30 \cdot |p_{vr}| \qquad (2.101b)$$

$$0.30 \cdot E_h(+)E_v \qquad \text{or} \qquad 0.30 \cdot |p_{tot}^h|(+)|p_{vr}| \qquad (2.101c)$$

where E_h and E_v represent the action effects due to the application of the seismic action along the horizontal and vertical axes respectively; the symbol (+) implies "to be combined with" and should be taken as the most unfavourable for the particular action effect under verification as described at Section 5.5 (Point 3.2 (3)P, EC8-4). An additional combination rule proposed at Point 5.3.2, ACI 350.3-06 and Point 4.10.2, IITK-GSDMA 2007 is:

$$|p_{tot}^h| = |p_{hs}| + \sqrt{(|p_i| + |p_w|)^2 + p_{c1}^2 + p_{vr}^2} \qquad (2.102)$$

similar to the rule suggested at Point C4.2, NZSEE-09 related to the method of combining actions.

The usual procedure followed by seismic codes in order to reduce by a certain amount the elastic spectrum depends on ductility, overstrength, redundancy of the structure, and the hysteretic energy dissipation mechanism. Significant differences can be highlighted in the strategies followed by the codes specified in Section 2.1 to reduce elastic seismic force. In the US standards, a detailed classification of tanks, as clearly stated in Table 2.4, is given and the corresponding response modification factor for each type of tank (steel, reinforced and prestressed tanks, above-ground and elevated), support (hinged, fixed, anchored or unanchored), and for each type of behaviour (impulsive and convective) is assigned. The NZSEE-09 recommendation at Point 3.2 also suggests a classification for tanks in order to evaluate the ductility and corresponding effective damping ratio (Section 6.2) to be used to determine the correction factor for the NZS 1170.5 elastic spectrum. It is important to note that EC8-4 at Points 4.4 and A.2.1.6 uses the behaviour factor q and explicitly suggests that it must be assumed equal to 1.0 for the convective mode, thereby implying that no reduction due to energy dissipation mechanism is considered. General anchored tanks should be either designed for elastic response (q up to 1.5), or, in properly justified cases, for inelastic response, provided that it is demonstrated that an inelastic response is acceptable. According to Point 4.4(5), EC8-4, steel cylindrical tanks above ground may be designed with a behaviour factor q greater than 1.5 but not larger than 2.0 in the case of unanchored tanks with a bottom base plate designed to allow uplift or sliding and with a thickness which should be less than the thickness of the lower part of the vertical wall (Section 3.2.4 and Points A.9.1 and A.9.5, EC8-4; Point C5.5.3, NZSEE-09). For ground-supported steel tanks with ductile anchored bolts, Point 4.4(5) in EC8-4 suggests a behaviour factor $q = 2.5$ if the anchors allow an increase in length without rupture equal to $\frac{R}{200}$ with R the radius of the tank. Elevated tanks designed for low dissipative structural behaviour (DCL as defined in UNI EN 1998-1:2004, Points 5–7) should have a behaviour factor q up to 1.5. For supporting structures designed according to a dissipative structural strategy (DCM or DCH), the behaviour factor q may be taken as being greater than 1.5 (Point 3.4(4), EC8-4). In particular, for skirt-supported or shaft-type elevated tanks (Point

Table 2.4 Response modification coefficients for impulsive R_i and convective R_c mode from USA documents.

	Support[a]	ASCE 7 R_i	ASCE 7 R_c	ACI 350.3 R_i	ACI 350.3 R_c	AWWA D110 R_i	AWWA D110 R_c	AWWA D115 R_i	AWWA D115 R_c
Reinforced[b] and prestressed concrete tanks (above ground)	Anchored flexible base	3.25	1.5	3.25	1.0	4.5	1.0	3.25	1.0
	Unanchored flexible base	1.5	1.5	1.5	1.0	2.0	1.0	1.5	1.0
	All other (fixed or hinged)	1.5	1.5	2.0	1.0	2.75	1.0	3.25	1.0

	Support	ASCE 7 R_i	ASCE 7 R_c	AWWA D100 R_i	AWWA D100 R_c	API 650 R_i	API 650 R_c
Steel tanks (above ground)	Anchored	3.0	1.5	3.0	1.5	4.0	2.0
	Unanchored	2.5	1.5	2.5	1.5	3.5	2.0

	Support	ASCE 7 R_i	ASCE 7 R_c	ACI 350.3 R_i	ACI 350.3 R_c	ACI 371.R R_i	ACI 371.R R_c	AWWA D100 R_i	AWWA D100 R_c
Elevated tanks	Shaft-type	2.0[c]	1.5	2.0	1.0	2.0	–	3.0[e]	1.5[e]
	Frame-type	3.0[d]	1.5	–	–	–	–	3.0[f]	1.5

Source: Adapted from [270].
a) Figure 2.17 for types of joints used between wall and foundation in concrete tanks.
b) See Table 1.2 for the list of various international codes, standards, guidelines and recommendations.
c) Used also in case of unbraced legs or asymmetrically braced legs for a frame-type tank.
d) Symmetrically braced legs.
e) Just for steel pedestal-type tanks.
f) The response modification coefficient R_i for cross-braced column-supported elevated tanks only applies to tanks with tension-only diagonal bracing.

4.4(4), EC8-4) the upper limit values of q defined in UNI EN 1998-1:2004, Points 5–7 for inverted pendulum structures may be applied. In the case of frame-type elevated tanks, the upper limit of q is defined for the corresponding structural system in UNI EN 1998-1:2004, Points 5–7 times a factor equal to 0.7 in the case of irregularity in elevation (Point 3.4(5), EC8-4). NZSEE-09 does not give any explicit value of the ductility factor for elevated tanks, and it merely states that a capacity design approach should be used to protect elevated tanks against failure (Section 4.1). However, NZSEE-09 suggests that if the staging is detailed to have significant ductility, then it may be designed to respond beyond the yield level under the design earthquake. Approximate methods to estimate the maximum inelastic displacement demands from the maximum displacement demand of a linear elastic single degree of freedom system are suggested at Section 4.4.

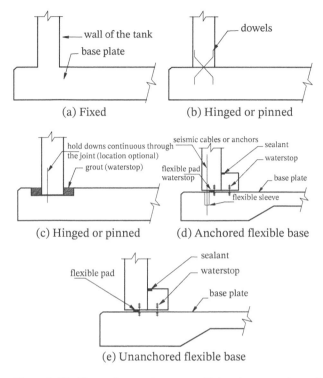

(a) Fixed

(b) Hinged or pinned

(c) Hinged or pinned

(d) Anchored flexible base

(e) Unanchored flexible base

Figure 2.17 Types of ground-supported joints between the wall and its foundation in case of liquid-containing concrete tanks (Point 4.1, AWWA D110-04, Point 2.1, ACI 350.3-06 and Point 8, IS:3370 (Part I)). Source: Adapted from [267].

2.2.10 Tank Forces and Stresses

With the maximum values of the hydrodynamic wall pressure determined using the rules defined in Section 2.2.9, the corresponding tank forces and stresses may be computed. For a vertical cylindrical tank, the maximum values of the following quantities are required for seismic design and safety evaluation (Figure 2.18):

- membrane action such as hoop N_θ and axial force N_y, and in plane shear force $N_{y\theta}$;
- bending action such as the out-of-plane shear Q_y and vertical bending moment M_y.

Equations (2.5) and (2.27) define the pressure distributions for impulsive and convective mode respectively (see also Section 2.2.9). These distributions vary up the vertical axis ζ and around the circumference θ. In the case of pressures due to an earthquake action, Priestley in [457] has shown that the pressure varies only slowly in the circumferential direction, hence the stress distribution at the cylinder generator of maximum pressure, corresponding to $\theta = 0$, can be calculated without significant loss of accuracy by assuming that the entire tank is subjected to an axisymmetric pressure distribution equal to the pressure distribution at $\theta = 0$ (Figure 2.19 and Point 4.9.3, IITK-GSDMA 2007). For design purposes, the maximum values of the dynamic forces and moments induced in the tank by the horizontal

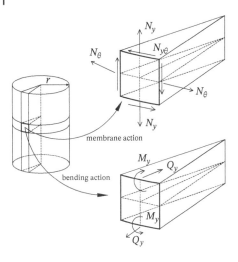

Figure 2.18 Membrane and bending stress resultants for the design of cylindrical tanks with a vertical axis.

and vertical components of the earthquake action should be combined by the square root of the sum of the squares rule and then superimposed on those induced by the hydrostatic pressure (Point 4.2, NZSEE-09, Chapter 4, ACI 350.3-06, Point E.6, API 650 2012 Edition, Point 13.5, AWWA D100-11, Point 4.3.1, AWWA D110-04 and D115-06) [267] or by adding the absolute values of the maxima (Point A.2.1.6, EC8-4). The resulting bending moment along the wall of the tank and across sections normal to the tank axis, using the SRSS rule, may be determined from (Point C4.2, NZSEE-09):

$$M_y = \sqrt{(M_0 + M_w)^2 + M_1^2 + M_V^2} \tag{2.103}$$

where the hydrostatic bending moment M_{hs} should be added; M_w is the bending moment due to the tank inertia (Equation (2.51b) or (2.52b)); M_0 is the hydrodynamic impulsive moment induced on a section of the tank wall immediately above the base (Equation (2.14) or (2.23b)) and M_1 the convective moment, arising from pressures on the wall, immediately above the tank base (Equation (2.35) or (2.42b)); M_V is the bending moment due to the linearly varying axisymmetric hydrodynamic wall pressure (Equation (2.46) or (2.47)) induced by the vertical component of the free-field ground acceleration.

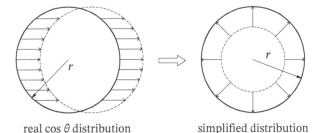

real cos θ distribution   simplified distribution

Figure 2.19 Simplified pressure distribution in circumferential direction on tank wall according to Priestley. Source: Adapted from [457].

If M_y in Equation (2.103) is the maximum value of the bending moment at a section immediately above the base, the maximum axial force in the tank wall is determined from [626]:

$$N_y = \frac{M_y}{Z} s \tag{2.104}$$

where the axial action due to the self-weight of the wall should be added; s is the uniform thickness of the wall and $Z = \pi R^2 s$ is the section modulus for the tank shell (Point C4.4.1, NZSEE-09). The corresponding total foundation moment (or overturning moment) may be expressed in the form (Point C4.1.1, NZSEE-09):

$$M_y' = \sqrt{(M_0' + M_w)^2 + (M_1')^2} \tag{2.105}$$

where M_0' is the rigid impulsive moment below the base plate (Equation (2.12) or (2.24)) and M_1' is the total convective moment at the base, including the base pressure component (Equation (2.33) or (2.43)). The earthquake actions from a vertical seismic excitation are rotationally symmetric, rendering them irrelevant with regard to the overturning foundation moment. As clearly written in Point A.2.1.6, EC8-4, the value of the moment M_y (Equation (2.103)) above the bottom of the base plate should be used for the calculation and verification of stresses in the wall of the tank, while the moment M_y' (Equation (2.105)) should be used for the calculation and verification of the supporting structure, base anchors, or foundation.

The resultant hoop force in the circumferential direction is given by (Point C4.2, NZSEE-09):

$$N_\theta = \sqrt{(N_{\theta i} + N_{\theta w})^2 + N_{\theta c1}^2 + N_{\theta V}^2} \tag{2.106}$$

where the hydrostatic hoop force $N_{\theta hs}$ should be added; $N_{\theta i}$ and $N_{\theta c1}$ are the impulsive and first convective mode hoop effects, respectively, produced by the horizontal component of the ground shaking while $N_{\theta V}$ is the effect induced by the vertical components of the excitation; $N_{\theta w}$ is the maximum value of the circumferential hoop force in the tank wall and is given by Equation (2.53).

The radial base shear applied at the bottom of the tank should be determined by:

$$Q_y = \sqrt{(Q_0 + Q_w)^2 + Q_1^2 + Q_V^2} \tag{2.107}$$

where the hydrostatic shear force Q_{hs} should be added; Q_V is the out-of-plane shear due to the linearly varying axisymmetric hydrodynamic wall pressure (Equation (2.46) or (2.47)) induced by the vertical component of the free-field ground acceleration; Q_0 is the hydrodynamic impulsive shear at the base (Equation (2.9) or (2.23a)), and Q_1 the convective first mode shear, arising from pressures on the wall, at the tank base (Equation (2.31) or (2.42a)); Q_w is the base shear due to the tank inertia (Equation (2.51a)). The corresponding value of the in the plane shearing force in circumferential direction is (Point C4.5, NZSEE-09 and Point R5.2.2, ACI 350.3-06) [68]:

$$N_{y\theta} = \frac{Q_y \sin \theta}{\pi R} \tag{2.108}$$

With the maximum value of the hydrodynamic wall pressures determined, impulsive p_i (Equation (2.5) or (2.18)/(2.19)), convective p_c (Equation (2.27) or (2.40)), and vertical p_{vr} (Equations (2.46) and (2.47)), the corresponding maximum values of the tank forces N_θ^{max} and M_y^{max} may be computed by:

- applying the vertical distribution of the pressure on a finite element model of the tank [214, 297, 506];
- applying the dimensionless design charts, for impulsive, convective, and hydrostatic (or vertical) pressures with rigid container assumptions, related to tank radius R, height H and wall thickness s for a

range of $\frac{H}{R} = 0.2, 0.5, 1.0, 2.0, 5.0$ and $\frac{R}{s} = 50, 100, 200, 500$ and 1000. The simple frame analogy proposed by Priestley in [457] has been used, in conjunction with the actual pressure distribution, to produce the dimensionless design charts in Appendix A for hoop forces $N_\theta^\star$ and vertical bending moments $M_y^\star$ in the walls of cylindrical tanks with a constant wall thickness (Points C4.3.1 and Appendix A, NZSEE-09) [459]. The hoop force N_θ^{max} and the vertical bending moment M_y^{max} are given by:

$$N_\theta^{max} = N_\theta^\star \cdot R \cdot p^{max} \tag{2.109a}$$

$$M_y^{max} = M_y^\star \cdot R \cdot s \cdot p^{max} \tag{2.109b}$$

The expanded form of the components (2.109) is then for hydrostatic ($z = 0$):

$$p_{hs}^{max} = \gamma_l H \tag{2.110a}$$

$$N_{\theta hs}^{max} = N_{\theta hs}^\star R p_{hs}^{max} \tag{2.110b}$$

$$\sigma_{\theta hs}^{max} = \frac{N_{\theta hs}^{max}}{s} \tag{2.110c}$$

$$M_{yhs}^{max} = M_{yhs}^\star R s p_{hs}^{max} \tag{2.110d}$$

$$\sigma_{yhs}^{max} = \frac{M_{yhs}^{max}}{Z} \tag{2.110e}$$

for impulsive (Equation (2.21) and $z = 0$):

$$p_i^{max} = q_0(0)S_e(T_0)\rho R \tag{2.111a}$$

$$N_{\theta i}^{max} = N_{\theta i}^\star R p_i^{max} \tag{2.111b}$$

$$\sigma_{\theta i}^{max} = \frac{N_{\theta i}^{max}}{s} \tag{2.111c}$$

$$M_{yi}^{max} = M_{yi}^\star R s p_i^{max} \tag{2.111d}$$

$$\sigma_{yi}^{max} = \frac{6M_{yi}^{max}}{s^2} \tag{2.111e}$$

for convective (Equation (2.41) and $z = H$):

$$p_c^{max} = q_1(z = H)S_e(T_1)\rho R \tag{2.112a}$$

$$N_{\theta c}^{max} = N_{\theta c}^\star R p_c^{max} \tag{2.112b}$$

$$\sigma_{\theta c}^{max} = \frac{N_{\theta c}^{max}}{s} \tag{2.112c}$$

$$M_{yc}^{max} = M_{yc}^\star R s p_c^{max} \tag{2.112d}$$

$$\sigma_{yc}^{max} = \frac{6M_{yc}^{max}}{s^2} \tag{2.112e}$$

for vertical ($z = 0$):

$$p_{vr}^{max} = S_{ve}(T_V)\rho R \tag{2.113a}$$

$$N_{\theta vr}^{max} = N_{\theta vr}^\star R p_{vr}^{max} \tag{2.113b}$$

$$\sigma_{\theta vr}^{max} = \frac{N_{\theta vr}^{max}}{s} \tag{2.113c}$$

$$M_{yvr}^{max} = M_{yvr}^{\star} R s p_{vr}^{max} \tag{2.113d}$$

$$\sigma_{yvr}^{max} = \frac{6 M_{yvr}^{max}}{s^2} \tag{2.113e}$$

As suggested by Appendix A in NZSEE-09, expressions (2.111a) and (2.113a), may be also used as a good approximation for flexible tanks, using the period's equations defined in Section 5.4 (with or without soil–structure interaction);

- applying an equivalent linear pressure distribution, where wall thickness does not vary excessively (Point C4.3.2, NZSEE-09; Point 4.9.4, IITK-GSDMA 2007; Point 5.3.1, ACI 350.3-06) [412], adopted from Point CA2.4, NZS 3106 (2009) and depicted in Figure 2.25 using Equations (2.135) and (2.137) in which 50% of the total impulsive Q_0 and convective Q_1 resultant base horizontal shear (in kN) must be assigned, according to Point C4.9.4, IITK-GSDMA 2007 (Point A2.4, NZS 3106 (2009)), to each side of the tank. In the following are the relations, for cylindrical tanks, substituting Equations (2.136) and (2.138):

$$p_0^{top} = \frac{Q_0(6h_0 - 2H)}{\pi R H^2} \tag{2.114a}$$

$$p_0^{bot} = \frac{Q_0(4H - 6h_0)}{\pi R H^2} \tag{2.114b}$$

regarding the impulsive component:

$$p_1^{top} = \frac{Q_1(6h_1 - 2H)}{\pi R H^2} \tag{2.115a}$$

$$p_1^{bot} = \frac{Q_1(4H - 6h_1)}{\pi R H^2} \tag{2.115b}$$

regarding the convective part.

For the hydrostatic stress resultants, the representative pressure p_{hs} may be used in combination with the dimensionless design tables given in "Circular concrete tanks without prestressing", Portland Cement Association, 1993 or in Part IV Indian Standard - IS:3370 (1967 and reaffirmed in 2004). In both cases, solutions are given for the ratio $0.4 \le \frac{H^2}{2Rs} \le 16$, with a fixed or hinged base when a uniform, triangular, or trapezoidal load is applied along the wall, and a shear force at the top in order to take into account a slab on the top of the tank to prevent free movement at that location.

2.2.11 Effects of Rocking Motion

As clearly explained previously, the analysis of the seismic response of a cylindrical rigid tank to horizontal ground shaking is normally carried out on the assumption that the tank base moves horizontally without any rotation. In reality, as observed during the Prince William Sound, Alaska, earthquake (1964) [213, 478], because of the flexibility of the supporting soils [221, 638], the tank base experiences a rigid base rocking component of motion, $\theta_g(t)$ (corresponding to an angular acceleration of the base $\ddot{\theta}_g(t)$ in Figure 2.1 (and shown later on flexible tanks in Figure 5.18) evaluated according to [204, 330, 337, 498, 593]), even under a purely translational free-field ground motion. The procedure suggested and recommended for use in practice by Veletsos and Tang [638] in the case of a rocking response of a rigid cylindrical tank subjected to an angular acceleration of the base $\ddot{\theta}_g(t)$, defines the following impulsive expressions:

$$p_i(z,\theta,t) = c_i^r(z)\rho_l R(H\ddot{\theta}_g(t))\cos\theta \tag{2.116a}$$

$$Q_i(t) = m_i^r(H\ddot{\theta}_g(t)) = m_i\frac{h_i'}{H}(H\ddot{\theta}_g(t)) \tag{2.116b}$$

$$M_i(t) = i_i mH^2\ddot{\theta}_g(t) \tag{2.116c}$$

$$M_i'(t) = i_i' mH^2\ddot{\theta}_g(t) \tag{2.116d}$$

and, consequently, the following convective expressions [638]:

$$p_c(z,\theta,t) = \rho_l R\cos\theta\sum_{n=1}^{\infty}c_n^r(z)a_{cn}^r(t) \tag{2.117a}$$

$$Q_c(t) = \sum_{n=1}^{\infty}m_{cn}^r a_{cn}^r(t) = \sum_{n=1}^{\infty}m_{cn}\frac{h_{cn}'}{H}a_{cn}^r(t) \tag{2.117b}$$

$$M_c(t) = \sum_{n=1}^{\infty}m_{cn}^r h_{cn}a_{cn}^r(t) \tag{2.117c}$$

$$M_c'(t) = \sum_{n=1}^{\infty}m_{cn}^r h_{cn}'a_{cn}^r(t) \tag{2.117d}$$

where ρ_l is the mass density of the liquid ($m = \rho_l\pi R^2 H$ is the total contained mass of the fluid). The impulsive mass m_i and height h_i' may be derived from relations (2.10) and (2.13); m_{cn}, h_{cn}' and h_{cn} come respectively from Equations (2.32), (2.34), and (2.36). The tank-liquid system investigated is depicted in Figure 2.1 (and also Figure 5.17) and it is presumed to be of uniform thickness and clamped to a rigid base. The impulsive pressure exerted against the rigid tank wall, Equation (2.116a), depends on the dimensionless function $c_i^r(z)$ that define, the heightwise variation of the pressure:

$$c_i^r(z) = 2\gamma\sum_{n=1}^{\infty}\frac{(-1)^{n+1}-\frac{2}{v_n'}}{(v_n')^2}\frac{I_1\left(\frac{v_n'}{\gamma}\right)}{I_1'\left(\frac{v_n'}{\gamma}\right)}\cos\left(v_n'\frac{z}{H}\right)+1-\frac{z}{H} \tag{2.118a}$$

$$v_n' = \frac{2n-1}{2}\pi \tag{2.118b}$$

$$\gamma = \frac{H}{R} \tag{2.118c}$$

and on $H\ddot{\theta}_g(t)$ that is the horizontal acceleration of the tank wall at the level of the still liquid surface [638]. The distribution of $c_i^r(z)$ is shown in Figure 2.20(a) for tanks with several different values of $\frac{H}{R}$. Note that for broad tanks, $c_i^r(z)$ increases monotonically from top $z = H$ to bottom $z = 0$, whereas for slender/tall tanks, it varies irregularly, attaining its maximum value away from the bottom of the tank itself. This trends are in sharp contrast to that reported in Figure 2.2(a) for a laterally excited tank, for which the pressure increases always monotonically from top to bottom irrespective of the values assumed for $\frac{H}{R}$.

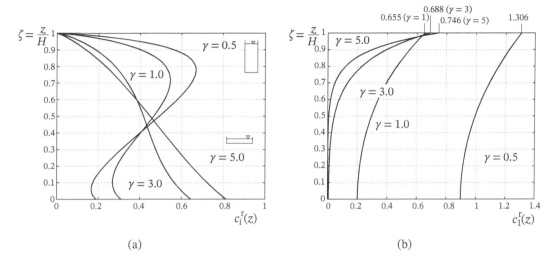

Figure 2.20 (a) Coefficient $c_i^r(z)$ for impulsive pressure variation in rigid (and flexible) tanks corresponding to four values as a function of the tank aspect ratio $\gamma = \frac{H}{R}$. Source: Aadapted from [638]. (b) Convective component $c_1^r(z)$ along the wall of the tank.

The dimensionless coefficients i_i and i'_i in expressions (2.116c) and (2.116d) may be expressed respectively as:

$$i_i = \frac{1}{6} + 2\gamma \sum_{n=1}^{\infty} \frac{1 - \dfrac{3(-1)^{n+1}}{v'_n} + \dfrac{2}{(v'_n)^2} I_1\left(\dfrac{v'_n}{\gamma}\right)}{(v'_n)^3} \, \frac{}{I'_1\left(\dfrac{v'_n}{\gamma}\right)} \tag{2.119a}$$

$$i'_i = i_i + \Delta i_i = i_i + \frac{1}{4\gamma^2} + 2\sum_{n=1}^{\infty} \frac{(-1)^{n+1} - \dfrac{2}{v'_n} I_2\left(\dfrac{v'_n}{\gamma}\right)}{(v'_n)^3} \, \frac{}{I'_1\left(\dfrac{v'_n}{\gamma}\right)} \tag{2.119b}$$

The maximum value of the impulsive pressure p_i, shear Q_i, and moments, M_i and M'_i, may then be determined by replacing $\ddot{\theta}_g(t)$ by its rocking spectral value S^r_e [243, 359, 458] using the impulsive natural frequency, given in Section 2.2.5 for tanks excited laterally (with or without soil–structure interaction) and an equivalent viscous damping value according to the expression proposed by Makris and Konstantinidis in [359]:

$$-0.34\ln(r) \tag{2.120}$$

where the value r of the coefficient of restitution is equal to [243, 458]:

$$r = \left(1 - \frac{3}{2}\sin^2\bar{\alpha}\right)^2 \tag{2.121}$$

and $\bar{\alpha}$ is the slenderness of the tank:

$$\bar{\alpha} = \tan^{-1}\left(\frac{2R}{H}\right) \tag{2.122}$$

The convective nondimensional component of the hydrodynamic pressure, (2.117a), for tanks excited in rocking is related to the same quantity for tanks excited laterally:

$$c_n^r(z) = q_n(z)\frac{h'_{cn}}{H} \tag{2.123}$$

where $q_n(z)$ may be derived from Equation (2.41); the distribution of 1st−sloshing nondimensional component of the hydrodynamic pressure, $c_1^r(z)$, is shown in Figure 2.20(b). The quantity $a_{cn}^r(t)$ represents the instantaneous value of the pseudo-acceleration induced by the prescribed base horizontal acceleration $H\ddot{\theta}_g(t)$ in an undamped single degree of freedom linear oscillator, with a prescribed natural frequency according to Equation (2.30) or (2.54) (Point C3.3.1, NZSEE-09) [635] for tanks excited laterally:

$$a_{cn}^r(t) = \omega_{cn}\int_0^t (H\ddot{\theta}_g(t))\sin\left[\omega_{cn}(t-\tau)\right]d\tau \tag{2.124}$$

The maximum value of the convective pressure p_c, shear Q_c, and moments, M_c and M'_c, may then be determined by replacing $a_{cn}^r(t)$ by its rocking spectral value S_e^r using the natural frequency according to Equation (2.30) or (2.54) (as already done in expressions (2.40)) (Point C3.3.1, NZSEE-09) [635].

2.3 Rectangular Tanks Fully Anchored at the Base

The procedure for the evaluation of the earthquake pressures, forces, and moments on a rigid rectangular tank of dimensions $2L \times 2B$ is quite similar to that described previously for vertical cylindrical tanks fully anchored at the base.

2.3.1 Impulsive and Convective Pressure Components

The spatial-temporal variation (z, t) of the rigid impulsive pressure on the walls acting perpendicular to the direction of the ground motion $a(t)$ follows the expression (Point A.4.1, EC8-4 and Eq. (C3.16) Point C3.3.1, NZSEE-09):

$$p_i(z, t) = q_0(z)\rho_l La(t) \tag{2.125}$$

where $a(t)$ is the free-field ground acceleration, with peak value denoted by a_g; L is the half-width of the tank parallel to the direction of the seismic action because, in the two-dimensional problem, the half-width of the tank is substituted for the tank radius R as the characteristic length used to form the dimensionless variables describing the equivalent mechanical model [247]; ρ_l (in kg/m^3) is the mass density of the liquid. The maximum value of the impulsive pressure p_i may then be determined by replacing $a(t)$ by its spectral value S_e using the natural frequency and damping value according to the terms already specified for cylindrical tanks in Equations (2.18) and (2.19) [635]. The dimensionless function $q_0(z)$, obtained from the potential flow theory [217] is used to give the variation of the impulsive pressure along the height H and the trend is very close to that of a cylindrical tank with $R = L$ (as clearly stated by comparing Figures 2.5(a) and 2.21(a)) [217]:

$$q_0(z) = \frac{H}{L} \sum_{n=1}^{\infty} \frac{8(-1)^{n+1}}{[(2n-1)\pi]^2} \tanh\left[(2n-1)\frac{\pi}{2}\frac{L}{H}\right] \cos\left[(2n-1)\frac{\pi}{2}\frac{z}{H}\right] \qquad (2.126)$$

Alternatively to Equation (2.126), the approximate pressure distribution derived by Housner [240, 241], which considers only the first term of the series, may be used (Point 5.2.7.2, IS:1893 (1894)):

$$q_0(z) = \frac{H}{L}\frac{\sqrt{3}}{2}\left[1 - \left(\frac{z}{H}\right)^2\right] \tanh\left(\sqrt{3}\frac{L}{H}\right) \qquad (2.127)$$

The two distributions are somewhat different and are compared in Figure 2.21(d) [217]; the peak value $q_0(0)$ of dimensionless impulsive pressures along the height z and for different ratio $\frac{H}{L}$ is depicted in Figures 2.21(b) and 2.21(c). However, the difference between the total lateral shear forces in the two formulations does not exceed 10% in the range $0.2 \leq \frac{H}{2L} \leq 1.4$. The shape and magnitude of the dimensionless pressure coefficient $q_0(z)$ vary with the slenderness $\gamma = \frac{H}{L}$ and along the height of the tank z as shown in Figures 2.21(a)–2.21(c). It is important to highlight that the pressure distribution $p_i(z,t)$ is constant along the direction $2B$ perpendicular to the direction of the seismic action.

The spatial-temporal variation (z, t) of the convective pressure, corresponding to the nth–sloshing mode of vibration of the liquid in the tank, can be obtained by the following relation [126, 127]:

$$p_{cn}(z,t) = q_n(z)\rho_l L a_{cn}(t) \qquad (2.128)$$

where $a_{cn}(t)$ represents the instantaneous value of the pseudo-acceleration induced by the prescribed free-field ground motion $a(t)$ in a single degree of freedom linear oscillator having a circular frequency ω_{cn} (Equation (2.30)) and a viscous damping ratio (Section 2.2.6), both equal to those of the nth–sloshing mode of vibration of the liquid in the tank. Considering just the dominant contribution of the fundamental mode of vibration, we obtain from Equation (2.128) the following expression corresponding to the 1st–sloshing mode of vibration (Point A.4.1, EC8-4 and Eq. (C3.17), Point C3.3.1, NZSEE-09) [126]:

$$p_{c1}(z,t) = q_1(z)\rho_l L a_{c1}(t) \qquad (2.129)$$

The maximum value of the convective pressure p_{c1} may then be determined by replacing $a_{c1}(t)$ by its spectral value S_e using the natural frequency and damping value according to the terms already defined in Equation (2.40) for cylindrical tanks [635]. The dimensionless convective pressure functions $q_n(z)$ are given by [247]:

$$q_n(z) = \frac{2}{\alpha_n^2} \frac{\cosh\left(\alpha_n \frac{z}{H}\right)}{\cosh\left(\alpha_n \frac{H}{L}\right)} \qquad (2.130)$$

and are shown in Figures 2.22(a) and 2.22(b) for the first two sloshing modes respectively ($\alpha_1 = \frac{\pi}{2}$ for the first mode and $\alpha_2 = \frac{3\pi}{2}$ for the second mode).

The impulsive and convective forces (Equations (2.9), (2.23a), (2.31), and (2.42a)) and moments (Equations (2.12), (2.14), (2.23b), (2.24), (2.33), (2.35), (2.42b), and (2.43)) given for cylindrical tanks

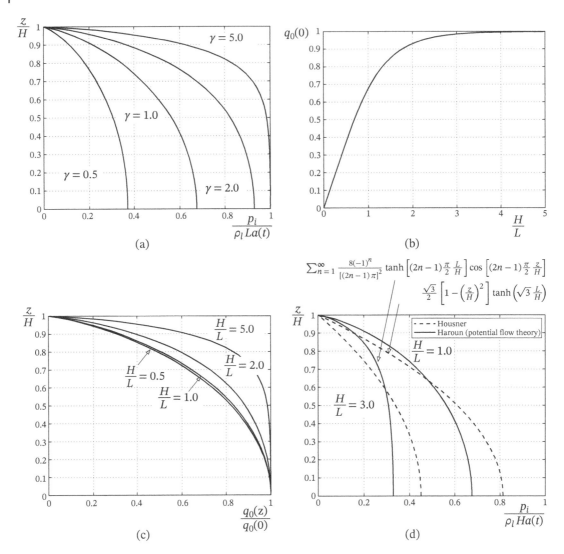

Figure 2.21 (a) Coefficient $q_0(z)$ for impulsive pressure variation, normalized to $\rho_l La(t)$, in rigid rectangular tanks: four values as a function of the tank aspect ratio $\gamma = \frac{H}{L}$; (b) peak value at the base ($z = 0$) of dimensionless impulsive pressures on rectangular wall perpendicular to direction of the earthquake; (c) shape and magnitude variation of the dimensionless ratio $\frac{q_0(z)}{q_0(0)}$ as a function of the geometric ratio $\gamma = \frac{H}{L}$ and the height of the tank; (d) difference between the impulsive hydrodynamic pressures in the two formulations by Housner [240, 241] and Haroun [217].

may be applied to rectangular containers using the following equivalent mechanical parameters in the case of reinforced and prestressed concrete tanks (Point 9.2, ACI 350.3-06):

$$\frac{m_0}{m} = \frac{\tanh\left(\sqrt{3}\frac{L}{H}\right)}{\sqrt{3}\frac{L}{H}} \tag{2.131a}$$

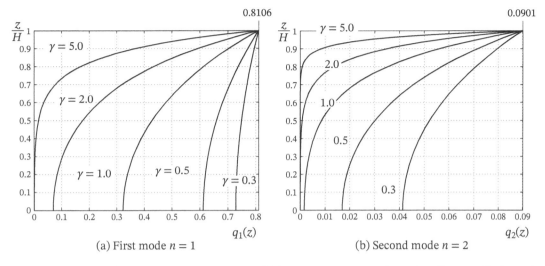

Figure 2.22 Dimensionless convective pressure coefficients $q_n(z)$ on rectangular tank wall perpendicular to the direction of the earthquake as a function of the tank aspect ratio $\gamma = \frac{H}{L}$.

$$\frac{m_1}{m} = \frac{1}{3}\sqrt{\frac{5}{2}\frac{L}{H}}\tanh\left(\sqrt{\frac{5}{2}\frac{H}{L}}\right) \tag{2.131b}$$

$$\frac{h_0}{H} = \frac{3}{8} \qquad\qquad \text{if} \quad \frac{2L}{H} \geq 1.333 \tag{2.131c}$$

$$\frac{h_0}{H} = 0.5 - 0.09375\frac{2L}{H} \qquad\qquad \text{if} \quad \frac{2L}{H} < 1.333 \tag{2.131d}$$

$$\frac{h_0'}{H} = \frac{\sqrt{3}\dfrac{L}{H}}{2\tanh\left(\sqrt{3}\dfrac{L}{H}\right)} - \frac{1}{8} \qquad\qquad \text{if} \quad \frac{2L}{H} \geq 0.75 \tag{2.131e}$$

$$\frac{h_0'}{H} = 0.45 \qquad\qquad \text{if} \quad \frac{2L}{H} < 0.75 \tag{2.131f}$$

$$\frac{h_1}{H} = 1 - \frac{\cosh\left(\sqrt{\dfrac{5}{2}\dfrac{H}{L}}\right) - 1}{\sqrt{\dfrac{5}{2}\dfrac{H}{L}}\sinh\left(\sqrt{\dfrac{5}{2}\dfrac{H}{L}}\right)} \tag{2.131g}$$

$$\frac{h_1'}{H} = 1 - \frac{\cosh\left(\sqrt{\dfrac{5}{2}\dfrac{H}{L}}\right) - 2.01}{\sqrt{\dfrac{5}{2}\dfrac{H}{L}}\sinh\left(\sqrt{\dfrac{5}{2}\dfrac{H}{L}}\right)} \tag{2.131h}$$

In the case of rigid rectangular steel tanks, the equivalent masses (m_i or m_0 and m_{c1} or m_1) and their heights above the base (h_i', h_i, h_{c1}' and h_{c1} or h_0', h_0, h_1' and h_1) may be assumed to be the same as for the circular cylindrical tank (Sections 2.2.1 and 2.2.2) with the half-length L of the rectangular tank used instead of the circular radius R (Table C 1, IITK-GSDMA 2007, Point 13.5, AWWA D100-11, Point 6.3, "Nuclear Reactors and Earthquakes", TID-7024). As explicitly written in Point C3.3.1, NZSEE-09 in most cases this approximation is expected to give base shears and moments at the base within 15% of the value from a more exact theory [317].

2.3.2 Periods of Vibration

The convective period of vibration of the nth–sloshing mode of vibration of the liquid in the tank can be obtained by the following relation (Eq. (C3.25), Point C3.6, NZSEE-09, Point 6.3, "Nuclear Reactors and Earthquakes", TID-7024, Point 4.3.2, IITK-GSDMA 2007; Point 4.4.1, HSE Research Report RR527):

$$T_n = \frac{2\pi}{\omega_{cn}} = \frac{2\pi\sqrt{\dfrac{L}{g}}}{\sqrt{\alpha_n \tanh\left(\dfrac{\alpha_n H}{L}\right)}} \tag{2.132}$$

where $\alpha_1 = \frac{\pi}{2}$ for the first mode of vibration and consequently $\alpha_2 = \frac{3\pi}{2}$ and $\alpha_3 = \frac{5\pi}{2}$ respectively for the second and third modes of vibration.

The horizontal T_0 and vertical T_V impulsive periods of vibration for rigid rectangular tanks, including soil–structure interaction, may be evaluated from relations (2.55) and (2.59) respectively, where R_b is the radius of an equivalent circular foundation. Alternatively to an equivalent foundation and alternatively to Equations (2.56), (2.57), and (2.60), Point D.3.1, EN 1998-5, 2020 Edition can be used and the following expressions adopted:

$$K_x = \frac{2}{2 - v_s} G_s B \left[1.2 + 3.3\left(\frac{L}{B}\right)^{0.65}\right] \alpha_x \tag{2.133a}$$

$$K_\theta = \frac{1}{1 - v_s} G_s B^3 \left[3.6\left(\frac{L}{B}\right)^{2.4}\right] \alpha_\theta \tag{2.133b}$$

$$K_V = \frac{2}{1 - v_s} G_s L \left[0.73 + 1.54\left(\frac{B}{L}\right)^{0.75}\right] \alpha_V \tag{2.133c}$$

where the rectangular tank has dimensions $2L \times 2B$, with $2L$ the tank foundation's longest length, parallel to the direction of the earthquake loading (x), and $2B$ is the foundation's shortest width.

In the case of a rigid tank ignoring the soil–structure interaction, the impulsive period T_0 is theoretically zero, and $S_e(T_0) = Sa_g$, where a_g is the peak value of the ground acceleration $a(t)$ and S the soil factor (Point 3.2.2.2, UNI EN 1998-1:2004). For the same conditions, the vertical period T_V is theoretically zero, and $S_{ve}(T_V) = a_{vg}$ (Section 2.2.3) where a_{vg} is the vertical component of the free-field ground acceleration (Point 3.2.2.3, UNI EN 1998-1:2004). Unless calculated on the basis of a more rigorous analysis, Point CA2.2.1, NZS 3106 (2009) suggests a ground-supported rectangular (or circular) tank's horizontal and vertical mode should assume $T_0 = 0.1$ sec. This puts the response within or near to the peak spectral response, consequently the peak response spectrum ordinates are required to be used, unless it can be shown that other ordinates are appropriate.

2.3.3 Convective Wave Displacement

The maximum vertical displacement of the convective sloshing wave from the at-rest level of the liquid for rectangular tanks is given by (Eq. (C3.37), Point C3.9, NZSEE-09; Point A2.5, NZS 3106 (2009)):

$$d_{\max} = L \frac{S_e(T_1)}{g} \tag{2.134}$$

where $S_e(T_1)$ is the spectral acceleration obtained from a horizontal elastic response spectrum corresponding to the convective period of the first mode (Equation (2.132)) and to the viscous damping ratio, as given in Section 2.2.6 and g is the gravitational acceleration. Equation (2.134) only includes the first mode displacement but generally higher mode effects can be ignored (Point C3.9, NZSEE-09; Point 4.11, IITK-GSDMA 2007).

2.3.4 Tank Forces and Stresses

When a rectangular tank is subjected to the action of an earthquake motion, tank walls experience convective p_{c1} and impulsive p_i hydrodynamic horizontal and vertical p_{vr} pressures and inertia pressure p_w, in addition to the hydrostatic pressure p_{hs} (Figure 2.23). The stress resultants required for the design of rectangular tanks of dimensions $2L \times 2B$ are depicted in Figure 2.24:

- membrane action such as vertical axial force N_y (where $Z = 4Ls\left(B + \frac{L}{3}\right)$ must be used in Equation (2.104)), horizontal axial force N_z, and in plane shear force N_{yz};
- vertical and horizontal shear/bending action such as the out-of-plane shear Q_y and the bending moments M_y and M_z.

For design purposes, the tank is divided into leading half and the trailing half portions as shown in Figure 2.24. It is usually assumed in Point CA2.4, NZS 3106 (2009) and in Point 5.3.1, ACI 350.3-06, that the impulsive and convective forces are equally resisted by the leading and the trailing walls perpendicular to the direction of the earthquake force. Thus, walls perpendicular to the ground motion being investigated in the leading (or trailing) half of the tank should be loaded perpendicular to their plane (dimension $2B$ in Figure 2.23) by the wall's own inertia pressure p_w, hydrostatic p_{hs} and vertical p_{vr}, one-half the impulsive pressure p_i and one-half the convective pressure p_{c1}.

Figure 2.23 Static and hydrodynamic pressure distribution acting on tank wall (Point 5.3, ACI 350.3-06). Source: Adapted from [217].

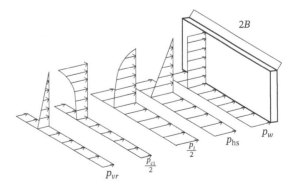

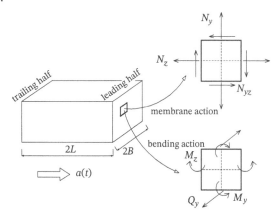

Figure 2.24 Membrane and bending stress resultants for design of rectangular tanks (Point C4.2, NZSEE-09).

With the vertical and horizontal distribution of the hydrostatic, inertia and hydrodynamic wall pressures determined (Figure 2.23), impulsive p_i (Equation (2.125)), convective p_{c1} (Equation (2.129)), and vertical p_{vr} (Equations (2.46) and (2.47)), the corresponding values of the tank forces may be computed by:

- applying the vertical distribution of the pressures on a finite element model of the tank;
- applying the dimensionless design charts, for hydrostatic p_{hs} (or vertical p_{vr}) or uniform p_w pressure with rigid container assumptions, related to a rectangular tank or an individual wall, given in "Rectangular concrete tanks", Portland Cement Association, 1998 [411] or in Part IV, Indian Standard - IS:3370 (1967 and reaffirmed in 2004). In both cases the solutions are given for the ratio $1.0 \leq \frac{2L}{H} \leq 4.0$ and $0.5 \leq \frac{2B}{H} \leq 3.0$, with a fixed or hinged base and free or hinged at the top;
- applying a reasonable approximation, where wall thickness does not vary excessively (Point C4.3.2, NZSEE-09; Point 4.9.4, IITK-GSDMA 2007; Point 5.3.1, ACI 350.3-06) [412], adopted from Point CA2.4, NZS 3106 (2009). As depicted in Figure 2.25, an equivalent linear pressure distribution with height, rather than a curved one for convective and impulsive components, may be assumed so as to give the same base shear and bending moment at the bottom of the tank wall. Analysis of the earthquake response of the tank should be based on distributions of the impulsive and convective equivalent pressure that vary linearly from p_0^{top} at the surface of the liquid to p_0^{bot} at the base (or p_1^{top} and p_1^{bot} for the convective components), applying translational and rotational equilibrium conditions (Figure 2.25(a)):

$$\frac{Q_0}{4B} = q_0^u + q_0^l = (p_0^{top} + p_0^{bot})\frac{H}{2} \tag{2.135a}$$

$$\frac{Q_0}{4B}h_0 = q_0^u\frac{H}{2} + q_0^l\frac{H}{3} = (2p_0^{top} + p_0^{bot})\frac{H^2}{6} \tag{2.135b}$$

and resulting in the two following main unknowns Point A2.4, NZS 3106 (2009):

$$p_0^{top} = \frac{Q_0(6h_0 - 2H)}{4BH^2} \tag{2.136a}$$

$$p_0^{bot} = \frac{Q_0(4H - 6h_0)}{4BH^2} \tag{2.136b}$$

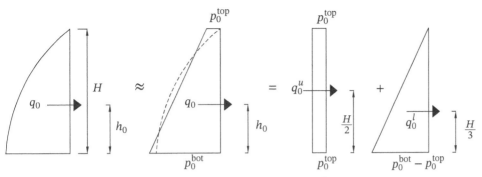

(a) Impulsive pressure: $q_0 = \frac{Q_0}{\pi R}$ for cylindrical and $q_0 = \frac{Q_0}{4B}$ for rectangular tanks

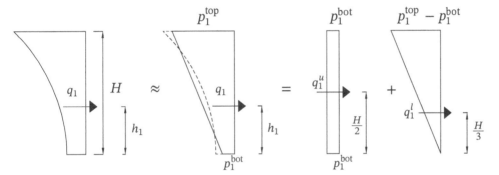

(b) Convective pressure: $q_1 = \frac{Q_1}{\pi R}$ for cylindrical and $q_1 = \frac{Q_1}{4B}$ for rectangular tanks

Figure 2.25 Equivalent linear distribution of impulsive (a) and convective (b) pressures on tank wall perpendicular to the earthquake force (Point C4.3.2, NZSEE-09; Point 4.9.4, IITK-GSDMA 2007; Point 5.3.1, ACI 350.3-06; Point CA2.4, NZS 3106 (2009)). Source: Adapted from [412].

The same expression may be written for the convective component (Figure 2.25(b)):

$$\frac{Q_1}{4B} = q_1^u + q_1^l = (p_1^{\text{top}} + p_1^{\text{bot}})\frac{H}{2} \tag{2.137a}$$

$$\frac{Q_1}{4B}h_1 = q_1^u\frac{H}{2} + q_1^l\frac{H}{3} = (2p_1^{\text{top}} + p_1^{\text{bot}})\frac{H^2}{6} \tag{2.137b}$$

where Point A2.4, NZS 3106 (2009):

$$p_1^{\text{top}} = \frac{Q_1(6h_1 - 2H)}{4BH^2} \tag{2.138a}$$

$$p_1^{\text{bot}} = \frac{Q_1(4H - 6h_1)}{4BH^2} \tag{2.138b}$$

Note that in Equations (2.135) and (2.137) 50% of the total impulsive Q_0 and convective Q_1 resultant base horizontal shear is assigned to each side of the tank in agreement with Point 5.3.1, ACI 350.3-06 and Point CA2.4, NZS 3106 (2009).

3

Above-Ground Unanchored Rigid Tanks

After reading this chapter you should be able to:

- Understand the equivalent mechanical model to analyse unanchored tanks
- Evaluate impulsive and convective modes and masses for rigid wall tanks
- Identify the critical components in a tank's design

3.1 Introduction

The tank wall in the anchored tanks described in Chapter 2 is effectively fixed to a foundation that must be sufficiently heavy to prevent uplift during an earthquake [251]. In practice, a large number of anchor bolts are required to be able to transmit the earthquake-induced vertical tension in the tank wall to the foundation. Thus, anchoring a tank is expensive, because an anchored tank must be provided with a massive foundation adequate to resist any uplift tendency in the shell and because improperly or poorly designed attachments can cause considerable damage to the shell wall (Figure 3.1(a)) and to the foundation (Figure 3.1(b)). As a result, it is far more common, particularly for large capacity broad tanks, to build the tank on a simple ringwall foundation, adequate to prevent differential settlement, without any attempt to restrain uplift of the shell in response to seismic load.

A number of analytical and experimental research studies have been reported in the literature on investigations of the seismic behaviour of unanchored tanks [220]. Experimental studies involving both static-tilt tests and dynamic shaking table tests have been reported by Clough [95], Niwa [427], Clough and Niwa [99], Cambra [67], Manos and Clough [381], Shih and Babcock [536, 537], Akiyama and Yamaguchi [7], Manos and Talaslidis [383], and Sakai et al. [497]. Clough [95] first investigated the uplift of the tank bottom plate by imposing an equilibrium between the overturning moment induced by the lateral forces of the tank–fluid systems and the restoring moment induced by the reaction forces in a given displaced configuration. Using this formulation, the uplift region of the tank bottom plate was identified and the response may be characterized by a rigid body rotation of the tank, which lifts off the foundation on one side and rocks up on its toe. The geometrical and material properties of the bottom plate and the tank's wall do not affect Clough's relation. An improvement on Clough's method

Seismic Design and Analysis of Tanks, First Edition. Gian Michele Calvi and Roberto Nascimbene.
© 2023 John Wiley & Sons, Inc. Published 2023 by John Wiley & Sons, Inc.

(a) Damage to the wall   (b) Damage to the foundation

Figure 3.1 Damaged tanks observed during the Emilia earthquake in Italy, 20 and 29 May 2012.

was defined first by Wozniak and Mitchell [672] (small deflection theory) and successively by Leon and Kausel [333], considering that the uplifting resistance of a base plate may be gained from the solution of the somewhat simpler problem of a uniformly loaded prismatic beam uplifted at one end. In the rigid–plastic beam model used by Wozniak and Mitchell, the effect of membrane forces on the base plate is ignored, with the result that the maximum load capacity for the beam is reached as soon as the two plastic hinges develop: one at the uplifted end, and the other at the section of the maximum moment within the beam. The Wozniak and Mitchell model has provided the basis of the method proposed in Appendix E, API 650 (2012) (relation E.6.2.1.1-1b) and Point 13.5.4, AWWA D100-11 for the design of unanchored welded steel tanks. The experimental tests by Clough and Niwa [99] and Manos and Clough [381] revealed the importance of membrane forces in the load-carrying mechanism and provided valuable information for analytical developments. An approximate solution, based on the theory of a semi-infinite beam on an elastic foundation [429], that incorporates the effects of the membrane forces, was proposed by Cambra [67] (large deflection theory) using, on the basis of experimental data, certain simplifying assumptions regarding the magnitude of the axial and shearing forces in the beam. A second-order beam theory that provides for the effects of membrane forces more accurately was used by Auli et al. [19, 165] considering the vertical restraining action of the base plate and the foundation modelled by non-linear Winkler springs. However, since these investigations have not been discussed from the perspective of the dynamics of rocking motion, the only overturning moment induced by the traditional translational tank–fluid system was used. Along these lines, Kobayashi and Ishida [255, 312] were the first to analyze the uplift behaviour of tanks from the rocking dynamics approach. Then, recently, the rocking dynamics of a tank was discussed through the introduction of the rock–translation interaction by Taniguchi in [594] and the phenomenon of the tank "walking" on its base [597], as observed during the Prince William Sound, Alaska, earthquake (1964) [213, 478].

A refined second-order beam formulation, providing for the effects of both material and geometric non-linearities, was presented by Malhotra and Veletsos in [373, 376] for a uniformly loaded, semi-infinite, prismatic beam resting on a rigid base and uplifting by a vertical force at one end, experiencing the effect of yielding away from the uplifted end [363]. Within the framework of the afore-mentioned formulation, an improvement was introduced by using the moderately large deflection plate theory for the uplifted region of the plate [374, 376], also considering the entire tank-liquid system [375]. More refined formulations accounting for a flexible soil foundation were proposed by Malhotra in [364, 365, 368]. A simplified non-linear pushover analysis for an unanchored tank was presented by Malhotra in [370] and then compared in [315] with the method proposed at Point A.9, EC8-4 for different values of the slenderness ratios. Based on the results from Malhotra and Veletsos in [373], Ahari et al. [6] presented results from a tapered beam model, compared to those of the constant prismatic beam model.

Several researchers have taken numerical approaches to calculate the seismic response: Haroun and Badawi [220], Lau et al. [328, 329], and Malhotra and Veletsos [376] using the classical Ritz method [104, 110], investigated the static tilt behaviour of an unanchored cylindrical tank; Peek and Jennings [448, 449, 452] solved the contact problem of the partially uplifted base plate and its interaction with the cylindrical shell, using the finite difference energy method [58, 59]; Natsiavas [419] by Hamilton's principle; then Peek and El-Bkaily were able to determine the extent of plastic buckling of the tank wall near the base (post buckling behaviour) based on static [450] and implicit dynamic time history analysis [142]; solutions for a partially uplifted tank using a fully non-linear, three-dimensional, finite element model were obtained by Barton and Parker [27], Vathi et al. [623], Zeiny in [684], applying the code developed in [149], and by Rammerstorfer, Fischer et al. for wind loads on a large liquid storage tank [467] and then extended to incorporate earthquake action in [166, 506] where many design charts corresponding to a range of slenderness $0.5 \leq \frac{H}{R} \leq 3.0$ have been produced in order to obtain the increase of the vertical membrane force in the wall due to uplift, with respect to the same force in the anchored case, as a function of a non-dimensional overturning moment (these charts are a part of Point A.9, EC8-4 [506]); an explicit integration procedure for the fluid tank system is employed by Ozdemir et al. [351, 434, 435], using an arbitrary Lagrangian Eulerian (ALE) formulation to model the coupling effects between fluid, structure, and soil; explicit dynamic analysis with the endurance time (ET) method [151] was examined by Alembagheri and Estekanchi in [9], using a general-purpose, finite element program developed in [300].

3.2 Vertical Cylindrical Tanks

The basic system investigated is shown in Figures 3.2 and 3.3 where it is depicted as a ground-supported, vertical, rigid flat-bottomed cylindrical tank of radius R, supported on a rigid base, and excited with an angular acceleration $\ddot{\theta}_g(t)$ acting on the pivoting bottom edge. The tank is filled with a fluid of mass density ρ_l (kg/m^3) to a level H. The upper surface of the liquid is considered to be free. The angular acceleration at any time t is denoted by $\ddot{\theta}_g(t)$ [204, 215, 221, 330, 337, 498, 593]. The locations of points in the liquid and the tank are univocally defined by the cylindrical coordinate system (r, θ, z), with the origin taken at the centre of the tank base and $\theta = 0$ taken in the direction of the horizontal excitation. Two main non-dimensional coordinates are then introduced as $\xi = \frac{r}{R}$ and $\zeta = \frac{z}{H}$. This system is considered under the assumption that the motion experienced by the tank is the same as the free-field ground motion $\ddot{\theta}_g(t)$. Furthermore, it is assumed that the effects of the sloshing motion on the fluid pressure are disregarded and that the liquid is incompressible, inviscid, and irrotational, and that all structural and liquid motions remain within the linearly elastic range of response (no separation or cavitation between the liquid and

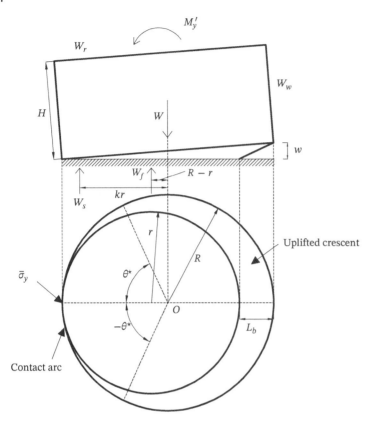

Figure 3.2 Response to overturning moment in plan and elevation (source: adapted from [95]): contact area at base and restoring moment system.

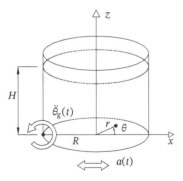

Figure 3.3 Unanchored tank-liquid geometry investigated: coordinate system (r, θ, z), ground acceleration $a(t)$ and rigid base rotation $\ddot{\theta}_g(t)$ acting at the shell-to-bottom head juncture [204, 330, 337, 596].

the tank). Under these assumptions, Taniguchi and Ando [596] analytically solved the Laplace's equation subject to the appropriate boundary conditions along the tank wall ($r = R$), base ($z = 0$), and the free liquid surface ($z = H$) and obtained the spatial-temporal variation (R, ζ, θ, t) of the rigid impulsive pressure on the wall at the height z and azimuth angle θ [596]:

$$p_{iw}(R, \zeta, \theta, t) = \rho_l R \ddot{\theta}_g(t) \left\{ H - z - 2 \sum_{n=1}^{\infty} \frac{\cos\theta}{\lambda_n^2 - 1} \left[R \left(\frac{1}{\lambda_n} + \frac{J_0(\lambda_n)}{J_1(\lambda_n)} \right) \sinh(\lambda_n \gamma \zeta) \right. \right.$$
$$\left. \left. + \left(\frac{H}{\cosh(\lambda_n \gamma)} - R \left(\frac{1}{\lambda_n} + \frac{J_0(\lambda_n)}{J_1(\lambda_n)} \right) \tanh(\lambda_n \gamma) \right) \cosh(\lambda_n \gamma \zeta) - z \right] \right\} \tag{3.1}$$

and on the bottom plate at the radius r and azimuth angle θ [596]:

$$p_{ib}(r, 0, \theta, t) = \rho_l R \ddot{\theta}_g(t) \left\{ H - 2 \sum_{n=1}^{\infty} \frac{J_1(\lambda_n \xi) \cos\theta}{J_1(\lambda_n)(\lambda_n^2 - 1)} \left[\frac{H}{\cosh(\lambda_n \gamma)} \right. \right.$$
$$\left. \left. - R \left(\frac{1}{\lambda_n} + \frac{J_0(\lambda_n)}{J_1(\lambda_n)} \right) \tanh(\lambda_n \gamma) \right] \right\} \tag{3.2}$$

where J_0 and J_1 are the Bessel functions of the first kind of order 0 and 1 respectively [561] and the numbers λ_n are the n-roots of the first derivative of J_1. The first ten of these roots are $\lambda_1 = 1.8412$, $\lambda_2 = 5.3314$, $\lambda_3 = 8.5363$, $\lambda_4 = 11.7060$, $\lambda_5 = 14.8636$, $\lambda_6 = 18.0155$, $\lambda_7 = 21.1644$, $\lambda_8 = 24.3113$, $\lambda_9 = 27.4571$, and $\lambda_{10} = 30.6019$ [23, 221, 596, 646]. Figure 3.4 shows the normalized fluid pressure distribution, employing the maximum tangential acceleration given by the product of the angular acceleration acting on the pivoting bottom edge and diagonals $\sqrt{H^2 + 4R^2}$ of the tank [596]: along the shell's wall and along the bottom plate, both corresponding to $\theta = 0°$ in Figures 3.4(a) and 3.4(b) respectively ($\theta = 0°$ corresponds to the right side in Figure 3.3); furthermore, the negative pressure distributions along the wall in the case of two different azimuth angles 120° and 180° (left side in Figure 3.3) are also depicted in Figures 3.4(c) and 3.4(d) respectively.

For the designer's convenience, the following describes the method presented by Clough in [95] and slightly modified by Clough and Niwa in [99], comparing the behaviour of the tank experimentally observed with the response predicted by the design procedures. This approach, included in NZSEE-09 at Point C4.4.2 [459], treats the container as if its wall were rigid, or constrained to move quasi-statically in a prescribed manner. The load assumed to be effective in producing a structural response is the total seismic overturning moment M'_y (Equation (2.105) for a rigid tank) which is based on the response of an equivalent anchored tank. Figure 3.2 shows a typical cylindrical tank, of radius R, filled with liquid to a depth H, subjected to a total overturning (or foundation) moment M'_y. Response is characterized by a rigid-body rotation of the shell, which lifts off the foundation on one side and rocks up on its toe. A crescent-shaped portion of the tank bottom is lifted off the foundation, and the portion which remains in contact is assumed to be circular, with radius r, and to be tangent to the tank shell at its point of contact with the foundation. Thus, the magnitude of uplift may be described by the non-dimesional ratio $\mu = \frac{r}{R}$. The unknown uplifted area is adjusted until the resisting moment M_R, due to the reactions, balances the applied peak overturning moment M'_y, applying the following iterative procedure [5]:

Step 1 Assume a value of radius ratio $\mu = \frac{r}{R}$ ranging from 1.0, corresponding to no uplift, down to 0.3, which corresponds to considerable uplift.

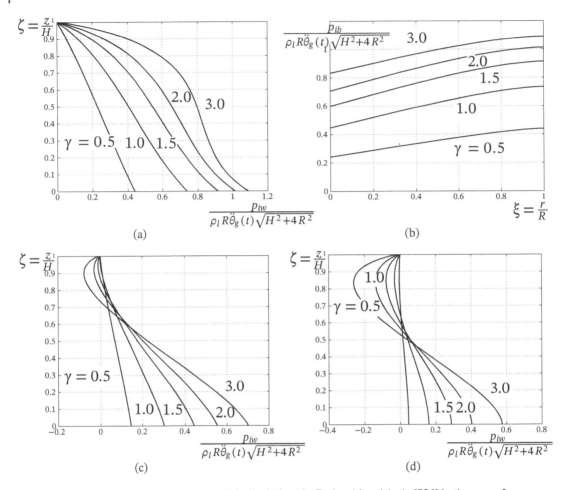

Figure 3.4 Fluid pressure distribution analytically derived by Taniguchi and Ando [596] in the case of an unanchored ground-supported, vertical, rigid flat-bottomed cylindrical tank: (a) along the wall ($\theta = 0°$), (b) the bottom plate ($\theta = 0°$); along the wall ($\theta = 120°$ and $\theta = 180°$) in (c) and (d) respectively.

Step 2 Further, we assume that the shell circumference remains in contact with the foundation over a circular arc, extending from $\theta^\star$ to $-\theta^\star$, as shown in Figures 3.2 and 3.5. It is also assumed that the axial compressive stress of the shell varies linearly from a maximum value $\bar{\sigma}_y$ on the excitation axis (at $\theta^\star = 0$) to zero at both ends of the contact arc (Figure 3.5). The half-angle of the contact sector may be derived by the following expression[1] [159, 165, 328]:

$$\theta^\star = \arctan\left(\frac{\mu}{1-\mu}\right) \qquad \text{(rad.)} \tag{3.3}$$

Step 3 The overturning moment M'_y will be resisted by the action of the weight of the liquid which remains directly supported by the foundation over the area of the base that does not uplift

---

1 A full derivation of Equation (3.3) is carried out in Appendix I of [95].

Figure 3.5 Contact area at base of an unanchored tank. Source: From [216]/American Society of Civil Engineers.

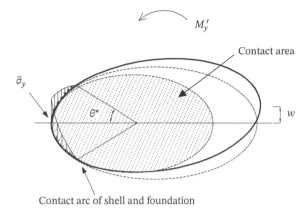

(radius r), $W_f = \pi r^2 H \gamma_l$ (in kN; γ_l is the unit weight of the liquid in kN/m³), and the reaction along the contact arc between the shell and the foundation in Figure 3.5, $W_s = W + W_w + W_r - W_f$, where W is the total weight of the liquid contents, W_w and W_r are the weight of the wall and the roof respectively; so $W - W_f$ could be considered the weight of the liquid acting on the uplifted portion of the tank floor. Reactions W_s and W_f, acting with moment arms kR and $(R - r)$ from the centre line of the tank, respectively, resist the total overturning moment M_y'. The linear distribution of W_s over the $2\theta^\star$ arc is depicted in Figure 3.5 [67]. The dimensionless constant k represents the effective moment arm of the reaction force W_s, and is given by [95]:

$$k = \frac{2}{\theta^{\star 2}} \left(1 - \cos \theta^\star \right) \tag{3.4}$$

Step 4 The resisting moment is equal to [5]:

$$M_R = W_s \cdot (kR) + W_f (R - r) \tag{3.5}$$

or as a function of the radius ratio μ:

$$M_R = RW \left[k \left(1 + \frac{W_w + W_r}{W} \right) + (1 - k)\mu^2 - \mu^3 \right] \tag{3.6}$$

Step 5 Step 1–4 are repeated until $M_R = M_y'$;

Experiences from past earthquakes, the Prince William Sound, Alaska (1964) [213, 478], San Fernando in 1971 [273], Miyagi-Ken-oki (Japan, 1978) [295, 678], Imperial County in 1979 [216, 331, 537], Livermore in 1980 [406], and Coalinga in 1983 [382], experimental [7, 67, 95, 97, 99, 381, 383, 427, 497, 536] and theoretical studies [19, 95, 256, 333, 672], have shown these systems to be prone to damage due to:

1) buckling of the tank wall caused by large axial compressive stress (Section 3.2.1) [481];
2) failure of the piping connections to the wall and other appurtenances connected to the tank, due to their inability to accommodate large base uplifts (Section 3.2.2);
3) loss of the tank contents due to extensive tearing of the tank bottom plate, as a consequence of the large uplifting displacements imposing large radial stresses on the base plate itself (Section 3.2.3) [432, 497];

4) rupture at the junction of the tank wall and base plate, caused by excessive plastic hinge rotation (Section 3.2.4).

A different methodology for the seismic evaluation of unanchored tanks is outlined in Appendix H, EPRI NP-6041-SL, "A Methodology for Assessment of Nuclear Power Plant Seismic Margin (Revision 1)" and BNL 52361 (Rev 10/95), "Seismic Design and Evaluation Guidelines for the Department of Energy High-Level Waste Storage Tanks and Appurtenances" [25, 112].

3.2.1 Axial Membrane Stress in a Shell Wall

Studies from Auli et al. [19], Natsiavas [419], Barton and Parker [27], and Ishida and Kobayashi [256], among others, have demonstrated that base uplifting may significantly influence the dynamic response of tanks and lead to axial membrane stresses in the tank wall, which are substantially higher than those induced in similarly excited fixed base systems. Two of the most widely used standards for the seismic design of unanchored liquid storage tanks are Appendix E, API 650 (2012) and Point C.4.4.2, NZSEE-09 [431]. Following the iterative procedure described in Section 3.2, the maximum axial stress in the shell is computed according to Equation C4.22 at Point C4.4.2, NZSEE-09 [459]:

$$\bar{N}_y = \frac{ABW_s}{\theta \star R} \quad \rightarrow \quad \bar{\sigma}_y = \frac{\bar{N}_y}{s} = \frac{ABW_s}{\theta \star Rs} \tag{3.7}$$

where A is a foundation stiffness factor equal to 1.0 for a rigid foundation and 0.5 for a flexible foundation; B is a calibration factor and the suggested value is 2.5 [99]; s is the uniform thickness of the wall and W_s is computed as:

$$W_s = W \left(1 + \frac{W_w + W_r}{W} - \mu^2 \right) \tag{3.8}$$

where $W_f = \pi r^2 H \gamma_l = W \mu^2$ has been substituted.

The Wozniak and Mitchell analysis [672] has provided the basis of the current guidelines for the design of unanchored welded steel tanks, according to Appendix E, API 650 (2012). The first parameter to be evaluated must be the anchorage ratio (Point E.6.2.1.1.1, API 650 and Point 13.5.4.1, AWWA D100-11):

$$J = \frac{M_y}{(2R)^2 \left[w_t (1 - 0.4 S_{vd}(T_V)) + w_a - 0.4 w_{\text{int}} \right]} \tag{3.9}$$

where M_y (in Nm) is the hydrodynamic moment induced on a section of the tank wall immediately above the base (Equation (2.103)); $2R = D$ is the nominal diameter of the tank in m [623]; $w_t = w_w + w_r$ (in N/m) is the total weight of the wall and roof acting at the base of the shell, where $w_w = \frac{W_w}{2\pi R}$ and $w_r = \frac{W_r}{2\pi R}$, including 10% of the specified snow load; $S_{vd}(T_V)$ is the design spectral acceleration for the vertical component of the seismic action (Point 3.2.2.5(5), EC8-4) corresponding to the period T_V with or without soil–structure interaction (Section 2.2.5 in the case of a rigid tank; while in the case of flexible tanks T_{Vf} (Equation (5.60)) or T'_{Vf} (Equation (5.57)) should be used) and to an equivalent viscous damping described in Section 6.2, ($S_{vd}(T_V)$ may be derived from $S_{ve}(T_V)$ using the appropriate behaviour factor at Section 2.2.9); $w_{\text{int}} = 0.25 D P_{\text{int}}$ (in N/m) [623] is the calculated design uplift load due to product pressure per unit circumferential length, where P_{int} is the tank internal design pressure above the liquid level (if the tank is an atmospheric tank, then it would be zero); w_a (in N/m) is the force-resisting uplift in the annual region at the bottom plate, or better the maximum weight of tank contents which may be used to resist the shell overturning [672]:

$$w_a = 99s_b\sqrt{f_y^b HG} \leq 201.1H(2R)G \tag{3.10}$$

In Equation (3.10) G is the design-specific gravity of the liquid to be stored (1.0 for water) and f_y^b is the minimum specified yield strength of the bottom plate (in MPa); H and R are in metres; s_b (in mm) is the thickness of the bottom plate, excluding corrosion allowance, that will extend radially at least the distance L_b from the inside of the wall. The maximum longitudinal shell membrane compression stress (in MPa) at the bottom of the tank wall, when there is no calculated uplift ($J \leq 0.785$), under the design seismic moment M_y (Nm), may be obtained by the formula (Point E.6.2.2.1, API 650 and Point 13.5.4.2.1, AWWA D100-11) [672]:

$$\bar{\sigma}_y = \left[w_t(1 + 0.4S_{vd}(T_V)) + \frac{1.273M_y}{(2R)^2}\right]\frac{1}{1000s_t} < \frac{83s_t}{2R} \qquad \text{for} \qquad \frac{GH(2R)^2}{s_t^2} \geq 44 \tag{3.11a}$$

$$\bar{\sigma}_y = \left[w_t(1 + 0.4S_{vd}(T_V)) + \frac{1.273M_y}{(2R)^2}\right]\frac{1}{1000s_t} < \frac{83s_t}{5R}$$

$$+ 7.5\sqrt{GH} < 0.5f_y^w \qquad \text{for} \qquad \frac{GH(2R)^2}{s_t^2} < 44 \tag{3.11b}$$

where s_t (in mm) is the thickness of the bottom course of the wall less corrosion allowance and f_y^w is the minimum specified yield strength of the shell course (in MPa). When the shell is uplifting, $0.785 < J \leq 1.54$, the unanchored tank is self-anchored and stable (Table E-6, API 650), which means that it uses its own self-weight (tank and stored product) to resist the overturning moment, providing the following shell compression requirements are satisfied (Point E.6.2.2.1, API 650 and Point 13.5.4.2.1, AWWA D100-11) [672]:

$$\bar{\sigma}_y = \left[\frac{w_t(1 + 0.4S_{vd}(T_V)) + w_a}{0.607 - 0.18667J^{2.3}} - w_a\right]\frac{1}{1000s_t} < \frac{83s_t}{2R} \qquad \text{for} \qquad \frac{GH(2R)^2}{s_t^2} \geq 44 \tag{3.12a}$$

$$\bar{\sigma}_y = \left[\frac{w_t(1 + 0.4S_{vd}(T_V)) + w_a}{0.607 - 0.18667J^{2.3}} - w_a\right]\frac{1}{1000s_t} < \frac{83s_t}{5R}$$

$$+ 7.5\sqrt{GH} < 0.5f_y^w \qquad \text{for} \qquad \frac{GH(2R)^2}{s_t^2} < 44 \tag{3.12b}$$

In the third case, the anchorage ratio is $J > 1.54$ that corresponds to the situation in which the tank is not stable and cannot be self-anchored. Usually mechanical anchorage should be added, paying attention to the attachments of the anchors to the shell to avoid the possibility of tearing the shell. As a second strategy, the bottom plate can be modified within the limit of s_b and L_b. To be clear, the thickness of the bottom annular ring, s_b, should not exceed the thickness of the bottom shell course, s_t, and where the bottom annular ring is thicker than the remainder of the bottom plate, the width L_b of the annular ring should be equal to or greater than (Point E.6.2.1.1.2, API 650 and Point 13.5.4.1.2, AWWA D100-11) [672]:

$$L_b = 0.01723s_b\sqrt{\frac{f_y^b}{HG_e}} \leq 0.07R \tag{3.13}$$

where $G_e = G(1 - 0.4S_{vd}(T_V))$ is the effective gravity, including vertical seismic effects. Furthermore, L_b (in m) is measured from the inside of the shell wall. Equation (3.13) says that the maximum width of the annulus for determining the resisting force is 3.5% of the tank diameter (Point E.6.2.1.1, API 650).

To provide only a useful indication of the buckling susceptibility of liquid storage tanks, Rinne [478] proposed the calculation of a dimensionless buckling resistance coefficient C_E [382, 448]:

$$C_E = \frac{2220 \cdot 10^6}{D\rho_l}\left(\frac{s}{H}\right)^2 > 0.44 \qquad \text{no buckle at the base} \qquad (3.14)$$

where s, $D = 2R$, and H are the thickness, diameter, and liquid's height of the tank (in feet), and ρ_l (in pounds per cubic foot) is the mass density of the liquid. Equation (3.14) has been derived by Rinne [478], assuming that the total weight of the tank and contents, $W + W_w + W_r$, is about 1.1 the total weight of the contents W, and that the overturning lateral force, producing M'_y, acts at a height of $0.4H$ above the base [448]. Furthermore, he assumed that the critical buckling stress is about 0.18 times the classical buckling stress derived independently by Timoshenko, Lorenz, and Southwell [54, 64, 148, 294, 357, 493, 612, 686]. The accuracy of Equation (3.14), compared with more refined formulation, based on experimental studies [381, 423, 428], has been discussed by Manos in [380].

3.2.2 Shell Uplift

It has been shown in several tank model tests [7, 67, 95, 99, 381, 383, 427, 497, 536] that horizontal earthquake action will cause unanchored tanks to lift off the ground [12]. This uplift behaviour is known to occur at even relatively low lateral acceleration and becomes more pronounced with increasing tank aspect ratio $\frac{H}{R}$. Examples from Alaska (1964) [213, 478], San Fernando in 1971 [273], and Livermore in 1980 [406], provide field verification of this experimental observation. According to Point C5.3.1, NZSEE-09, the minimum peak ground velocity[2] v_g (m/sec), corresponding to the velocity component $v(t)$ of the ground motion $a(t)$, required to cause overturning is given by:

$$v_g^2 \geq \frac{4R^2}{H} \qquad \text{or} \qquad R \leq \frac{v_g^2}{4}\frac{H}{R} \qquad (3.15)$$

where R and H are in metre units. Equation (3.15) comes from an approximation by Ishiyama [257, 258]. To evaluate the inherent stability or overturning resistance of an unanchored tank to seismic loading, it is of primary importance to evaluate the uplift displacement of the tank bottom. An empirical procedure, taken from Oden and Ripperger [429], was outlined by Cambra in [67], and then modified by Equation C4.25 at Point C4.4.2, NZSEE-09, to describe the kinematic behaviour of bottom plating when tanks experience a lateral earthquake. The method proposed by Cambra considers bending, catenary stiffness of the bottom plate to represent the uplifted region, and the semi-infinite beam on an elastic foundation to simulate the plate region in contact with the ground. Cambra's original formula [67] has been modified in order to include a foundation stiffness factor A and to assume that the base material will yield at the base-shell joint [5]:

$$w = \frac{1}{A}\left[\frac{f_y^b s_b^2}{6N_x} + \frac{p_0 L_b}{N_x}\left(\frac{L_b}{2} - \sqrt{\frac{\bar{E}s_b^3}{12N_x}}\right)\right] \qquad (3.16)$$

where $\bar{E} = \frac{E}{1-\nu^2}$, E and ν are the Young's modulus and Poisson's ratio of the tank bottom plate, whose thickness is s_b; $p_0 = \gamma_l H$ is the hydrostatic pressure on the base in which γ_l is the unit weight of the liquid (in kN/m^3); $L_b = 2R(1 - \mu)$ is the radial uplift separation length where $\mu = \frac{r}{R}$ must be determined by the iterative procedure in Section 3.2; the axial membrane bottom plate force is $N_x = \sigma_x s_b$ where σ_x may be derived from Equations (3.18) or (3.19); f_y^b is the yield strength of the bottom plate (Tab. 3.1, Point 3, UNI

---

2 According to Ishiyama [257, 258], this is the lower limit of the maximum velocity to overturn the tank.

ENV 1993-1-1:2005 and UNI EN 1993-1-3:2007); A (Equation (3.7)) is a foundation stiffness factor equal to 1.0 for a rigid foundation and 0.5 for a flexible foundation.

The tank uplift (in mm) estimated following Point E.7.3.1 API 650 and Point A.13.6.1 AWWA D100-11 is [431]:

$$w = 12.10 \frac{f_y^b L_b^2}{s_b} \qquad (3.17)$$

where the width L_b (in m) of the annular ring comes from Equation (3.13) and s_b (in mm) is the thickness of the bottom plate less corrosion allowance (f_y^b in MPa).

3.2.3 Radial Membrane Stress at Base

An estimate of the radial membrane stress σ_x on the base plate as a result of uplift has been derived by Cambra in [67] (Equation C4.24, Point C4.4.2 NZSEE and Point A.9.4 EC8-4):

$$\sigma_x = \frac{1}{s_b} \left[\frac{2\bar{E}s_b p_0^2 R^2 (1 - \mu)^2}{3} \right]^{\frac{1}{3}} \qquad (3.18)$$

where by substituting $L_b = 2R(1 - \mu)$ we obtain [5]:

$$\sigma_x = \frac{1}{s_b} \left[\frac{\bar{E}s_b p_0^2 L_b^2}{6} \right]^{\frac{1}{3}} \qquad (3.19)$$

or in a more convenient form, according to Vathi et al. [623] and assuming a Poisson's ratio of the tank bottom plate $v = 0.3$:

$$\sigma_x = 0.5679E \left[\frac{p_0 L_b}{Es_b} \right]^{\frac{2}{3}} \qquad (3.20)$$

where the symbols are the same as applied in Equation (3.16). The analysis determines first the value of the radius ratio $\mu = \frac{r}{R}$ by the iterative procedure Steps 1–5 at Section 3.2, then the amount of uplift of the tank base is estimated through the uplifted length $L_b = 2R(1 - \mu)$ and finally the radial membrane stress in base σ_x associated with the uplift mechanism is determined.

3.2.4 Plastic Rotation at Base

According to Point A.9.1, EC8-4, flexural yielding is assumed to take place in the base plate. It is recommended, by Point E.6.2.1.1.2, API 650, Point 13.5.4.1.2, AWWA D100-11, Point A.9.5 EC8-4 and Point C5.5.3 NZSEE-09, the bottom annular ring (Equation (3.13)) should be designed with a thickness s_b less than the wall thickness s_t, so as to avoid flexural yielding at the base of the wall. For a membrane action to develop in the tank (Point 11.4 (5), UNI EN 1993-4-2), as described at Sections 3.2.1 and 3.2.3, a plastic hinge must develop in the tank bottom base at the junction with the wall (Point C5.5.3.2, NZSEE-09) [414]. The rotation θ_p in Figure 3.6 must be limited to a maximum value θ_p^{max} to not cause fracture [34, 35, 623]:

$$\theta_p^{max} = \left(\frac{0.05}{\frac{s_b}{2}} \right) 2s_b = 0.20 \text{ rad} \qquad (3.21)$$

where it has been assumed a maximum allowable steel strain equal to 0.05 and a plastic hinge length equal to twice the base plate thickness s_b. From Figure 3.6, the required plastic rotation θ_p for an

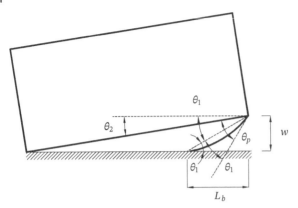

Figure 3.6 Plastic rotation of base plate of uplifting tank.

uplift w (Equation (3.16)), should be compared with the maximum rotation [5]:

$$\theta_p = 2\theta_1 - \theta_2 = \left(\frac{2w}{L_b} - \frac{w}{2R}\right) \leq \theta_p^{max} = 0.20 \text{ rad} \tag{3.22}$$

If θ_p exceeds the limit, $\theta_p > 0.20$ rad, it can be reduced by increasing the base plate thickness within the limit of s_b and L_b (Equation (3.13); Point E.6.2.1.1.2, API 650; Point 13.5.4.1.2, AWWA D100-11 and Point C5.5.3.2, NZSEE-09). If the base plate thickness so calculated exceeds the thickness of the bottom shell course, s_t, it is necessary to mechanically anchor the tank.

3.3 Rectangular Tanks

The expressions for the evaluation of the earthquake pressures on a rigid rectangular tank of dimensions $2L \times 2B$ that undergoes uplift motion due to an angular acceleration $\ddot{\theta}_g(t)$ pivoting at its left bottom edge (contact line in Figure 3.8) have been analytically derived from Taniguchi and Ando [595]. Assuming that the effects of the sloshing motion on the fluid pressure is disregarded and a perfect fluid and velocity potential, Taniguchi and Ando [595] analytically solved the Laplace's equation, subject to the appropriate boundary conditions, and obtained the spatial-temporal variation (x, z, t) of the rigid impulsive pressure on the side walls (left side $x = 0$ and right side $2L$ in Figure 3.8) along the height z [595]:

$$p_{iw}^{left}(0, z, t) = -\rho_l L \ddot{\theta}_g(t) \left\{ z - H + \frac{8}{\pi^2} \sum_{n=1,3,5,\cdots}^{\infty} \frac{1}{n^2} \left[z - \frac{4L}{n\pi} \sinh \frac{n\pi z}{2L} \right. \right.$$

$$\left. \left. + \left(\frac{4L}{n\pi} \tanh \frac{n\pi H}{2L} - \frac{H}{\cosh \frac{n\pi H}{2L}} \right) \cosh \frac{n\pi z}{2L} \right] \right\} \tag{3.23}$$

$$p_{iw}^{right}(2L, z, t) = -\rho_l L \ddot{\theta}_g(t) \left\{ z - H - \frac{8}{\pi^2} \sum_{n=1,3,5,\cdots}^{\infty} \frac{1}{n^2} \left[z - \frac{4L}{n\pi} \sinh \frac{n\pi z}{2L} \right. \right.$$

$$\left. \left. + \left(\frac{4L}{n\pi} \tanh \frac{n\pi H}{2L} - \frac{H}{\cosh \frac{n\pi H}{2L}} \right) \cosh \frac{n\pi z}{2L} \right] \right\} \tag{3.24}$$

and on the bottom plate ($z = 0$) at the distance x [595]:

$$p_{ib}(x,0,t) = -\rho_l L \ddot{\theta}_g(t) \left[-H + \frac{8}{\pi^2} \sum_{n=1,3,5,\cdots}^{\infty} \frac{1}{n^2} \left(\frac{4L}{n\pi} \tanh \frac{n\pi H}{2L} - \frac{H}{\cosh \dfrac{n\pi H}{2L}} \right) \cosh \frac{n\pi x}{2L} \right] \qquad (3.25)$$

Figures 3.7(a–c) depict the values of the fluid pressures on the right side, the left side wall, and the bottom plate, respectively corresponding to five values of aspect ratio $\gamma = \frac{H}{L} = 0.5, 1.0, 1.5, 2.0$ and 3.0. The normalized fluid pressure distribution employing the maximum tangential acceleration given by the product

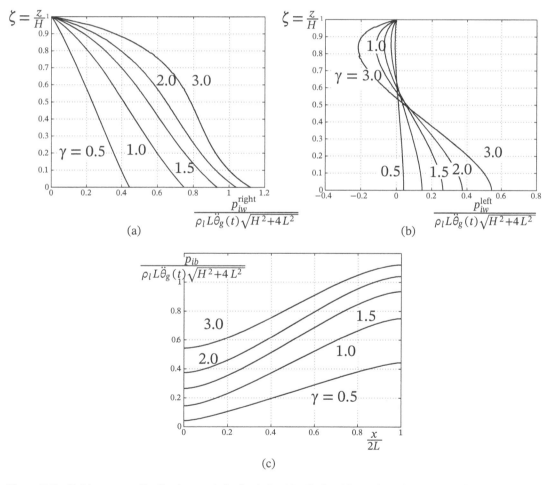

(a)

(b)

(c)

Figure 3.7 Fluid pressure distribution analytically derived by Taniguchi and Ando in the case of an unanchored ground-supported, rigid rectangular tank: (a) along the right side wall ($x = 2L$ and Equation (3.24)), (b) along the left side wall ($x = 0$ and Equation (3.23)), and (c) along the bottom plate ($z = 0$ and Equation (3.25)). Source: Adapted from [596].

of the angular acceleration acting on the pivoting bottom edge and diagonals $\sqrt{H^2 + 4L^2}$ of the tank, is shown [596].

For the designer's convenience, the following introduces the same procedure presented in Section 3.2 for cylindrical tanks, for the analysis of uplifting rectangular tanks which is based upon Equations (3.5), (3.7), (3.16), and (3.18) slightly modified. This approach considers a rectangular container, with dimensions $2L \times 2B$, uplifted as depicted in Figure 3.8. Using the mechanism in Figure 3.8 and writing the rotation equilibrium about the compression reaction W_s, the resisting moment M_R is now given by (Point C4.4.2, NZSEE-09):

$$M_R = (W_w + W_r)L + W(1 - \mu)\,[2L\mu + L(1 - \mu)] = (W_w + W_r)L + WL(1 - \mu^2) \tag{3.26}$$

that can be solved directly to obtain the non-dimesional ratio μ:

$$\mu = \sqrt{1 - \left(\frac{M_R}{WL} - \frac{W_w + W_r}{W}\right)} \tag{3.27}$$

The maximum axial stress in the wall $(2B)$, perpendicular to the direction of the ground motion, is computed according to Equation C4.28 an Point C4.4.2, NZSEE-09:

$$\bar{N}_y = B'\left[\frac{W_w + W_r + W(1 - \mu)}{2B}\right] \quad \rightarrow \quad \bar{\sigma}_y = \frac{\bar{N}_y}{s} \tag{3.28}$$

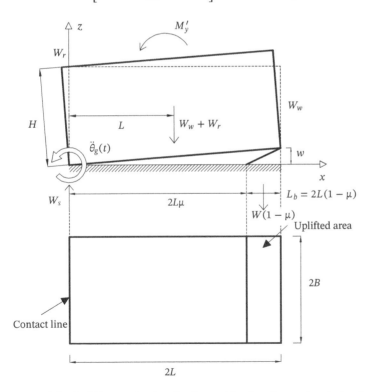

Figure 3.8 Uplifting mechanism, in plan and in elevation, and restoring forces on a rectangular tank.

where $B' = 2.5$ [99]. An estimate of the radial membrane stress σ_x in the base plate as a result of uplift has been derived by Priestley et al. in Point C4.4.2, NZSEE-09 (Equation C4.24):

$$\sigma_x = \frac{1}{s_b} \left[\frac{2\bar{E}s_b p_0^2 L^2 (1 - \mu)^2}{3} \right]^{\frac{1}{3}}$$

(3.29)

where all terms are defined similar to their use in Equation (3.18). The tank uplift w may be derived using Equation (3.16) substituting $L_b = 2L(1 - \mu)$.

4

Elevated Tanks

After reading this chapter you should be able to:

- Understand the equivalent mechanical model to analyse elevated tanks
- Evaluate impulsive and convective modes and masses for elevated tanks
- Identify components related to the performance of elevated tanks

4.1 Introduction

Elevated storage tanks should remain functional in the post-earthquake period, even after a major earthquake, to ensure water supply to control the fires, which cause a great deal of damage and loss of lives [343]. Nevertheless, several elevated tanks sustained moderate to severe damage during past earthquakes [233, 408] (Figure 4.1): Kern County, California, earthquake, 21 July 1952 (Figure 4.1(a)), the Chile earthquake, 22 May 1960 [15, 564] (Figures 4.1(b), 4.1(c) and 4.1(d)), El Asnam, Algeria, earthquake, 10 October 1980 (Figure 4.1(e)), Bihar-Nepal earthquake, 21 August 1988 [233, 269], Jabalpur (Indian state of Madhya Pradesh) earthquake, 22 May 1997 [461], Bhuj (state of Gujarat in India) earthquake, 26 January 2001 [233, 461–463] (Figure 4.1(f)), Maule earthquake, 27 February 2010 [141], Van earthquake, Turkey, 23 October 2011 [619]. Such performance reveals a complex behaviour mainly due to the presence of two primary components: the tank, which contains the liquid, and its supporting structure. Even if, in past earthquakes, a number of elevated tanks suffered damage, however, this damage was primarily located in the supporting structures. A wide variety in the configuration of elevated tanks can be found in civil engineering applications [20, 222, 344] and may be classified into three main categories: (1) frame elevated tanks; (2) axisymmetrical pedestal elevated; and (3) composite elevated tanks (Figure 4.2), as explicitly highlighted in FEMA 450 at Point 14.4.7.9 [20, 222, 343, 344, 395].

There have been several studies [75, 145, 146, 170, 222, 253, 408, 409, 465, 532, 574] in which many simplified procedures used for a straightforward estimate of the seismic hazard of elevated tanks have been

Seismic Design and Analysis of Tanks, First Edition. Gian Michele Calvi and Roberto Nascimbene.
© 2023 John Wiley & Sons, Inc. Published 2023 by John Wiley & Sons, Inc.

proposed. However, most of them refers to three main simplified models depending on the simplifications adopted [343, 408, 462]:

1) a single lumped-mass model in which the elevated tank has to be analysed as a single degree of freedom system;
2) a two uncoupled mass model in which two masses are assumed to be uncoupled and the earthquake forces on the support are estimated by considering two separate single degree of freedom systems;
3) a the more refined one based on the two coupled masses model.

4.1.1 Frame Elevated Tanks

Frame elevated tanks have a steel cross-braced supporting tower [533] (Figures 4.1(a) and 4.2(a)) or a reinforced concrete multi-column assembly [50] (Figures 4.1(b), 4.1(c), 4.1(e) and 4.2(b)). The small-capacity (less than 0.76 ML[1]) steel elevated tank usually has a cylindrical sidewall, an ellipsoidal bottom and roof; medium-capacity from 0.76 ML to 1.9 ML uses torus bottom and ellipsoidal roof and large-capacity tanks (larger than 1.9 ML) may have a diameter from 11 to 20 m and a capacity from 750 to more than 3500 m^3 (Figure 4.3). Elevated tanks on reinforced concrete frame stagings [136, 356, 444, 445] may have circumferential beams (Figures 4.1(c) and 4.4(a)), radial beams with a central column (Figures 4.1(b) and 4.4(b)) and both radial and circumferential beams, with or without a central column (Figures 4.1(e) and 4.4(c)). The moment-resisting-frame-type polygonal braced (or unbraced) stagings used for elevated tanks cannot be simply treated as plane frames and should strictly be analysed as space frames [501]. It is well established from Dutta et al. [133–136] that small accidental eccentricity may cause considerably amplified torsional response, even under horizontal ground shaking, in any structure if it has torsional-to-lateral time period ratio, τ, in the critical range of 0.7–1.25 [133]:

$$\tau = \varrho \sqrt{\frac{N_p + \frac{(N_p-1)K_r}{\cos^2\left(\frac{\pi}{N_c}\right)}}{0.0025N_p(4N_p^2 - 1) + N_p + 2(N_p - 1)K_r}} \quad (4.1)$$

where N_c and N_p are the number of columns and panels respectively [445]; the relative flexural stiffness K_r of columns with respect to that of beams is:

$$K_r = \frac{(E_cI_c)/h}{(E_bI_b)/L} \quad (4.2)$$

in which E_c is the modulus of elasticity of the column material, I_c the moment of inertia of the column cross-section, and h the panel height; E_b is the modulus of elasticity of the beam, I_b the moment of inertia of the beam, and L is the span; ϱ is the ratio of effective radius of gyration $R_{\text{e-gyr}}$ of the mass of the tank container and impulsive mass of water to the radius of staging R_s (by considering vertical columns resting on the perimeter of a circle [133, 500]):

$$\varrho = \frac{R_{\text{e-gyr}}}{R_s} \quad (4.3)$$

---

1 ML means megalitre (1 million litres) and is equal to 10^6 L.

(a) (b) (c)

(d) (e) (f)

Figure 4.1 A wide range of configuration of elevated tanks damaged during earthquake events: (a) Kern County, California, earthquake (21 July 1952, Magnitude 7.7); (b)–(d) two elevated reinforced concrete water tank and an elevated all-steel inverted pendulum-type water tank (Chile earthquake, 22 May 1960, Magnitude 8.5); (e) failure of the reinforced concrete supporting tower of an elevated reinforced concrete water tank (El Asnam, Algeria, earthquake, 10 October 1980, Magnitude 7.7); (f) tank container with hollow circular shaft support (Bhuj earthquake, 26 January 2001, Magnitude 7.7). Source: (a) to (d) from Karl V. Steinbrugge Collection; (e) from William G. Godden (Vol. 4) Collection, University of California, Berkeley; and (f) [463]/Springer Nature.

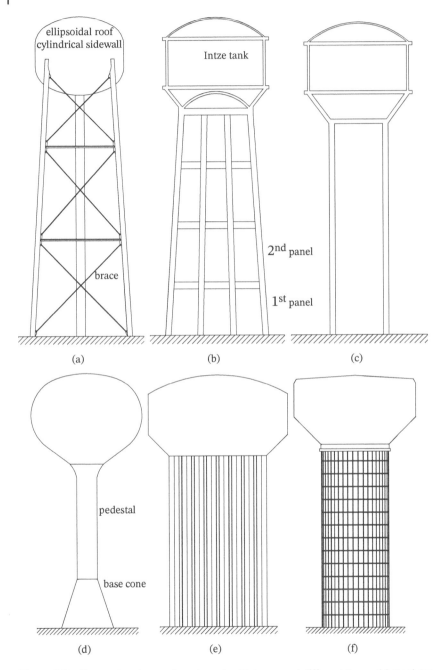

(a) (b) (c)

(d) (e) (f)

Figure 4.2 Elevated tanks may be categorized into several different types: (a) steel and (b) concrete frame elevated tanks; (c) elevated tanks on uniform cylindrical shaft-type tower or (d) on a flared base; (e) fluted column elevated storage tank and (f) composite tanks.

Figure 4.3 Large-capacity water elevated steel tank near Urbana, Illinois. Source: William G. Godden Collection, University of California, Berkeley.

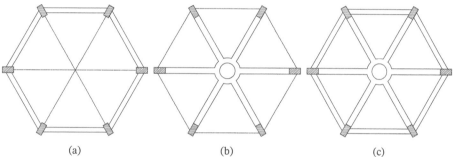

(a)   (b)   (c)

Figure 4.4 Staging configurations to be used in case of frame-supported reinforced concrete tanks: (a) circumferential and (b) radial beams; (c) staging with both types of beams in radial and circumferential directions.

where:

$$R_{\text{e-gyr}} = R_{\text{gyr}} \sqrt{\frac{m_r + m_w + m_b}{m_r + m_w + m_b + m_i}} \tag{4.4}$$

In Equation (4.4) $m_r + m_w + m_b$ (see Figure 2.10) is the mass of the tank structure alone composed of the mass of the roof m_r, wall m_w, and base m_b; m_i is the impulsive mass of the liquid which is included in the lateral impulsive mode of vibration and may be derived from Equation (2.10) (or (2.25a)–(2.25b)) for cylindrical tanks and from expression (2.131a) for rectangular tanks; R_{gyr} is the usual radius of gyration of the tank when empty.

4.1.2   Axisymmetrical Tanks

Axisymmetrical pedestal elevated tanks are supported by a single circular steel or concrete tower (capacity less than 0.76 ML; Figures 4.1(d), 4.1(f) and 4.2(c)) or a single cylindrical support pedestal with a flared conical base (capacity from 0.76 ML to 7.6 ML; Figure 4.2(d)) used to contain pumping units

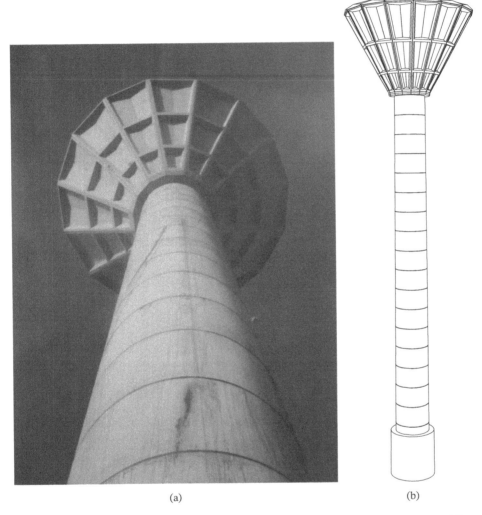

(a)   (b)

Figure 4.5 Single-pedestal tank built by Adriano Rivoli S.p.A. from Monopoli in the town of Cancello Arnone in 1991: Italian seismic zone 2 according to D. Min. LL. PP. 16 Gennaio 1996. Source: Courtesy of Ing. Vincenzo Di Bernardo from Teramo.

and other operating equipment. A modified single-pedestal tank is an elevated welded steel storage tank supported by a single large diameter corrugated steel support column, to give structural rigidity, and is usually denoted as a fluted (or folded-plate) column elevated storage tank (capacity greater than 1.9 ML; Figure 4.2(e)). The support column may be constructed of steel (Figure 4.1(d)), reinforced concrete (Figure 4.2(c)) or assembled using precast elements (Figure 4.5). The container on the top of the supporting tower may be spherical or various combinations of cones and cylinders (such as the classical Intze type of tanks in Figures 4.1(c), 4.1(e) and 4.2(c)).

4.1.3 Composite Elevated Tanks

Composite elevated tanks (Figure 4.2(f)) are comprised of a welded steel tank for watertight containment at the top and a single pedestal concrete support structure [393]. These tanks are also sometimes referred to as "concrete pedestal elevated tanks" (ACI 371R-98 and Point 14.4.7.9.6, FEMA 450) and use the best design features of steel and concrete: the steel tank provides a proven, watertight container while the reinforced concrete support column provides a cost-effective, structurally robust pedestal with minimal maintenance [393]. They are commonly built to store between 2.8–7.6 ML.

4.2 Single Lumped-Mass Model

According to the single lumped-mass model suggested by Chandrasekaran and Krishna in [75] (Figure 4.6), the Indian seismic code Point 5.2, IS:1893 (1984) on "Elevated Tower-Supported Tanks" [461, 464], Point 4.7.5.2, ACI 371R-98 and Point A5.1.2.8.2.1.5, ACI 371R-08 (for concrete towers), Point 13.4, AWWA D100-11 (for steel towers), allows elevated tanks to be analysed as a single degree of freedom system. Also Point 15.7.10.2, ASCE 7 assumes that the material stored should be considered as a single rigid mass acting at the volumetric centre of gravity. For tanks which are closed and fully filled with water (this prevents the vertical motion of water sloshing) or are completely empty, the behaviour of the

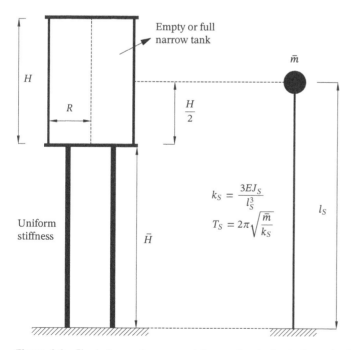

$$k_S = \frac{3EJ_S}{l_S^3}$$

$$T_S = 2\pi\sqrt{\frac{\bar{m}}{k_S}}$$

Figure 4.6 Single lumped-mass model according to the Indian seismic code Point 5.2, IS:1893 (1984), Point 4.7.5.2, ACI 371R-98, Point A5.1.2.8.2.1.5, ACI 371R-08 (for concrete tower) and Point 13.4, AWWA D100-11 (for steel tower).

tank may be well estimated as a one-mass system [391, 408]. It must be stated that this can be a realistic assumption for long and narrow tank containers with a height-to-radius ratio $\frac{H}{R} \geq 4$ (Point 14.4.7.5.1, FEMA 450), because all fluid mass participates in the impulsive mode of vibration and moves with the container wall in an impulsive fashion [343, 461]. Furthermore, as noted before, elevated tanks can have different types of supporting structures, which could be in the form of a steel (Figures 4.1(a) and 4.2(a)) or reinforced concrete frame (Figures 4.1(b), 4.1(c), 4.1(e) and 4.2(b)) and concrete (Figures 4.1(f), 4.2(f) and 4.5) or steel pedestal (Figures 4.1(d), 4.2(c) and 4.2(e)). In a single-mass lumped model, it is assumed that the supporting structure acts as a cantilever of uniform rigidity along the height [343] and this assumption is more suitable for a steel or reinforced concrete shell supporting structure as schematically represented in Figures 4.2(c)–4.2(f). Hence, the single-mass model is a reasonable approximation for an elevated water tank where the total water weight W (in kN) is concentrated near the top of the structure and accounts for the largest portion of the total gravity weight $W_G = W + W_S + W_L$ (Point A5.1.2.8.2.1.5, ACI 371R-08), such that $W \geq 80\% \, W_G$ [464] (Points 4.7.5.2 and A4.7.5.2, ACI 371R-98), where W_S is the total weight of the empty container W_C and of the supporting structure W_P, whereas W_L should be $\geq 25\%$ of the floor live load in areas used for storage along the height of the staging. The fundamental structure period of vibration commonly used is (Point 4.7.5.2, ACI 371R-98 and Point A5.1.2.8.2.1.5, ACI 371R-08; Point 4.3.1.3, IITK-GSDMA 2007 and Point 5.2.3, IS:1893 (1984)):

$$T_S = 2\pi \sqrt{\frac{\overline{m}}{k_S}} \tag{4.5}$$

where $\overline{m} = \frac{\overline{W}}{g}$ and $\overline{W} = W_C + \frac{2}{3}W_P + W$ (according to Point 4.7.5.2(b), ACI 371R-98, it should be used as a maximum of two-thirds (66%) the self-weight W_P of steel/concrete support) or $\overline{W} = W_C + \frac{1}{3}W_P + W$ (according to Point 4.3.1.3, IITK-GSDMA 2007 and Point 5.2.3, IS:1893 (1984); this assumption considers staging acting like a lateral spring and comes from classical results obtained by Tse et al. in [617]); Point C3.11, NZSEE-09 uses the total weight of the supporting structure such that $\overline{W} = W_C + W_P + W$. The lateral flexural stiffness of the supporting structure k_S is determined by the deflection of the supporting structure acting as a cantilever beam of length l_S, subjected to a concentrated end load (Point A4.7.5.2, ACI 371R-98) [343, 500]:

$$k_S = \frac{3EJ_S}{l_S^3} \tag{4.6}$$

where J_S is the moment of inertia of the steel section or gross concrete section about the centroidal axis and ignoring reinforcement (Point A4.7.5.2, ACI 371R-98 and Point A5.1.2.8.2.1.5, ACI 371R-08) and E is the modulus of elasticity of the material of the supporting structure.

4.3 Two Uncoupled Mass Model

A more satisfactory method, compared to the single lumped-mass model, is the two mass model ($m_i + \Delta m, m_{c1}$) suggested by Housner in [242] and inserted at Point C3.11, NZSEE-09, Point A.6, EC8-4 and

Point 4.2.2.4, IITK-GSDMA 2007 [267–269]. In this approach, the two masses are assumed to be uncoupled and the earthquake forces on the support are estimated by considering two separate single degree of freedom systems, as depicted in Figure 4.7. The first mass m_{c1} is representing the sloshing of the convective mass in a tank assumed to be rigidly connected to the ground; the second mass is representing a sum between the impulsive component m_i of the fluid and the structure masses Δm behaving as an inverted pendulum with a stiffness equal to that of the supporting staging. In this model, the liquid and the supporting structure may be accounted for by considering the following masses:

- an impulsive mass m_i (Equation (2.10) or (2.25a)–(2.25b) for cylindrical tanks and expression (2.131a) for rectangular tanks) rigidly connected to the tank's wall and located at a height h_i (Equation (2.15) or (2.25c)–(2.25d), (2.26a) and (2.131c)–(2.131d)) or h_i' (Equation (2.13) or (2.25e)–(2.25f), (2.26b)–(2.26c) and (2.131e)–(2.131f)) above the base plate respectively. In the evaluation of the quantities h_i and h_i', the height of the staging $\overline{H}$ (Figures 4.6 and 4.7) should not be included. For tanks of other shapes (different from cylindrical or rectangular), an equivalent circular tank is to be considered. Joshi [278] has shown that such an approach gives satisfactory results for Intze tanks. Similarly, for tanks of truncated inverted conical shape, Point A.6, EC8-4 suggests considering it in the model as an equivalent cylinder of the same volume of liquid and an equivalent radius R_{eq} equal to that R_T of the cone at the level of the liquid (Point A.6, EC8-4):

$$H_{eq} = \frac{\text{volume of the fluid}}{\pi R_{eq}^2} \tag{4.7a}$$

$$R_{eq} = R_T \text{ at liquid level} \tag{4.7b}$$

$$s_{eq} = s \text{ thickness of the wall} \tag{4.7c}$$

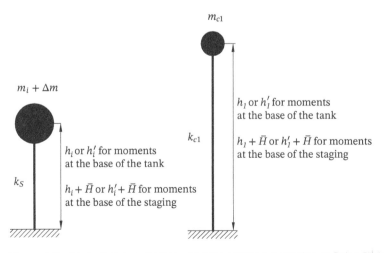

Figure 4.7 Equivalent two masses uncoupled system, according to Point C3.11, NZSEE-09 and Point A.6, EC8-4 [242, 267, 268]: two masses are assumed to be uncoupled and the earthquake forces on the support are estimated by considering two separate single degree of freedom systems (Point 4.2.2.4, IITK-GSDMA 2007).

A different way is to predict the behaviour of the conical tank by an equivalent geometry of a cylinder having the same thickness and projected perpendicular to the longitudinal axis of the cone [21, 277]:

$$H_{eq} = \frac{H}{\cos \theta_V} \tag{4.8a}$$

$$R_{eq} = \frac{2R_B + H \tan \theta_V}{2 \cos \theta_V} \tag{4.8b}$$

$$s_{eq} = s \text{ thickness of the wall} \tag{4.8c}$$

where R_B is the base radius of the conical tank and θ_V the vertical inclination angle of the tank walls. Alternatively, the refined equivalent mechanical models proposed by El Damatty, Sweedan et al. in [143–147, 575, 577–579] and Gavrilyuk et al. in [174, 175] may be used. An experimental shaking table test of an elevated conical tank has been reported by Diaconu et al. in [117];

- an added structural mass $\Delta m = \dfrac{\overline{W}}{g}$ where $\overline{W} = W_C + \frac{2}{3}W_P$ (according to Point 4.7.5.2(b), ACI 371R-98, it should be used a maximum of two-thirds (66%) the self-weight W_P of the concrete/steel staging) or $\overline{W} = W_C + \frac{1}{3}W_P$ (according to Point C4.2.2.3, IITK-GSDMA 2007 and Point 5.2.3, IS:1893 (1984)); Point C3.11, NZSEE-09 uses the total weight of the supporting structure such that $\overline{W} = W_C + W_P$. Mass Δm should be added to the impulsive mass m_i and hence connected to the tank's wall and located at a height h_i or h_i' above the base plate;

- a convective mass m_{c1} (where the first convective mode has been used in Equation (2.32) as suggested by Haroun et al. in [222, 223] and by Livaoğlu and Doğangün in [342]) or m_1 (Equations (2.44a) and (2.131b)) connected to the tank walls by springs of stiffness [88, 100, 646] (Equation (1.37)):

$$k_{c1} = \omega_{c1}^2 m_{c1} = 4\pi^2 \frac{m_{c1}}{T_1^2} \tag{4.9}$$

where ω_{c1} comes from Equation (2.30) and T_1 from expression (2.54). The convective mass m_{c1} should be located, above the base plate of the tank, at a height h_{c1}' and h_{c1} (Equations (2.34) and (2.36)) or h_1' and h_1 (Equations (2.44c) or (2.45) and (2.44b); Equations (2.131g) and (2.131h)). In order to evaluate impulsive and convective masses and heights the simplified Malhotra method [377] could be also used, according to Point A.3.2.2, EC8-4, and described in Section 5.6.

This method will be satisfactory for design purposes, if the ratio of the period of the two uncoupled systems exceeds 2.5 [268] as indicated in Point C3.11, NZSEE-09 and Point C4.2.2.4, IITK-GSDMA 2007; in ASCE 7 (2010 Edition) in Point 15.7.10.2(a), the same ratio has to be assumed to be equal to 3 in order to include the effects of fluid–structure interaction in determining the forces, the effective period, and the mass centroids of the system. If impulsive and convective time periods are not well separated, then a coupled two degrees of freedom system will have to be solved using elementary structural dynamics, as clearly explained in Section 4.4.

4.4 Two Coupled Masses Model

A two coupled masses model is suggested in Point 6.6 of the Technical Report from the United States Atomic Energy Commission titled "Nuclear Reactors and Earthquakes", Reactor technology TID-4500 (22nd Edition, August 1963, TID-7024). The resulting models (Figures 4.8 and 4.9) is a two degrees

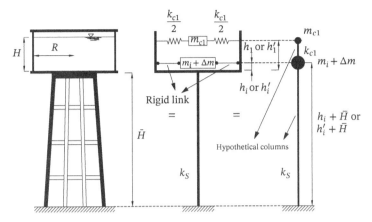

Figure 4.8 Two coupled masses dynamic model for elevated fluid container (Point 6.6, "Nuclear Reactors and Earthquake" from TID-4500).

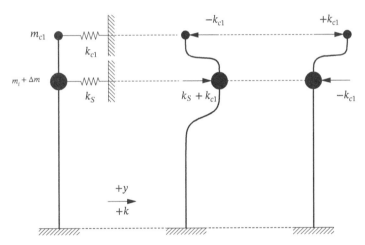

Figure 4.9 Equivalent two degrees of freedom system corresponding to the two coupled masses dynamic model for elevated fluid container depicted in Figure 4.8.

of freedom system, to which the equations previously described in Section 4.3 and regarding masses $(m_i, \Delta m, m_{c1})$ and heights $(h_i, h'_i, h_{c1}, h'_{c1})$ apply. The fictitious springs joining m_{c1} to the tank walls in Figure 4.8 have been replaced with a single hypothetical column (Figure 4.9) of the same stiffness, k_{c1} (Equation (4.9)), forming a direct coupling between $m_i + \Delta m$ and m_{c1}. The weight $m_i + \Delta m$ is connected to the ground through a similar hypothetical column representing the support structure and having a spring constant equal to k_S (Figure 4.9). The procedure for calculating frequencies ω_n, periods T_n of vibrating n−mode shapes ϕ_n, and maximum forces and deflections of the system can be found by following the procedure illustrated in detail below:

Step 1 Evaluation of the masses $(m_i, \Delta m, m_{c1})$ and heights $(h_i, h'_i, h_{c1}, h'_{c1})$ according to the procedure described in Section 4.3.

Step 2 Calculation of the convective spring constant k_{c1} from Equation (4.9), and computation of the spring staging constant k_S from a deflection analysis of the supporting structure. Spring constant k_S (Figure 4.9) is defined as the horizontal shear in the hypothetical column joining $m_i + \Delta m$ to the ground, when $m_i + \Delta m$ is translated horizontally through a unit distance with m_{c1} restrained against translation. Eventually, a closed-form expression was derived for the lateral stiffness of the frame staging by the method outlined in [500] (and also [133–136]) and assuming equal heights for all panels.

Step 3 Determination of the n–natural vibration frequencies, ω_n, of the system in Figure 4.9 in terms of the restraining forces, (k_S, k_{c1}), and the masses, $(m_i, \Delta m, m_{c1})$, are determined from the following relation:

$$\omega_n^2 = \frac{1}{2}\left[\frac{k_S + k_{c1}}{m_i + \Delta m} + \frac{k_{c1}}{m_{c1}} \pm \sqrt{\left(\frac{k_S + k_{c1}}{m_i + \Delta m} - \frac{k_{c1}}{m_{c1}} \right)^2 + 4\frac{k_{c1}^2}{(m_i + \Delta m)m_{c1}}} \right] \tag{4.10}$$

where $n = 1$ for the first mode and $n = 2$ for the second mode. The first two periods of vibration, $(T_1 > T_2)$ (corresponding respectively to the two natural vibration frequencies ω_1 and ω_2), are given by:

$$T_n = \frac{2\pi}{\omega_n} \tag{4.11}$$

Step 4 Evaluation of the eigenvectors ϕ_n at the level of the mass $m_i + \Delta m$, which corresponds to the mode shape of the freely vibrating system using either of the following expressions:

$$\phi_n = \frac{-\dfrac{k_{c1}}{m_i + \Delta m}}{\dfrac{k_S + k_{c1}}{m_i + \Delta m} - \omega_n^2} = \frac{-\dfrac{k_{c1}}{m_{c1}} + \omega_n^2}{\dfrac{-k_{c1}}{m_{c1}}} \tag{4.12}$$

assuming the eigenvector at the level of the mass m_{c1} to be unity ($\bar{\phi} = 1$) and the positive signs in Figure 4.9.

Step 5 Determination of the maximum displacements $\bar{y}_{0n}$ and $\bar{y}_{1n}$, corresponding to the masses $(m_i + \Delta m)$ and m_{c1} respectively:

$$\bar{y}_{01} = \frac{K_1 S_{ev}(T_1)}{\omega_1}\phi_1 = \bar{y}_{11}\phi_1 \tag{4.13a}$$

$$\bar{y}_{11} = \frac{K_1 S_{ev}(T_1)}{\omega_1} \tag{4.13b}$$

for the first mode ($n = 1$) and:

$$\bar{y}_{02} = \frac{K_2 S_{ev}(T_2)}{\omega_2}\phi_2 = \bar{y}_{12}\phi_2 \tag{4.14a}$$

$$\bar{y}_{12} = \frac{K_2 S_{ev}(T_2)}{\omega_2} \tag{4.14b}$$

for the second mode ($n = 2$). The maximum response of the elevated tank in Figure 4.8, given by Equations (4.13) and (4.14), is obtained for each of the two modes in terms of the following participation factors [100]:

$$K_1 = \frac{(m_i + \Delta m)\phi_1 + m_{c1}}{(m_i + \Delta m)\phi_1^2 + m_{c1}} \tag{4.15a}$$

$$K_2 = \frac{(m_i + \Delta m)\phi_2 + m_{c1}}{(m_i + \Delta m)\phi_2^2 + m_{c1}} \tag{4.15b}$$

Response (Equations (4.13) and (4.14)) is to be based on the velocity response spectrum S_{ev} (m/sec) corresponding to period T_1 and T_2 and an appropriate viscous damping ratio (Section 2.2.6; Point 2.3.3.2, EC8-4, Points 3.2 and 3.7, NZSEE-09, Point E.1, API 650 2012 Edition, and Point 15.7.2, ASCE 7) for the two modes of vibration. The damping associated with the sloshing action is usually quite low, for water in the range of 0.5%, whereas that associated with the motion of the supporting structure may be considerably larger. Usually the first mode involves sloshing almost entirely, whereas the second mode is primarily motion of the supporting structure. Consequently it is reasonable to use the damping associated with the fluid in the tank for the first mode and a value for the second mode appropriate to the type of staging structure involved (steel/concrete and framing/pedestal).

Step 6 Calculation of the maximum horizontal inertia forces for each mode acting on each mass. These forces can be expressed in terms of the spring constants, k_S and k_{c1}, and the maximum deflections, $\bar{y}_{0n}$ and $\bar{y}_{1n}$, as follows:

$$F_{01} = -k_{c1}\bar{y}_{11} + (k_S + k_{c1})\bar{y}_{01} \tag{4.16a}$$

$$F_{11} = k_{c1}\bar{y}_{11} - k_{c1}\bar{y}_{01} \tag{4.16b}$$

for the first mode ($n = 1$) and:

$$F_{02} = -k_{c1}\bar{y}_{12} + (k_S + k_{c1})\bar{y}_{02} \tag{4.17a}$$

$$F_{12} = k_{c1}\bar{y}_{12} - k_{c1}\bar{y}_{02} \tag{4.17b}$$

for the second mode ($n = 2$). The maximum tower horizontal shear is:

$$Q = F_{01} + F_{11} + F_{02} + F_{12} \tag{4.18}$$

Step 7 Determination of the maximum vertical sloshing wave displacement, $d_{\max}$, of the fluid surface for cylindrical tanks with radius R as depicted in Figure 4.10 (it should be noted that the following relations hold also in case of above-ground storage tanks) [241]:

$$\theta_h^1 = 1.534\frac{\bar{y}_{11} - \bar{y}_{01}}{R}\tanh\left(\sqrt{\frac{27}{8}\frac{H}{R}}\right) \tag{4.19a}$$

$$d_{\max}^1 = \frac{\frac{2}{3}\sqrt{\frac{3}{8}}R\coth\left(\frac{27}{8}\frac{H}{R}\right)}{\dfrac{g}{\omega_1^2\theta_h^1 R} - 1} \tag{4.19b}$$

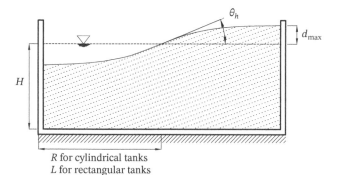

Figure 4.10 Displacement d_{max} and angular amplitude θ_h of free oscillations at the fluid surface.

for the first mode ($n = 1$) and:

$$\theta_h^2 = 1.534\frac{\bar{y}_{12} - \bar{y}_{02}}{R}\tanh\left(\sqrt{\frac{27}{8}\frac{H}{R}}\right)$$ (4.20a)

$$d_{max}^2 = \theta_h^2 R$$ (4.20b)

for the second mode ($n = 2$). The maximum vertical displacement, d_{max}, of the fluid surface is obtained by summing the contributions of the two modes:

$$d_{max} = d_{max}^1 + d_{max}^2$$ (4.21)

Step 8 In the case of a rectangular tank of dimension $2L \times 2B$ the half-length L of the rectangular tank is used instead of the circular radius R [241]:

$$\theta_h^1 = 1.58\frac{\bar{y}_{11} - \bar{y}_{01}}{L}\tanh\left(\sqrt{\frac{5}{2}\frac{H}{L}}\right)$$ (4.22a)

$$d_{max}^1 = \frac{0.527L\coth\left(\sqrt{\frac{5}{2}\frac{H}{L}}\right)}{\dfrac{g}{\omega_1^2\theta_h^1 L} - 1}$$ (4.22b)

for the first mode ($n = 1$) and:

$$\theta_h^2 = 1.58\frac{\bar{y}_{12} - \bar{y}_{02}}{L}\tanh\left(\sqrt{\frac{5}{2}\frac{H}{L}}\right)$$ (4.23a)

$$d_{max}^2 = \theta_h^2 L$$ (4.23b)

for the second mode ($n = 2$). The maximum vertical displacement, d_{max}, of the fluid surface may be derived from Equation (4.21).

Most structures will experience inelastic deformations when subjected to moderate to severe ground motions [453]. If the staging is detailed to have significant ductility μ, then it may be designed to respond beyond the yield level under the design earthquake. Approximate methods to estimate the maximum inelastic displacement demands from the maximum displacement demand of a linear elastic single degree of freedom system are particularly useful in these situations as consistently highlighted in many published research works [28, 89, 152, 400, 402, 401, 523] or recommendations (ATC-32, ATC-40, FEMA 273 and 274). In all of these approximate methods a key step is the estimation of the maximum inelastic displacement demand of single degree of freedom systems:

$$\ddot{x} + 2\xi_0\omega_0\dot{x} + \frac{F(x)}{m} = -\ddot{x}_g \tag{4.24}$$

from the maximum displacement demand of an equivalent linear elastic single degree of freedom systems [28]:

$$\ddot{x}_{eq} + 2\xi_{eq}\omega_{eq}\dot{x}_{eq} + \omega_{eq}^2 x_{eq} = -\ddot{x}_g \tag{4.25}$$

Equation (4.24) describes the motion in term of displacement, velocity and acceleration $(x(t), \dot{x}(t), \ddot{x}(t))$ of an idealized elastoplastic oscillator, with associated mass m, where $F(x)$ is the restoring force of the yielding system subject to a ground acceleration $\ddot{x}_g(t)$; ω_0 and ξ_0 are the circular frequency of vibration and damping ratio, respectively, of the non-linear oscillator. Using equivalent linearization methods, the maximum response of the system (whose exact solution is computed with Equation (4.24)) is approximated with the maximum response of an equivalent linear system whose response x_{eq} is computed with Equation (4.25) [402]. The concept of equivalent viscous damping ξ_{eq} was proposed by Jacobsen in [262] to obtain approximate solutions of the steady forced vibration of damped single degree of freedom systems with linear force–displacement relationships. Years later, Jacobsen in [264] extended the concept of equivalent viscous damping to yielding single degree of freedom systems by considering simultaneously an equivalent viscous damping ratio and a period shift T_{eq} (or ω_{eq}). As clearly highlighted by Hadjian in [210], if the equal energy dissipation principle is employed, different methods of treating the period shift are the reasons for the different equivalent viscous damping ratios given by many authors, such as Rosenblueth and Herrera in [479], Gulkan and Sozen in [201], and Iwan in [259]. Hence, the three parameters of the linear equivalent system $(\omega_{eq}, T_{eq}, \xi_{eq})$ in Equation (4.25) may be derived using one of the following relations [28, 402, 683]:

1) Rosenblueth and Herrera's equivalent linearization method [479] was historically the first to propose using the secant stiffness k_{sec} at maximum deformation as the basis for selecting the period shift. In the case of a bilinear system with harmonic loading and a post-yield stiffness of α' times the initial stiffness, the circular frequency of vibration ω_{eq}, the ratio between the period of vibration of the equivalent system T_{eq} to that of the original system T, and the viscous damping ratio in the equivalent linear elastic system ξ_{eq}, are given by:

$$\omega_{eq} = \sqrt{\frac{k_{sec}}{m}} = \frac{2\pi}{T_{eq}} \tag{4.26a}$$

$$\frac{T_{eq}}{T} = \sqrt{\frac{\mu}{1 - \alpha' + \alpha'\mu}} \tag{4.26b}$$

$$\xi_{eq} = \xi_0 + \frac{2}{\pi}\left[\frac{(1 - \alpha')(\mu - 1)}{\mu - \alpha'\mu + \alpha'\mu^2}\right] \tag{4.26c}$$

In the case of elastoplastic system ($\alpha' = 0$), relations (4.26b) and (4.26c) reduce to:

$$\frac{T_{eq}}{T} = \sqrt{\mu} \tag{4.27a}$$

$$\xi_{eq} = \xi_0 + \frac{2}{\pi}\left(1 - \frac{1}{\mu}\right) \tag{4.27b}$$

2) In later research, Gulkan and Sozen [201] noted that under earthquake loading, instead of a harmonic loading, usually the displacement would be significantly smaller than the maximum response. This means that the equivalent damping ratio evaluated with expressions (4.26c) or (4.27b) would result in an overestimation of the equivalent viscous damping and hence would lead to an underestimation of the response [402]. Using model by Takeda et al. [583] with degrading stiffness hysteretic response and experimental shake table results, Gulkan and Sozen developed the following empirical equation to compute the equivalent damping ratio:

$$\xi_{eq} = \xi_0 + \frac{1}{5}\left(1 - \frac{1}{\sqrt{\mu}}\right) \tag{4.28}$$

The substitute structure method, developed by Shibata and Sozen in [534, 535], is the extension of the empirical method, proposed by Gulkan and Sozen [201], to multi degrees of freedom systems;

3) Using a hysteretic model [94], derived from a combination of linear elastic and Coulomb slip elements [173, 443], Iwan and Gates [259–261] suggested the following empirical relations to estimate the period shift and equivalent damping ratio (Equations C3.43/C3.44, Point C3.114 NZSEE-09):

$$\frac{T_{eq}}{T} = 1 + 0.121(\mu - 1)^{0.939} \tag{4.29a}$$

$$\xi_{eq} = \xi_0 + 5.868(\mu - 1)^{0.371} \tag{4.29b}$$

5

Flexible Tanks

After reading this chapter you should be able to:

- Understand the equivalent mechanical model to analyse flexible tanks
- Evaluate impulsive and convective modes and masses for flexible tanks
- Identify components related to the performance of flexible tanks

5.1 Introduction

After the extensive damage observed during the Prince William Sound, Alaska, earthquake in 1964 [213, 478] and many other earthquakes till 1983 [424, 469], the importance of the influence of wall flexibility on the seismic tank's response became evident [114]. Several analytical studies have been conducted to assess the effects of wall flexibility: Edwards [140] was the first investigator to consider tank flexibility in establishing the hydrodynamic forces exerted on ground-supported, cylindrical tanks subject to horizontal earthquake ground motions; Veletsos and Yang [624, 645, 646, 680], Shaaban and Nash [524], Balendra et al. [23, 24]; Haroun and Housner in [223, 224] extended Housner's mechanical model [240–242] for rigid tanks to include the tank wall deformability by considering only the fundamental mode of vibration of the tank wall and then Haroun [214] modelled the tank shell by using harmonic ring- shaped finite elements and the fluid by added mass through an analytical solution from continuum mechanics (Fischer [158], Kana [286], Tedesco and Kostem [599]). It has been observed, mainly by Kana and Dodge [287] and by Balendra et al. [23], that for tanks with realistic flexibility, the pressures and forces are considerably different from those in rigid tanks (Point A.3.1, EC8-4). The walls of a rigid tank experience the same motion of the ground ($a(t)$ as in Equation (2.5)), whereas those for a flexible tank generally move differently [626]. To be more clear, the maximum acceleration in a given direction of a point on a deformable wall may be equal to, greater than, or smaller than the maximum ground acceleration, depending on the location of the point and the flexibility of the tank, whereas for a rigid tank it is the same as the corresponding maximum ground acceleration. As a consequence, these differences also are reflected in the magnitudes and distributions of the hydrodynamic wall pressures

Seismic Design and Analysis of Tanks, First Edition. Gian Michele Calvi and Roberto Nascimbene.
© 2023 John Wiley & Sons, Inc. Published 2023 by John Wiley & Sons, Inc.

and associated tank forces. Furthermore, the flexibility of the tank wall affects almost exclusively the impulsive component of the response because the convective effects are associated with oscillations of the much longer period than those characterizing the impulsive effects (Point 4.3.3, HSE Research Report RR527) [635]. Veletsos and Yang in [645, 646] were the first to argue that the convective effects could not be affected considerably by the flexibility of the tank wall and recommended for design purposes evaluating these effects by considering the tank to be rigid. Later, the adequacy of this approach has been confirmed by the results obtained by Haroun and Housner [223, 225].

The hydrodynamic fluid pressure exerted on the wall of a flexible tank due to a ground motion $a(t)$ is given by the following contributions [158, 164, 224, 408, 419] (Point A.3, EC8-4):

1) The impulsive fluid pressure component $p_i(\xi, \zeta, \theta, t)$ which varies synchronously with the horizontal ground acceleration $a(t)$ and therefore remains unaffected in the rigid case. It is called "rigid impulsive";

2) The long period component contributed by the convective fluid pressure $p_c(\xi, \zeta, \theta, t)$ driven by $a_{cn}(t)$ [215]. Because the convective components of the response are more or less insensitive to variations in wall flexibility, they may be considered to be the same as those obtained for rigid tanks [635];

3) The short period fluid pressure $p_f(\zeta, \theta, t)$ caused by the wall deformation relative to the base circle due to the deformability of the tank wall (called the "flexible impulsive") [223, 624, 626, 646]. The difference in the dominant period of vibration of the convective and flexible impulsive response weakens the coupling between the two components and decreases the sensitivity of one component to changes in the characteristics of the other. Based on this reasoning, the coupling between the wall displacement and sloshing is not included in the usual design approaches (Point 4.3.3, HSE Research Report RR527);

4) The fluid pressure $p_v(\zeta, t)$ due to the vertical motion of the tank. It is composed of the hydrodynamic rigid pressure $p_{vr}(\zeta, t)$ or $p_{vr}(z)$ and of the flexible wall component $p_{vf}(\zeta, t)$ induced by the vertical component of the ground acceleration $a_v(t)$.

The information in the following sections refers exclusively to the impulsive effects $p_f(\zeta, \theta, t)$ and $p_{vf}(\zeta, t)$ respectively given in Sections 5.2 and 5.3.

Wall deformation effects have been studied extensively in the case of vertical and horizontal cylinders and rectangular tanks, without giving any criteria about a possible subdivision or classification for the different cases. According to Table 1.2, only two codes introduce two different formulations and mechanical models for rigid and flexible tanks: UNI EN 1998-4:2006 titled *Design of Structures for Earthquake Resistance. Part 4: Silos, Tanks and Pipelines* and NZSEE-09 *Seismic Design of Storage Tanks: Recommendations of a Study Group of the New Zealand National Society for Earthquake Engineering*. Generally, concrete tanks are considered to be tanks with rigid wall, while steel tanks are considered to be tanks with a flexible wall. However, it is well known that in industrial applications, some steel vessels are rather thick (and consequently rigid enough) to resist high internal pressure [439] and, on the other hand, the impulsive terms in tall reinforced concrete tanks may be consistently affected by the flexibility of the walls [183]. This is the main reason why we suggest using Table 5.1 as a simple guide to select the behaviour of the tank using just the geometrical slenderness $\frac{H}{R}$ and the thickness slenderness $\frac{R}{s}$.

Table 5.1 Distinction between rigid (R) and flexible (F) tanks with respect to the geometrical slenderness $\frac{H}{R}$ and the thickness slenderness $\frac{R}{s}$.

$\frac{R}{s}$	$\frac{H}{R}$									
	0.50	0.75	1.00	1.50	2.00	2.50	3.00	4.00	5.00	6.00
100	R	R	R	R	R	R	R	R	R	R
125	R	R	R	R	R	R	R	R	R	R
150	R	R	R	R	R	R	R	R	R	F
250	R	R	R	R	R	R	R	R	F	F
400	R	R	R	R	R	R	R	F	F	F
600	R	R	R	R	R	R	F	F	F	F
800	R	R	R	R	R	F	F	F	F	F
1000	R	R	F	F	F	F	F	F	F	F
1500	F	F	F	F	F	F	F	F	F	F
2000	F	F	F	F	F	F	F	F	F	F

5.2 Impulsive Pressure Component

This section provides information on seismic analysis procedures for flexible tanks subjected to horizontal and vertical action, mainly using EC8-4, NZSEE and USA Standards, having the following characteristics:

- upright cylindrical shape (Section 5.2.1) and rectangular cross-section (Section 5.2.2);
- rigid foundation (Sections 5.2.1 and 5.2.2) and flexible foundation or soil–structure interaction (Section 5.2);
- fully anchored at the base (Sections 5.2.1 and 5.2.2) and under the effect of a rigid base rocking motion (Section 5.7).

The analysis procedures presented in the following consider a theoretical and analytical approach to handle the seismic design of liquid storage flexible containers, based on the seminal papers by Fischer, Rammerstorfer, and Scharf in [166, 468, 470, 507], Veletsos et al. in [591, 624, 626, 641, 645, 646], Haroun et al. in [219, 226] and Malhotra in [370]. Although such provisions represent a significant advantage in the design of tanks, Annex A, Point A.1, EC8-4 explicitly highlight that the dynamic interaction between the fluid, the flexible tank walls, and the underlying foundation soil, is a problem of considerable complexity usually requiring high computational resources and software such as LS-DYNA [351] using the Arbitrary Lagrangian Eulerian (ALE) method [132, 650].

5.2.1 Vertical Cylindrical Tanks

Following Section 2.2 and Figure 2.1, the tank under consideration is a circular cylindrical, flexible liquid container of radius R, length $\overline{L}$, and thickness s. The tank is filled with liquid (homogeneous, incompressible, inviscid, and initially at rest) to a height H and a cylindrical coordinate system (r, θ, z) is used with the centre of the base being the origin. The analysis of liquid motion requires the evaluation of a velocity potential function, $\Phi(r, \theta, z, t)$ (irrotational flow field), which satisfies the Laplace equation (2.1) in the region occupied by liquid ($0 \leq r \leq R$, $0 \leq \theta \leq 2\pi$ and $0 \leq z \leq H$) [215]. Since the velocity vector of the liquid is the gradient of the velocity potential, the liquid container boundary conditions can be expressed as follows [203, 214, 215, 221, 250, 637]:

- at the tank bottom, $z = 0$, the liquid velocity in the vertical direction must be the same as the vertical velocity of the rigid base (let us suppose that the vertical component of the liquid velocity must be zero for a horizontally excited tank):

$$\frac{\partial \Phi}{\partial z}(r, \theta, 0, t) = 0 \tag{5.1}$$

- at the quiescent liquid free surface, ($z = H$), the pressure is zero:

$$\frac{\partial \Phi}{\partial t}(r, \theta, H, t) = 0 \tag{5.2}$$

- the radial velocity of the liquid adjacent to the wall of the tank shell, ($r = R$), must be the same as that of the flexible shell:

$$\frac{\partial \Phi}{\partial r}(R, \theta, z, t) = \left[v(t) + \frac{\partial w}{\partial t}(\theta, z, t) \right] \cos \theta \tag{5.3}$$

where $w(\theta, z, t)$ is the shell radial displacement of the deformable wall and $v(t)$ is the velocity corresponding to the ground motion $a(t)$.

The solution to Equation (2.1) in order to evaluate the pressure $p_f(\zeta, \theta, t)$, together with its appropriate boundary conditions (5.1)–(5.3), is mainly based on the analysis reported by Veletsos et al. in [624, 626, 641, 645, 646], which utilizes concepts and hypotheses employed by Chopra [86], in his studies on the effects of fluid–structure interaction for reservoir–dam systems, and by Haroun, Housner et al. in [214, 221, 223, 224, 226, 227]:

1) It is assumed that the systems, schematically depicted in Figure 5.1(a), responds as a cantilever system with a deflection configuration $f(\zeta)$ of prescribed form (first, second and third mode, $k = 1,2$, and 3, in Figure 5.1(a)) at any time during the seismic motion. The dimensionless continuous function $f(\zeta)$ defines the assumed mode of deformation for the tank wall such that $|f(\zeta)| \leq 1.0$ [161, 162, 646]. Usually, as clearly explained by Fischer and Rammerstorfer in [164], tall tanks with the ratio $\frac{H}{R} > 1$ often show a more or less linear variation of the deformation over the tanks' height (Equation (5.9b)), whilst broad tanks with a ratio $\frac{H}{R} < 1$ often show a typical "concave" or "convex" wall deformation shape which can be approximated by relations (5.9a) and (5.9c);
2) It is further assumed that the cylindrical cross-section of the tank does not change shape during the deformation exerted on the container's wall by the action of an earthquake. This means that only the set of modes for which the circumferential variation of radial displacements is proportional to the first mode $\cos \theta$ with $m = 1$ need be considered (Figure 5.1(b)) [161, 163, 166, 214, 215, 284,

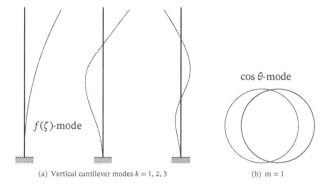

(a) Vertical cantilever modes $k = 1, 2, 3$ (b) m = 1

Figure 5.1 (a) Vertical nodal pattern corresponding to a beam-type mode of vibration and (b) circumferential nodal pattern corresponding to the first mode $\cos\theta$ with $m = 1$.

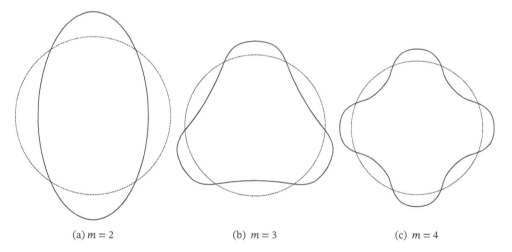

(a) $m = 2$ (b) $m = 3$ (c) $m = 4$

Figure 5.2 Circumferential nodal pattern corresponding to the first three mode $\cos m\theta$ with $m = 2,3$ and 4 [67, 412]: two, three and four lobe deformation pattern.

416, 470]. Furthermore we are investigating in this analytical approach, suggested at Point A.3.1, in EC8-4, perfectly circular cylindrical tanks for which a horizontal base excitation activates basic radial modes corresponding to wave number $m = 1$ only [164, 166]. Clough D.P. and Clough R.W. in [95, 96, 98] as well as Kana [285, 286], shaking table experiments from Niwa [427] and vibration tests on full-scale tanks by Housner and Haroun in [245], confirmed by extensive measurements that the deformed shape of cylindrical tanks subjected to horizontal earthquake excitation also contains modes with circumferential wave numbers $m > 1$ [158] mainly due to non-circular imperfections of the tank's wall. Circumferential wave numbers $m > 1$ (Figure 5.2) have been considered analytically by Fischer in [158] and numerically by Wu et al. in [673].

As a consequence of the two main hypotheses above, the natural free lateral vibration modes of a circular cylindrical tank can be classified as $\cos\theta$-type modes in the circumferential direction and a beam-type mode in the axial/radial direction [655], because the tank behaves like a vertical cantilever. So the impulsive flexible pressure $p_f(\zeta, \theta, t)$ caused by the wall deformation behaves in the following fashion depending on the modes of vibration (Point A.3.1, EC8-4):

$$f(\zeta)\cos\theta \tag{5.4}$$

Assuming the modes (5.4) known (first $f^1(\zeta)$, second $f^2(\zeta)$, and third $f^3(\zeta)$ mode, corresponding to $k = 1, 2$, and 3, in Figure 5.1(a) through $f(\zeta)$), the spatial-temporal variation (ζ, θ, t) of the flexible impulsive pressure on the wall is (Point A.3.1, EC8-4 and Point 4.3.3, HSE Research Report RR527) [158, 160, 164, 336, 408, 419, 507, 645]:

$$p_f(\zeta, \theta, t) = \left[\sum_{k=1}^{\infty} \left(\sum_{n=0}^{\infty} d_n(\zeta)\psi(\zeta)\cos(v_n\zeta) \right) a_{fk}(t) \right] \rho_l H \cos\theta \tag{5.5}$$

where [164, 645]:

$$\psi(\zeta) = \cfrac{\displaystyle\int_0^1 f(\zeta) \left[\overbrace{\frac{\rho_s}{\rho_l}\frac{s(\zeta)}{H}}^{\text{structural mass}} + \overbrace{\sum_{n=0}^{\infty} b'_n \cos(v_n\zeta)}^{\text{liquid mass}} \right] d\zeta}{\displaystyle\int_0^1 f(\zeta) \left[\underbrace{\frac{\rho_s}{\rho_l}\frac{s(\zeta)}{H}f(\zeta)}_{\text{structural mass}} + \underbrace{\sum_{n=0}^{\infty} d'_n \cos(v_n\zeta)}_{\text{liquid mass}} \right] d\zeta} \tag{5.6a}$$

$$b'_n = \frac{(-1)^n}{v_n^2} \frac{I_1(v_n/\gamma)}{I'_1(v_n/\gamma)} \quad \text{and} \quad d'_n = \frac{d_n}{2} \tag{5.6b}$$

$$d_n(\zeta) = 2 \frac{I_1(v_n/\gamma)}{I'_1(v_n/\gamma)} \frac{\displaystyle\int_0^1 f(\zeta)\cos(v_n\zeta)d\zeta}{v_n} \tag{5.6c}$$

$$v_n = \frac{2n+1}{2}\pi \tag{5.6d}$$

In Equations (5.5)–(5.6), ρ_l and ρ_s are the mass density of the liquid and of the tank wall's material (in kg/m^3), respectively; $s(\zeta)$ is the variable wall thickness and $a_{fk}(t)$ is the response acceleration, relative to the base, of a single degree of freedom linear oscillator induced by the prescribed free-field ground motion $a(t)$; the single degree of freedom system should have a circular frequency ω_{fk} (or a period $T_{fk} = \frac{2\pi}{\omega_{fk}}$; Equations (5.35)–(5.40)) and a damping ratio for the interaction during a common vibration of the tank wall and the liquid [161, 166] (for a system steel shell containing water it could be 2% [162, 166, 223, 297] or 1% [161, 162]; for a system concrete shell containing water it could be 3% [229] or 5% (Point R9.5, ACI 350.3-06 and [80, 183])), both corresponding to the kth–beam-type mode of vibration (Figure 5.1(a)). In the case of soil–structure interaction, the frequency and damping ratio should be selected according to the terms defined with reference to Equation (5.19). The coefficient ψ is a mode participation factor [297], depending on the ratio between the mass of the structure and of the containing liquid [164, 624, 645]; more specifically, the upper part of the relation represents the effective mass of the structure–fluid system for a rigid body motion of the tank and the lower part of Equation (5.6a) represents the effective mass of the system when vibrating in a deflection configuration specified by the function $f(\zeta)$ [645]. It is of interest to note that in, case of a rigid body motion ($f(\zeta) = 1$ [624]) of a rigid tank ($\psi = 1$) with $a_{fk}(t) \rightarrow a(t)$, Equation (5.6c) reduces to the following expression:

$$d_n = 2\frac{I_1(\nu_n/\gamma)}{I_1'(\nu_n/\gamma)}\frac{\int_0^1 f(\zeta)\cos(\nu_n\zeta)d\zeta}{\nu_n} \quad \rightarrow \quad 2\frac{I_1(\nu_n/\gamma)}{I_1'(\nu_n/\gamma)}\frac{(-1)^n}{\nu_n^2} \tag{5.7}$$

and consequently relation (5.5) reduces to Equation (2.5) and coefficient $d_n\cos(\nu_n\zeta) \rightarrow C_i(\xi = 1, \zeta)$ obtained from expression (2.6a).

A procedure which seems justified for many practical applications [27, 161, 166, 652] is to take into account only the fundamental beam-type mode of vibration, $k = 1$ ($a_{fk}(t) = a_f(t)$), as clearly required by Point A.3.1, in EC8-4. Hence, in Equations (5.5)–(5.6) the mode index $k = 1$ and the summation $\sum_{k=1}^{\infty}$ over all modal contribution may be dropped, obtaining:

$$p_f(\zeta, \theta, t) = \rho_l H \cos\theta \left(\sum_{n=0}^{\infty} d_n(\zeta)\psi(\zeta)\cos(\nu_n\zeta) \right) a_f(t) \tag{5.8}$$

where $f(\zeta)$ represents the dimensionless first mode deformation for the tank wall. The maximum value of the pressure p_f may then be determined by replacing $a_f(t)$ by its spectral value S_e using the natural frequency and damping value according to Equation (5.19) (Point C3.3.2, NZSEE-09) [635]. In deriving the flexible impulsive pressure $p_f(\zeta, \theta, t)$ (Equations (5.6) and (5.8)), the first eigenmode $f(\zeta)$ of vibration of the cylindrical wall of the container filled with liquid, may be derived following different strategies:

- Using a finite element analysis for determination of mode shapes of an elastic circular, cylindrical shell filled to an arbitrary depth with an incompressible and inviscid fluid. The problem of the coupled system has been investigated analytically and numerically by Wu et al. in [524, 673] and then using the computer program TANKFREQ [410].
- Applying the simplified analytical formulation proposed by Gupta [202, 203] for the evaluation of the eigenvalues and eigenvectors of partially or completely liquid-filled ground-supported cylindrical tanks.
- Following an iterative method for the determination of the first mode shape $f(\zeta)$ proposed by Fischer and Rammerstorfer in [158, 161, 163] (Point A.3.1, EC8-4):
  Step 1 Assume a vibration mode $f^i(\zeta)$ (trial configuration), where ^{i} is the number of the ith cycle in the iteration process. Three different types of assumed shape of wall deformation may be used according to Veletsos and Yang [624, 645], corresponding to a half-sine curve, a linear function and a versed-sine expression:

$$f(\zeta) = \sin\left(\frac{\pi}{2}\zeta\right) \tag{5.9a}$$

$$f(\zeta) = \zeta \tag{5.9b}$$

$$f(\zeta) = 1 - \cos\left(\frac{\pi}{2}\zeta\right) \tag{5.9c}$$

The trial configuration $f(\zeta)$ depends on the relative magnitudes of the flexural and shearing deformations of the structure as well as on the dimension of the tank and on the relative weight (and type) of roof system and of the contained liquid. However, as pointed out by Fischer and Rammerstorfer in [164] and by Virella et al. in [652], tall tanks with the ratio $\frac{H}{R} > 1$ often show a more or less linear variation of the deformation over the tanks' height (Equation (5.9b)), while broad tanks with a ratio $\frac{H}{R} < 1$ often show a typical "concave" or "convex" wall deformation shape which can be approximated by relations (5.9a) and (5.9c). After measurement and analysis, the distribution suggested by Fischer in [158] is given by wall deformation (5.9b), while the one used by Haroun and Abou-Izzeddine in [218] to simulate the tank wall vibration is given by Equation (5.9a).

Step 2 Evaluate a fictitious axisymmetric mass density $\rho_{\star}^{i}(\zeta)$ to be added to the mass density ρ_s of the tank wall [158, 160] (Point A.3.1, EC8-4) in order to determine an effective mass density $\rho^i(\zeta)$:

$$\rho^i(\zeta) = \rho_{\star}^{i}(\zeta) + \rho_s = \frac{p_f^i(\zeta)}{2gs(\zeta)f^i(\zeta)} + \rho_s \tag{5.10}$$

where ^{i} is the number of the ith cycle in the iteration process. In the "added mass" concept [27, 236, 652], a "dry shell" can act as an approximate substitute of the liquid-filled shell as long as vibrations, with an eigenmode approximated by $f(\zeta)$, are considered [163]. In Equation (5.10) $p_f^i(\zeta)$ may be derived from expression (5.8) normalized to a horizontal excitation of $1 \cdot g$ and assuming only the fundamental beam-type mode of vibration, $k = 1$ ($a_{fk}(t) = a_f(t)$), such that $\cos \theta a_f(t) = g$ [166, 297, 470]. The singularity associated with $\rho^i(\zeta)$ at $\zeta = 0$ may be treated numerically using a procedure based on the Rayleigh quotient [106, 109, 470] or using a simplified approach proposed by Habenberger in [206, 207]. A simplified procedure using a linear fictitious additional mass density variation has been described by Fischer in [159].

Step 3 In order to take the fluid pressure into account, the "dry shell" is attributed, along the wet height $0 \leq z \leq H$ (or $0 \leq \zeta \leq 1$), with the axisymmetric effective mass distribution $\rho^i(\zeta)$, obtained from expression (5.10). The solution of the eigenvalue problem of free vibration of this "dry shell" substitute, usually following a numerical procedure [283, 673], renders an eigenmode $f^{i+1}(\zeta)$ (corresponding to the first mode $\cos \theta$ with $m = 1$ [297]) that can be used for an improved assumption of the first mode shape of the liquid-filled tank [161, 163, 166].

Step 4 The new mode shape $f^{i+1}(\zeta)$ can now be used in Equation (5.10) to calculate a new effective mass density $\rho^{i+1}(\zeta)$. Then the iteration scheme given by the repetition of Steps 2–3 continues until a proper break–off criterium is attained. The following is suggested by Fischer and Rammerstorfer in [163]:

$$\left\| \left| \frac{\int_0^1 f^{i+1}(\zeta)d\zeta}{\int_0^1 f^i(\zeta)d\zeta} \right| - 1 \right\| < \epsilon \tag{5.11}$$

and this by Scharf in [506]:

$$\left| f^{i+1}(\zeta) - f^i(\zeta) \right| < \epsilon \tag{5.12}$$

where ϵ is a fixed tolerance value.

The first mode dimensionless distribution of the flexible impulsive pressure $\frac{p_f}{\rho_l R a_f(t)}$ is depicted in Figure 5.3(a) using two different values of thickness and as a function of the tank aspect ratio $\gamma = \frac{H}{R}$. Note that for broad tanks with values of $\frac{H}{R} < 1$, both the distributions and magnitudes of $\gamma \sum_{n=0}^{\infty} d_n(\zeta)\psi(\zeta)\cos(v_n\zeta)$ from Equation (5.8) are quite similar to those for the corresponding rigid tanks shown in Figure 2.2(a) [635]. The differences in shape between the impulsive pressure distributions for the first mode calculated considering flexible and rigid tanks (Figure 5.3(b)) increase with the aspect ratio γ of the tank [652]. Furthermore, despite the differences in the values of $\frac{s}{R}$, the results are not significantly different. The relative insensitivity of the pressure distribution functions to the $\frac{s}{R}$ ratio is further demonstrated by Veletsos and Shivakumar in [635] in the range of 0.0005 to 0.002. The practical implication of this behaviour is that the pressure relation (5.8) may be used with good accuracy for tanks with varying wall thickness as well. The insensitivity referred to above is for the first mode dimensionless distribution of the flexible impulsive pressure, but not for the associated pressure p_f and the following tank forces, shear and moments, in Equations (5.13), (5.15), and (5.17). The pressure and forces also depend on the response acceleration $a_f(t)$ that in turn depend on the natural frequencies of the tank-liquid system (Equation (5.34) or (5.59) excluding or including the effects of soil–structure interaction, respectively), which are relatively sensitive to variation in $\frac{s}{R}$ [635].

At the end of the iteration process, Steps 1–4, by integrating the converged flexible impulsive pressure $p_f(\zeta, \theta, t)$, over the vertical wetted surface of the tank, we can derived the corresponding resultant base horizontal shear $Q_f(t)$ [158, 160]:

$$Q_f(t) = \int_0^{2\pi} \int_0^H p_f(z, \theta, t) R \cos\theta \, dz \, d\theta = m_f a_f(t) \tag{5.13}$$

where [158, 160]:

$$m_f = m\psi\gamma \sum_{n=0}^{\infty} \frac{(-1)^n}{v_n} d_n \tag{5.14}$$

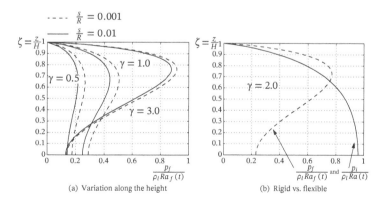

(a) Variation along the height

(b) Rigid vs. flexible

Figure 5.3 Coefficient $\gamma \sum_{n=0}^{\infty} d_n(\zeta)\psi(\zeta)\cos(v_n\zeta)$ (Equation (5.8)) [206] for the first mode flexible impulsive pressure variation, normalized to $\rho_l R a_f(t)$ as a function of the tank aspect ratio $\gamma = \frac{H}{R}$ and using $\cos\theta = 1$ (a); (b) comparison between the impulsive pressure distribution for the first mode of a flexible tank (Equation (5.8)) with $\frac{s}{R} = 0.001$ and a rigid tank (Equation (2.6a) and Figure 2.2(a)). Source: Adapted from [206].

It is of interest to note that in the case of rigid body motion ($f(\zeta) = 1$ [624]) of a rigid tank ($\psi = 1$), the flexible impulsive mass (5.14) reduces to the rigid impulsive mass (Equation (2.10)) thanks to the validity of relation (5.7).

The moment $M_f(t)$, immediately above the base, due only to the pressures exerted on the tank wall, may be evaluated according to the following expression [158]:

$$M_f(t) = m_f h_f a_f(t) \tag{5.15}$$

where h_f represents the height at which the flexible impulsive component of the base shear $Q_f(t)$ must be applied to yield the moment [158]:

$$h_f = H \frac{\left[\gamma \sum_{n=0}^{\infty} d_n \frac{(-1)^n v_n - 1}{v_n^2} \right]}{\gamma \sum_{n=0}^{\infty} d_n \frac{(-1)^n}{v_n}} \tag{5.16}$$

The total foundation moment $M_f'(t)$, immediately below the tank bottom, which also incorporates the contribution of the pressure on the tank base, is given by [158]:

$$M_f'(t) = m_f h_f' a_f(t) \tag{5.17}$$

where [158]:

$$h_f' = H \frac{\left[\gamma \sum_{n=0}^{\infty} d_n \frac{(-1)^n v_n - 1}{v_n^2} + \sum_{n=0}^{\infty} \frac{d_n}{v_n} \frac{I_2(v_n/\gamma)}{I_1(v_n/\gamma)} \right]}{\gamma \sum_{n=0}^{\infty} d_n \frac{(-1)^n}{v_n}} \tag{5.18}$$

in which $I_2(x) = I_0(x) - \frac{2I_1(x)}{x}$ denotes the modified Bessel function of the first kind of order 2 (Equations (2.7)) [561]. It is again interesting to highlight that in the case of rigid body motion ($f(\zeta) = 1$ [624]) of a rigid tank ($\psi = 1$), by substituting relation (5.7) into Equations (5.16) and (5.18), h_f and h_f' reduce to h_i (Equation (2.15)) and h_i' (Equation (2.13)), respectively. As stated in Section 1.7 and Figure 1.20, it is often advantageous to replace the liquid, conceptually, by an equivalent mechanical model based on a system of rigid bodies, spring, masses, and stiffness [119, 646]. The analytical formulation above suggested by Rammerstorfer, Fischer, and Scharf in [166, 468, 506] can be depicted in Figure 5.4 using a spring–mass analogy. A few approximate approaches for cylindrical tanks are presented in detail at Section 5.6.1.

According to Point C3.3.2, NZSEE-09 Recommendations for flexible tanks with a ratio $\frac{H}{R}$ in the range of 0.25 to 1.5, the maximum flexible impulsive component on the walls of a tank can be written as a spatial function of (z, θ) (Equation (C3.22), Point C3.3.2, NZSEE-09) replacing in relation (2.18) the impulsive period T_0 with the corresponding period of vibration of the first horizontal tank-liquid mode with or without the effects of soil–structure interaction [5]:

$$p_f(z, \theta) = q_0(z) \frac{S_e(T_f)}{g} \gamma_w R \cos\theta \qquad 0.25 \le \frac{H}{R} \le 1.5 \tag{5.19}$$

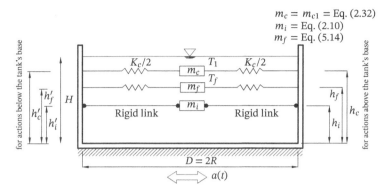

$$m_c = m_{c1} = \text{Eq. (2.32)}$$
$$m_i = \text{Eq. (2.10)}$$
$$m_f = \text{Eq. (5.14)}$$

Figure 5.4 Equivalent mechanical model for a flexible tank full of water under seismic excitation according to the analytical formulation suggested by Rammerstorfer, Fischer, and Scharf in [166, 468, 506].

where $q_0(z)$ is the dimensionless impulsive pressure functions of the height variable z and may be obtained from Equation (2.21); $S_e(T_f)$ is the spectral acceleration obtained from a horizontal elastic response spectrum corresponding to the impulsive period of vibration of the first flexible tank-liquid mode with ($T_f^\star$ in Equation (5.59)) or without (T_f in Equation (5.34)), the effects of soil–structure interaction and to the equivalent viscous damping level for horizontal mode consistent with the limit state (Point 2.3.3.1, EC8-4 and Point 4.1.3(1), EN 1998-2:2005; Point 3.4.3, ASCE/SEI 43-05)[25], tank type (bolted and welded steel; Point 4.1.3(1), EN 1998-2:2005) and interaction with the soil (Section 6.2). As clearly highlighted by Priestley et al. at Point C.3.2, NZSEE-09 and by Kettler in [297], Equation (5.19) gives conservative values for the flexible impulsive pressure below mid-height and under-estimates the same pressure above mid-height (Figure 5.3(b)).

5.2.2 Rectangular Tanks

Even though there have been numerous studies done on the fluid–structure interaction effects in flexible liquid containers, most of them are concerned with cylindrical tanks [140, 158, 164, 223, 224, 286, 408, 419, 524, 599, 624, 645, 646, 680]. Early investigation of dynamic analysis of fluid in flexible rectangular containers subjected to acceleration was conducted by Housner in [241]. An analytical solution based on three-dimensional modelling of the rectangular containers has been obtained by Kim et al. in [302] by applying the Rayleigh–Ritz method using assumed vibration modes of rectangular plate with suitable boundary conditions as admissible functions; in their study, only a pair of walls, orthogonal to the direction of the applied ground motion is assumed to be flexible, while the other pair remains rigid. Doğangün et al. in [124] investigated the seismic response of liquid-filled rectangular storage tanks using numerical methods based on a Lagrangian approach with a liquid assumed to be linear-elastic, inviscid, and compressible; more recently, Chen and Kianoush [79, 80] used the sequential method to calculate hydrodynamic pressure in two-dimensional rectangular tanks including wall flexibility effects. The effect of wall flexibility on the overall dynamic response of shallow and tall rectangular models has been thoroughly investigated by Ghaemmaghami and Kianoush [183] by comparing the results between rigid and flexible models. At Point C3.3.2 in the New Zealand recommendations, NZSEE-09 originally edited by Professor Priestley [459], it is recommended that the approximation for cylindrical tanks with flexible walls may be used for flexible rectangular tanks, highlighting that the order of the error is unknown.

An approximation for design purposes, suggested at Point A.4.2, EC8-4, is to use the same distribution (2.125) as for rigid walls but to replace the prescribed free-field ground motion $a(t)$ with the response acceleration $a_f(t)$ (assuming only the fundamental beam-type mode of vibration, $k = 1$ with $a_{fk}(t) = a_f(t)$), relative to the base, of a single degree of freedom linear oscillator induced by the prescribed free-field ground motion $a(t)$:

$$p_f(z, t) = q_0(z)\rho_l L a_f(t) \tag{5.20}$$

The single degree of freedom system should have a circular frequency ω_f (or a period $T_f = \frac{2\pi}{\omega_f}$; Equation (5.49) without or Equation (5.59) with soil–structure interaction) and a damping ratio for the interaction during a common vibration of the tank wall and the liquid (for a system concrete shell containing water it could be 3% [229] or 5% (Point R9.5, ACI 350.3-06 and [80, 183]); for a system steel shell–water it could be used 2% [162, 166, 223, 297] or 1% [161, 162]), both corresponding to the 1st–beam type mode of vibration (Figure 5.1(a)); in the case of soil–structure interaction, the damping ratio should be selected according to the terms defined with reference to Equation (5.19). The maximum value of the pressure p_f may then be determined by replacing $a_f(t)$ by its spectral value S_e using the natural frequency and damping value as previously outlined.

As stated in Section 1.7 and Figure 1.20, it is often advantageous to replace the liquid, conceptually, by an equivalent mechanical model based on a system of rigid bodies, spring, masses, and stiffness [119, 646]. The model suggested by Hashemi et al. in [229] and depicted in Figure 5.5, related to a rectangular tank of height H and dimension $2L \times 2B$, with $2L$ the tank length parallel to direction of earthquake loading (x), considers three response modes of the contained liquid (Point 3.3.2.1 and Figure 3.5, NZSEE-09) (Figure 5.6):

1) the rigid body impulsive response in which the portion of the liquid, represented by the mass $m_0 = m_d - m_f$ placed at a height $h_0 = \frac{m_d h_d - m_f h_f}{m_0}$ (or $h_0' = \frac{m_d h_d' - m_f h_f'}{m_0}$), accelerates with the tank walls synchronously with the horizontal ground acceleration $a(t)$:

$$m_d = 2m \sum_{n=0}^{\infty} \frac{(-1)^n \tanh(v_n/\gamma) \sin(v_n)}{v_n^3/\gamma} \tag{5.21a}$$

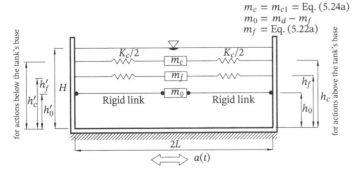

Figure 5.5 Equivalent mechanical model of rectangular flexible tanks as suggested by Hashemi et al. in [229]: rigid body impulsive response m_0, flexible impulsive response m_f and 1st–sloshing mode of the convective response m_{c1}. Source: Data from [229].

$$h_d = H \frac{\sum_{n=0}^{\infty}\left[\frac{(-1)^n \tanh(v_n/\gamma)\left[\cos(v_n)+v_n\sin(v_n)-1\right]}{v_n^4}\right]}{\sum_{n=0}^{\infty}\left[\frac{(-1)^n \tanh(v_n/\gamma)\sin(v_n)}{v_n^3}\right]} \tag{5.21b}$$

$$h_d' = H \frac{\sum_{n=0}^{\infty}\frac{(-1)^n}{v_n^3}\left[\frac{\tanh(v_n/\gamma)\left(\cos(v_n)+v_n\sin(v_n)-1\right)-\tanh(v_n/\gamma)}{v_n}+\frac{1}{\gamma}\right]}{\sum_{n=0}^{\infty}\left[\frac{(-1)^n \tanh(v_n/\gamma)\sin(v_n)}{v_n^3}\right]} \tag{5.21c}$$

where v_n comes from Equation (5.6d) and $\gamma = \frac{H}{L}$; h_d and h_d' refer to the hydrodynamic moments above and below the base plate, respectively;

2) the flexible impulsive response associated with the wall deformation relative to the base and due to the deformability of the tank wall:

$$m_f = m\frac{\tanh\left(\sqrt{3}\frac{L}{H}\right)}{\sqrt{3}\frac{2L}{H}} \tag{5.22a}$$

$$h_f = H\left[0.58 - 0.12\tanh\left(2.5\left(\frac{L}{H}-0.25\right)\right)\right] \tag{5.22b}$$

$$h_f' = \begin{cases} 0.58H & \frac{L}{H}\le 0.6 \\ H\dfrac{0.77\frac{L}{H}}{\tanh\left(\sqrt{3}\frac{L}{H}\right)} & \frac{L}{H}>0.6 \end{cases} \tag{5.22c}$$

where h_f should be used to evaluate the hydrodynamic moment induced on a section of the tank wall immediately above the base, whereas h_f' to evaluate the hydrodynamic moment immediately below the tank bottom, induced on the foundation. If the ratio $\frac{B}{H}<0.65$, a correction factor c_f should be applied to mass m_f in Equation (5.22a):

$$c_f = 1.25 - 0.71\left(\frac{B}{H}-0.3\right) \tag{5.23}$$

3) the nth–sloshing mode of the convective response supposed insensitive to variations in wall flexibility:

$$m_{cn} = m\frac{2\tanh(\alpha_n\gamma)}{\alpha_n^3\gamma} \tag{5.24a}$$

$$h_{cn} = H\left[\frac{\text{csch}(\alpha_n\gamma)-\coth(\alpha_n\gamma)}{\alpha_n\gamma}+1\right] \tag{5.24b}$$

$$h_{cn}' = H\left[\frac{2\,\text{csch}(\alpha_n\gamma)-\coth(\alpha_n\gamma)}{\alpha_n\gamma}+1\right] \tag{5.24c}$$

where $\alpha_1 = \frac{\pi}{2}$ for the first mode of vibration and consequently $\alpha_2 = \frac{3\pi}{2}$ and $\alpha_3 = \frac{5\pi}{2}$ respectively for the second and third modes of vibration; furthermore, h_{cn} should be used to evaluate the

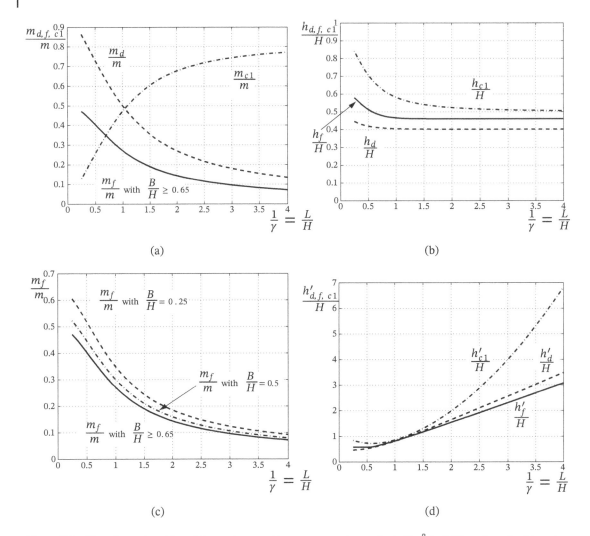

Figure 5.6 Non-dimensional impulsive and convective masses m_d/m, m_f/m (with $\frac{B}{H} \geq 0.65$) and m_{c1}/m (a), and heights h_d/H, h_f/H and h_{c1}/H (b) as a function of $\frac{L}{H}$; (c) application of the correction factor c_f to the mass m_f in case $\frac{B}{H} \geq 0.65$ and $\frac{B}{H} < 0.65$; (d) non-dimensional heights h'_d/H, h'_f/H and h'_{c1}/H.

hydrodynamic moment induced on a section of the tank wall immediately above the base, whereas use h'_{cn} to evaluate the hydrodynamic moment immediately below the tank bottom, induced on the foundation (Figure 5.6).

As previously highlighted with Equation (5.20), many standards and codes, such as the New Zealand recommendations NZSEE-09, EC8-4 (Point A.4.2), as well as ACI 350.3-06, do not consider the variation of the impulsive pressure, over the rectangular wall surface of the tank (i.e. $2B \times H$), in both vertical (z) and horizontal directions (y). The impulsive pressure distribution (intended here as the sum between the rigid and flexible impulsive components), suggested by Hashemi et al. in [229] and depicted in Figure 5.7, may be derived by solving the following linear system in terms of $p_i^{\max}$, p_{i1}^{base} and p_{i2}^{base}:

$$\tilde{h}B\left(p_i^{\max} + p_{i1}^{\text{base}}\right) + h_i^{\max} B\left(p_{i2}^{\text{base}} + 2p_{i1}^{\text{base}} + p_i^{\max}\right) = Q_y \tag{5.25a}$$

$$\tilde{h}B\left(p_i^{\max} + p_{i1}^{\text{base}}\right)\left(h_i^{\max} + \frac{\tilde{h}}{3}\right) + \left[\frac{(h_i^{\max})^2 \cdot B}{3}\left(p_{i2}^{\text{base}} + 3p_{i1}^{\text{base}} + 2p_i^{\max}\right)\right] = M_y \tag{5.25b}$$

$$\frac{B^2}{12}\left[\left(p_i^{\max} + 2p_{i1}^{\text{base}}\right)\tilde{h} + \left(p_{i2}^{\text{base}} + 4p_{i1}^{\text{base}} + p_i^{\max}\right)h_i^{\max}\right] = T_y \tag{5.25c}$$

where $\tilde{h} = H - h_i^{\max}$ and the height $h_i^{\max}$, at which maximum pressure $p_i^{\max}$ on the centreline of the walls perpendicular to the earthquake direction ($a(t)$) occurs, is given by:

$$\frac{h_i^{\max}}{H} = \begin{cases} 0.793 - 0.13\frac{L}{H} & \frac{L}{H} < 1.1 \\ 0.65 & \frac{L}{H} \geq 1.1 \end{cases} \tag{5.26}$$

In Equations (5.25), Q_y and M_y come from Equations (5.77a) and (5.77b) respectively, while T_y is the maximum value of the torsion action on the middle of the foundation if the impulsive hydrodynamic pressure on only half of one wall that is perpendicular to the earthquake direction is considered [229]:

$$T_y = \frac{1}{4}\sqrt{\left[m_f b_f S_e(T_f)\right]^2 + \left[\left(\frac{m_d B}{2} - m_f b_f\right)a_g\right]^2} \tag{5.27}$$

where

$$\frac{b_f}{H} = \begin{cases} 0.46 - 0.07\left(\frac{B}{H} - 0.3\right) & \frac{B}{H} \leq 1 \\ 0.41 & \frac{B}{H} > 1 \end{cases} \tag{5.28}$$

Once the linear system (5.25) is solved, numerical values for $p_i^{\max}$, p_{i1}^{base} and p_{i2}^{base} are derived and the flexible impulsive pressure distribution can be drawn easily (Figure 5.7); furthermore the effect of convective pressure is not considered.

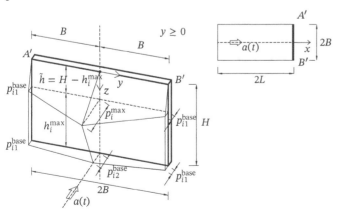

Figure 5.7 Pressure distribution over the rectangular wall surface $2B \times H$ of the tank (plane $y - z$) perpendicular to the earthquake $a(t)$ along x direction, represented using $p_i^{\max}$, p_{i1}^{base} and p_{i2}^{base} and some geometrical quantities such as $h_i^{\max}$ and $\tilde{h}$.

5.3 Effects of the Vertical Component of the Seismic Action

Past studies on the dynamic response of liquid storage tanks [78, 226, 227, 320, 353, 384, 627] have revealed that the hydrodynamic effects induced by a vertical component of ground shaking may be quite important, reaching the level of the hydrostatic effects [626] for a ground motion with a peak acceleration of about one-third the acceleration due to gravity [637, 639]. The dynamically activated pressure caused by the vertical earthquake excitation of a cylindrical tank is axisymmetrically distributed and does not contribute to the overturning moment. However, it must not be ignored in the strength and stability analyses in both senses [469]: on the one hand, the hydrodynamic pressure caused by a vertical motion is jointly responsible for plastic buckling and, on the other hand, as we will appreciate later in Section 5.5, it has a stabilizing effect with regard to the imperfection sensitivity of the axially loaded tank wall. Furthermore, vertical accelerations of circular cylindrical tanks and their liquid contents increase the maximum hoop forces in the tank walls.

Compared with the huge amount of research data dealing with the effects of a horizontal earthquake motion, much less interest has been spent in studying the dynamic of vertically excited tanks. The first paper published on this topic dates back to Bleich in [46]. A comprehensive treatment was presented by Yang in [680] and by Veletsos and Yang in [645]. Haroun and Tayel [226, 227] published numerical and analytical studies for calculating the natural frequencies and corresponding mode shapes of vertically excited tanks and then Haroun and Abou-Izzeddine [219] conducted a parametric study on numerous factors affecting the response of tanks under vertical excitations following the same procedure presented in the companion paper dealing with the response under horizontal excitations [218].

Principally, two procedures are currently used for estimating the hydrodynamic pressure due to a vertical earthquake excitation in a flexible cylindrical tank:

1) The first is used for estimating the dynamic loads of a flexible cylindrical tank resting on rigid soil and it was published in the guidelines of the ASCE.[1] It is based on the findings of Kumar [320] and Veletsos and Kumar [627] and a similar procedure is also used in Point C3.3.2, NZSEE-09, originally edited by Professor Priestley [459]. According to Veletsos and Kumar [626, 627], a generally conservative approximation may be obtained by assuming that a tank's flexibility affects only the magnitude and temporal variation of the hydrodynamic wall pressure but not its spatial variation. Under these assumptions, the pressure $p_v(z)$ is uniformly distributed in the circumferential direction and increases linearly from top to bottom, hence it is perfectly equal to that of a rigid tank (Equation (2.47)) (Point C3.3.2, NZSEE-09):

$$p_v(z) = \gamma_l H \left(1 - \frac{z}{H}\right) \frac{S_{ve}(T'_{Vf})}{g} \tag{5.29}$$

where S_{ve} is the spectral acceleration obtained from a vertical elastic response spectrum corresponding to the period T'_{Vf} (Section 5.4 and Equations (5.55), (5.56) or (5.57) for cylindrical tanks) of the vertical mode of vibration without soil-structure interaction or to the period T_{Vf} (Section 5.4

---

1 ASCE (American Society of Civil Engineers), *Guidelines for the Seismic Design of Oil and Gas Pipeline Systems*, Committee on Gas and Liquid Fuel Lifelines, 1984 [626].

and Equation (5.60)) of the vertical mode of vibration with soil-structure interaction; the damping factor is described in Section 6.2 in the case of soil–structure interaction while for a rigidly supported tank it could be 2% [627] or 1–3% [637].

2) The second procedure has been proposed by Sakai et al. [498] and Rammerstorfer et al. [470] superposing two contributions, one from the vertical rigid body motion (p_{vr} in Equations (2.46) or (2.47)) and the other one, p_{vf}, from the fluid–structures' interaction vibration (flexible contribution) [166, 468, 469, 506, 507]:

$$p_v(\zeta, t) = \sqrt{\left[p_{vr}(\zeta, t)\right]^2 + \left[p_{vf}(\zeta, t)\right]^2} \tag{5.30}$$

where, according to Point A.3.3, EC8-4, the square root of the sum of the squares (SRSS) rule has been used. The pressure p_{vf} may be estimated, according to Luft [353], considering a tank, rigidly supported on ground, that is free to expand radially at the base [297, 506]:

$$p_{vf}(\zeta, t) = 0.8\rho_l H \cos\left(\frac{\pi}{2}\zeta\right) a_{vf}(t) \tag{5.31}$$

Later Veletsos and Tang [637] examined the response of rigidly supported tanks on the assumption that the tank wall was either clamped or hinged at the base and demonstrated that Equation (5.31) can be used irrespective of the degree of base constraint.

Alternatively, a parametric study has been conducted by Fischer et al. in [166, 167, 468, 469, 470] leading to the following more accurate result:

$$p_{vf}(\zeta, t) = 0.815 f(\gamma)\rho_l H \cos\left(\frac{\pi}{2}\zeta\right) a_{vf}(t) \tag{5.32}$$

using a correction dimensionless factor $f(\gamma)$ depending on the slenderness $\gamma = \frac{H}{R}$:

$$f(\gamma) = 1.0 \qquad\qquad 0 < \gamma \leq 0.8 \tag{5.33a}$$

$$f(\gamma) = 1.078 + 0.274 \ln \gamma \qquad\qquad 0.8 < \gamma \leq 4.0 \tag{5.33b}$$

In Equations (5.31) and (5.32), $a_{vf}(t)$ is the response acceleration of a single degree of freedom linear oscillator induced by the vertical component of the free-field ground acceleration $a_{vg}(t)$; the single degree of freedom system should have a circular frequency ω'_{Vf} corresponding to the fundamental axisymmetric breathing mode of vibration without (Equations (5.55), (5.56) or (5.57)) or with soil–structure interaction (Equation (5.60)) and a damping ratio equal to that of the fundamental axisymmetric mode of vibration of the tank-liquid system (for rigidly supported tanks, it could be 2% [627] or 1–3% [637], while in the case of soil–structure interaction the damping factor described in Section 6.2). Following Equations (5.31) and (5.32) the hydrodynamic pressure on the wall of a flexible tank induced by the vertical component of ground shaking is uniformly distributed in the circumferential direction and may be considered to increase from top to bottom approximately as a quarter-sine curve [168]. The maximum value of the pressure p_v corresponds to the maximum value of both p_{vr} (Equations (2.46) and (2.47)) and p_{vf}. Hence, the maximum value of p_{vf} may then be determined by replacing $a_{vf}(t)$ by its spectral value S_{ve} using the natural frequency and damping value, according to Equation (5.29) [635].

In the case of rectangular flexible tanks, we suggests using the same expressions (5.29) or (5.30) using an equivalent cylinder of the same volume of liquid with an equivalent radius (as done in Section 4.3 for conical tanks); furthermore period T'_{Vf} (and T_{Vf}) have to be obtained from Equations (5.55)–(5.57), where R should be substituted by L.

A refined numerical analysis for the response of concrete rectangular liquid storage tanks subjected to vertical ground acceleration is investigated by Kianoush and Chen in [301] using the sequential method procedure [80]; further refinements including the effects of wall flexibility, the damping properties of the liquid, both horizontal and vertical components of an earthquake, and sloshing motion are taken into account by Ghaemmaghami and Kianoush in [183].

5.4 Periods of Vibration

The periods of vibration of the first impulsive tank-liquid horizontal mode excluding soil–structure interaction are given by the following.

For a vertical circular cylindrical tank of height H (wet height in metre) (Eq. (C3.26) Point C3.6, NZSEE-09 and Point 7.1, COVENIN 3623, 2000 Edition) [431]:

$$T_f = \frac{5.61\pi H}{K_h}\sqrt{\frac{\gamma_l}{Eg}} = \frac{5.61\pi H}{K_h}\sqrt{\frac{\rho_l}{E}} \tag{5.34}$$

or the circular frequency [344]:

$$\omega_f = \frac{2\pi}{T_f} \tag{5.35}$$

where γ_l is the unit weight of the liquid (in N/m^3; or ρ_l in kg/m^3 is the mass density of the liquid) and g is the gravitational acceleration; E is the Young's modulus of the wall material (in N/m^2); K_h is a dimensionless period coefficient which depends on the liquid height and thickness to radius ratio as clearly depicted in Figure 5.8 [223, 626]. Originally Equation (5.34) was derived by Haroun and Housner in [223] for a steel roofless cylindrical tank (Poisson's ratio for the tank material equal to 0.3) with uniform thickness and completely filled with water. Furthermore, Equation (5.34) is based on the assumption that the mass of the tank itself is negligible in comparison to that of the liquid [626]. According to this condition, normally satisfied in practice, Veletsos in [626] extended the possibility to use expression (5.34) to other liquids [5] by assuming a ratio between the density of the water ρ_w and the density of the wall ρ_s equal to $\frac{\rho_w}{\rho_s} = 0.127$ and hence $5.61 = 2\sqrt{\frac{1}{0.127}}$. As well as from Figure 5.8, a few analytical values of the coefficient K_h may be obtained from Table 5.2 for a steel roofless cylindrical tank (Poisson's ratio for the tank material equal to 0.3) with uniform thickness and completely filled with water in the case of $\frac{s}{R} = 0.0005, 0.001$ and 0.002 [626, 635]. It is interesting to note that API 650 2012 Edition does not recognize two impulsive modes (rigid and flexible wall) and at Point E.4.5.1 suggests a unique expression to compute the impulsive horizontal period of the tank [431]:

$$T_f^{\text{API}} = \frac{\overline{C}_i H}{\sqrt{\frac{s}{R}}}\sqrt{\frac{\rho_l}{E}} \tag{5.36}$$

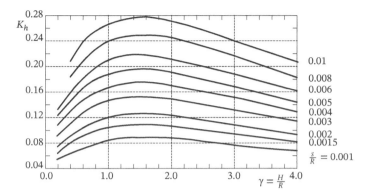

Figure 5.8 Dimensionless coefficient K_h for a steel roofless cylindrical tank in order to evaluate T_f for horizontal mode excluding soil–structure interaction Source: graphically adapted from [223, 626] and Point C3.6, NZSEE-09 and Point 7.1, COVENIN 3623, 2000 Edition.

Figure 5.9 Coefficient $\overline{C}_i$ used for the evaluation of the impulsive horizontal period T_f^{API} of a cylindrical steel tank according to Point E.4.5.1, API 650 2012 Edition.

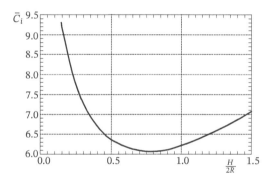

where ρ_l (in kg/m³) is the mass density of the liquid and E (in N/m²) is the Young's modulus of the material of the tank wall of height H (in metre); the dimensionless coefficient $\overline{C}_i$ is depicted in Figure 5.9. Note that expressions (5.34) and (5.36) are equivalent. For instance, consider a height to radius ratio of 2 and a shell thickness to radius ratio of 0.002, both Equations (5.34) and (5.36) give respectively:

$$T_f = 140.5H\sqrt{\frac{\rho_l}{E}} \tag{5.37a}$$

$$T_f^{API} = 138.6H\sqrt{\frac{\rho_l}{E}} \tag{5.37b}$$

This comparison confirms that API 650 considers that all the impulsive mass is concentrated in one mode of vibration and this corresponds to the first impulsive tank-liquid horizontal mode [431].

Expression (5.34) has been developed for the case of a steel tank and then extended by Veletsos and Shivakumar [635] to a concrete tank (Poisson's ratio for the tank material equal to 0.17):

$$T_f = \frac{3.16\pi H}{K_h}\sqrt{\frac{\gamma_l}{Eg}} \tag{5.38}$$

where the ratio between the density of the water ρ_w and the density of the wall ρ_s has been assumed to be equal to $\frac{\rho_w}{\rho_s} = 0.4$ and hence $3.16 = 2\sqrt{\frac{1}{0.4}}$. Table 5.2 shows the coefficient K_h for a reinforced concrete roofless cylindrical tank with uniform thickness and completely filled with a liquid in the case of $\frac{s}{R} = 0.01$ [635]. The first mode period of tanks of non-uniform wall thickness may be determined by using the average thickness and usually this assumption will give a good approximation up to $\frac{H}{R}$ ratios of about 4 (Point C3.6, NZSEE-09).

Table 5.2 Dimensionless coefficient K_h in order to evaluate the first impulsive period T_f for horizontal mode excluding soil–structure interaction: steel and reinforced concrete tank [626, 635].

	Steel tanks			Concrete tanks
$\frac{H}{R}$	$\frac{s}{R} = 0.0005$	$\frac{s}{R} = 0.001$	$\frac{s}{R} = 0.002$	$\frac{s}{R} = 0.01$
0.3	0.0421	0.0600	0.0856	0.1222
0.4	0.0468	0.0667	0.0946	0.1293
0.5	0.0506	0.0719	0.1019	0.1366
0.6	0.0538	0.0762	0.1079	0.1431
0.7	0.0564	0.0799	0.1129	0.1488
0.8	0.0587	0.0829	0.1170	0.1537
0.9	0.0605	0.0855	0.1204	0.1578
1.0	0.0620	0.0875	0.1231	0.1612
1.1	0.0632	0.0891	0.1252	0.1638
1.2	0.0641	0.0903	0.1268	0.1658
1.3	0.0647	0.0911	0.1278	0.1671
1.4	0.0650	0.0916	0.1284	0.1679
1.5	0.0652	0.0918	0.1286	0.1681
1.6	0.0652	0.0917	0.1284	0.1679
1.7	0.0650	0.0914	0.1280	0.1673
1.8	0.0647	0.0910	0.1273	0.1664
1.9	0.0642	0.0904	0.1265	0.1652
2.0	0.0637	0.0896	0.1254	0.1638
2.2	0.0625	0.0879	0.1230	0.1604
2.4	0.0611	0.0859	0.1202	0.1565
2.6	0.0596	0.0838	0.1172	0.1522
2.8	0.0580	0.0815	0.1140	0.1478
3.0	0.0563	0.0792	0.1108	0.1434

Alternatively to expression (5.34), the fundamental circular frequency of the tank–fluid system may be evaluated by means of the following approximate relation derived by Rammerstorfer et al. in [166, 468, 506] (Point A.3.1, EC8-4):

$$\omega_f = \frac{\pi}{R} \frac{\sqrt{\dfrac{Es(\zeta)}{\rho_l H}}}{(0.157\gamma^2 + \gamma + 1.49)} \tag{5.39}$$

where $\zeta = \frac{z}{H} = \frac{1}{3}$; ρ_l (in kg/m^3) is the mass density of the liquid, γ can be derived from Equation (2.6c) and E is Young's modulus of the material of the tank wall. A quite similar expression may be found at Point 4.3.1.1, IITK-GSDMA 2007 and at Point A.3.1, Eurocode 8, Part 4, Edition 2003 [206, 297]:

$$\omega_f = \frac{\pi}{R} \sqrt{\frac{Es(\zeta)}{\rho_l H}} (0.01675\gamma^2 - 0.15\gamma + 0.46) \tag{5.40}$$

As clearly pointed out by Kettler in [297], Equations (5.39) and (5.40) give approximately the same fundamental circular frequency due to the fact that:

$$0.01675\gamma^2 - 0.15\gamma + 0.46 \approx \frac{1}{0.157\gamma^2 + \gamma + 1.49} \tag{5.41}$$

Expressions (5.34), (5.36) and (5.39) are applicable to only those circular steel tanks in which the wall is rigidly attached to a base slab. In some liquid-containing concrete tanks, various types of ground-supported joints between the wall and its foundation are used (Figures 2.17). Hence, for the basic configurations of ground-supported, liquid-containing structures, Point 9.3.4, ACI 350.3-06 suggests using two different expressions to evaluate the period of vibration of the first impulsive tank-liquid mode depending on the boundary conditions. For tanks in Figures 2.17(a) to 2.17(c), the expression suggested has been adapted from Veletsos [626] (Equation (5.34)) and Veletsos and Shivakumar [635] (Equation (5.38)):

$$T_f = \frac{2\pi H}{C_w \sqrt{\dfrac{s}{10R}}} \sqrt{\frac{\gamma_c}{10^3 Eg}} \tag{5.42}$$

where γ_c is the weight of the concrete material of the tank's wall (in kN/m^3) and E (in MPa) is the Young's modulus of the wall's material of the tank of height H (in metre), radius R (in metre) and thickness s (in mm); g is the gravitational acceleration (in m/sec^2). The non-dimensional coefficient C_w may be derived from the relation suggested in ACI 350.3-06:

$$\begin{aligned} C_w = 9.375 \cdot 10^{-2} + 0.2039 \left(\frac{H}{2R}\right) - 0.1034\left(\frac{H}{2R}\right)^2 - 0.1253\left(\frac{H}{2R}\right)^3 \\ + 0.1267\left(\frac{H}{2R}\right)^4 - 3.186 \cdot 10^{-2}\left(\frac{H}{2R}\right)^5 \end{aligned} \tag{5.43}$$

valid for $\frac{2R}{H} > 0.667$. For tanks in Figures 2.17(d) and 2.17(e) (Point 9.3.4, ACI 350.3-06), the expression suggested has been adapted from Point 4.3.1.5, AWWA D115, version 1995 (First Edition):

$$T_f = \sqrt{\frac{8\pi \left(W_w + W_r + W_i\right)}{2gRk_a}} \tag{5.44}$$

where W_w and W_r are the weight of the wall and roof respectively (in kN) and $W_i = \frac{m_i g}{1000}$ (in kN) is the weight of the impulsive component of the stored liquid (m_i, in kg, may be derived from Equation (2.25a) as stated at Point 9.3.1, ACI 350.3-06; this suggests that ACI 350.3-06 (as well as API 650) considers that all the impulsive mass is concentrated in one mode of vibration and this corresponds to the first impulsive tank-liquid horizontal mode); R in metre. The period T_f from expression (5.44) should not exceed 1.25 seconds. The spring constant of the tank wall support system k_a (in N/m^2) is[2]:

$$k_a = 10^6 \left[\left(\frac{A_{steel} E_{steel} \cos^2 \alpha_s}{L_c S_c} \right) + \left(\frac{2 G_p w_p L_p}{t_p S_p} \right) \right]$$ (5.45)

where A_{steel} (in mm^2) and E_{steel} (in MPa) are, respectively, the cross-sectional area and modulus of elasticity of the base cable, wire, strand (whose angle with horizontal is equal to α_s in degree) or ordinary reinforcement; L_c (in mm) is the effective length of base cable or strand taken as the sleeve length plus 35 times the strand length and S_c (in mm) is the centre-to-centre distance between individual base cable loops; G_p (in MPa), w_p (in mm), L_p (in mm) and S_p (in mm) are respectively, the shear modulus, width, length, and centre-to-centre spacing of elastomeric bearing pads at the base (for a tank with a continuous bearing pad $L_p = 1.0$ and $\frac{L_p}{S_p} = 1.0$, Point 4.3.1.5, AWWA D115, version 1995 (First Edition); for a tank with a series of bearing pads having lengths 150 mm spaced at 300 mm on centres, $\frac{L_p}{S_p} = 0.5$); t_p (in mm) is the thickness of elastomeric bearing pads.

The influence of the roof mass on the period of vibration may be obtained using Dunkerley's expression (Equation (5.46)) as outlined by Veletsos in [626] (Point C3.6, NZSEE-09). If the mass of the tank's roof is sufficiently important to warrant its consideration, the natural frequency of the tank–roof–liquid system [215, 626, 651], ω_f', may be approximately determined using Dunkerley's expression [205, 349, 350, 505, 565]:

$$\frac{1}{\omega_f'^2} = \frac{1}{\omega_f^2} + \frac{1}{\omega_F^2} + \frac{1}{\omega_T^2}$$ (5.46)

where ω_f is the fundamental frequency of the tank-liquid system without the roof and may be derived from expressions (5.35), (5.36), (5.39) and (5.40) for steel tanks and from (5.38), (5.42), and (5.44) for concrete tanks; ω_F and ω_T are the natural frequencies for an empty tank of the same geometry of the actual tank but for which the only structural mass is the roof mass m_r [626]. Particularly, ω_F represents the natural frequency of an empty tank assumed to be acting as a cantilever flexural beam of flexural stiffness k_F:

$$\omega_F = \sqrt{\frac{k_F}{m_r}}$$ (5.47a)

$$k_F = \frac{3EJ}{\overline{L}^3} = 3\pi E s \left(\frac{R}{\overline{L}} \right)^3$$ (5.47b)

2 The original expression has been modified according to ERRATA as of 3 December ACI 350.3-06, *Seismic Design of Liquid-Containing Concrete Structures and Commentary* (ACI 350.3-06), first to seventh editions, Reported by ACI Committee 350.

whereas ω_T represents the natural frequency of an empty tank assumed to be acting as a cantilever shear beam of shearing stiffness k_T.

$$\omega_T = \sqrt{\frac{k_T}{m_r}} \tag{5.48a}$$

$$k_T = \kappa \frac{GA_s}{\overline{L}} = \frac{\pi E s}{2(1+\nu)} \left(\frac{R}{L}\right) \tag{5.48b}$$

In Equations (5.47b) and (5.48b), $\overline{L}$ is the height of the tank's wall, $J = \pi R^3 s$ is the moment of inertia of the cross–sectional area of the tank, assumed with a uniform wall thickness s, $A_s = 2\pi Rs$; κ is a shear coefficient which, for the ring section A_s considered, equals $\frac{1}{2}$; the shear modulus of elasticity for the tank material is $G = \frac{E}{2(1+\nu)}$.

The periods of vibration of the first impulsive tank-liquid horizontal mode, excluding soil-structure interaction are given for a rectangular tank of height H (wet height in meter) and dimension $2L \times 2B$, with $2L$ the tank length parallel to the direction of the earthquake loading (Eq. (C3.28), Point C3.6, NZSEE-09 and Point A.4.2, EC8-4) [302, 341]:

$$T_f = 2\pi \sqrt{\frac{d_f}{g}} \tag{5.49}$$

where d_f is the deflection of the tank wall (perpendicular to the ground motion) on the vertical centre–line at the height h_f (Equation (5.22b)) corresponding to the impulsive mass m_f (Equation (5.22a)), when loaded by a uniformly distributed load $\overline{p}$, of magnitude $\frac{m_f g}{4BH}$, in the direction of the ground motion; in the case of a concrete tank the contribution to the uniformly distributed load given by the inertia of the walls $\frac{m_w g}{(2B+2L)H}$ could also be added. For roofless tanks in the range $0.6 \leq \frac{H}{2B} \leq 1.5$ (if the ground motion is supposed to run in $2L$ direction), the deflection d_f may be calculated assuming the wall to be free at the top and fully fixed on the other three sides (Figure 5.10) using the following relations [613]:

$$f_1 = c_{f1} \frac{\overline{p}a^4}{\mathscr{B}} \tag{5.50a}$$

$$f_2 = c_{f2} \frac{\overline{p}a^4}{\mathscr{B}} \tag{5.50b}$$

$$m_{x1} = c_{x1}\overline{p}a^2 \qquad m_{x2} = c_{x2}\overline{p}a^2 \qquad m_{x3} = -c_{x3}\overline{p}a^2 \tag{5.50c}$$

$$m_{x4} = -c_{x4}\overline{p}a^2 \qquad m_{y2} = c_{y2}\overline{p}a^2 \qquad m_{y5} = -c_{y5}\overline{p}a^2 \tag{5.50d}$$

where coefficients c_{f1}, c_{f2}, c_{x1}, c_{x2}, c_{x3}, c_{x4}, c_{y2} and c_{y5} may be derived from Table 5.3; $\mathscr{B}$ is the flexural rigidity (or bending stiffness) of the shell/wall $\frac{Es^3}{12(1-\nu^2)}$ where E and ν are, respectively, the Young's modulus and Poisson's ratio of the tank. In the case of a shallow rectangular tank with $\frac{H}{2B} < 0.6$, the behaviour of the wall, along the centre–line, may be similar to that of a cantilever of unit section ($1 \times s$ where s is the thickness of the tank's wall) and height h_f, fully fixed at the base and free at the top loaded by $\frac{m_f g}{4B}$

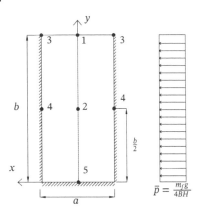

Figure 5.10 Notations for the solution in term of deflections, bending moments, and reactions of a uniformly loaded rectangular plate with three edges fully fixed and the fourth one free at the top. Source: From [613]/McGraw Hill.

Table 5.3 Dimensionless coefficients c_{f1}, c_{f2}, c_{x1}, c_{x2}, c_{x3}, c_{x4}, c_{y2} and c_{y5} for deflections, bending moments, and reactions of uniformly loaded rectangular plate with three edges built in and fourth edge free (Poisson's ratio for the material equal to $\frac{1}{6}$).

b/a	0.6	0.7	0.8	0.9	1.0	1.25	1.5
c_{f1}	0.00271	0.00292	0.00308	0.00323	0.00333	0.00345	0.00335
c_{f2}	0.00129	0.00159	0.00185	0.00209	0.00230	0.00269	0.00290
c_{x1}	0.0336	0.0371	0.0401	0.0425	0.0444	0.0467	0.0454
c_{x2}	0.0168	0.0212	0.0252	0.0287	0.0317	0.0374	0.0402
c_{x3}	0.0745	0.0782	0.0812	0.0836	0.0853	0.0867	0.0842
c_{x4}	0.0365	0.0439	0.0505	0.0563	0.0614	0.0708	0.0755
c_{y2}	0.0074	0.0097	0.0116	0.0129	0.0138	0.0142	0.0118
c_{y5}	0.0554	0.0545	0.0535	0.0523	0.0510	0.0470	0.0418

Source: Adapted from [613].

(the inertia of the walls $\frac{m_w g}{2B+2L}$ can also be added). Thus, for a tank with a wall of uniform thickness, one can obtain d_f as follows:

$$d_f = \frac{\frac{m_f g}{4B} h_f^3}{3EJ^\star} \quad \text{or} \quad d_f = \frac{\left(\frac{m_f g}{4B} + \frac{m_w g}{2B+2L}\right) h_f^3}{3EJ^\star} \tag{5.51}$$

where E (in MPa) is the Young's modulus of the wall's material of the tank and $J^\star = s^3/12$ is the moment of inertia of the cross-sectional area of the unit section of the wall $(1 \times s)$.

An alternative expression to (5.49) is proposed in Point 9.2.4, ACI 350.3-06, for a single wall $2B \times \overline{L}$ (if the ground motion is supposed to run in $2L$ direction):

$$T_f = 2\pi \sqrt{\frac{m_f^{\text{lin}} + m_w^{\text{lin}}}{k_w^{\text{lin}}}} \tag{5.52}$$

where the wall mass m_w^{lin} and the flexible impulsive mass m_f^{lin} (both in kg/m) are given by the following dimensional expressions[3] (g in m/sec^2):

$$m_w^{\text{lin}} = \overline{L}s\frac{\gamma_c}{g} \tag{5.53a}$$

$$m_f^{\text{lin}} = \frac{m_f}{m}LH\frac{\gamma_l}{g} \cdot 10^3 \tag{5.53b}$$

In expressions (5.53) the height of the tank's wall $\overline{L}$, the tank half-length L parallel to the direction of earthquake loading, the wet height H is in metre; the thickness of the wall s in mm; γ_l (the unit weight of the liquid) and γ_c (the weight of the concrete material of the tank's wall) in kN/m^3 and $\frac{m_f}{m}$ comes from Equation (5.22a). For fixed-base, free-top cantilever walls, such as in open-top tanks (as depicted in Figure 5.10), flexural stiffness for a unit width of wall k_w^{lin} (in N/m) may be approximated using the following expression from Point 9.2.4, ACI 350.3-06:

$$k_w^{\text{lin}} = \frac{E}{4 \cdot 10^3}\left(\frac{s}{h^{\text{lin}}}\right)^3 \tag{5.54a}$$

$$h^{\text{lin}} = \frac{\frac{\overline{L}}{2}m_w^{\text{lin}} + h_f m_f^{\text{lin}}}{m_w^{\text{lin}} + m_f^{\text{lin}}} \tag{5.54b}$$

Flexural stiffness expressions, different from (5.54a), may be developed for other wall support conditions. Such spring constants will generally fall within the low period range (less than about 0.3 sec) for tanks of normal proportions.

The fundamental period of vibration, corresponding to a vertical, axisymmetric and breathing mode of oscillation of the tank–liquid system, in the case of cylindrical and rectangular flexible tanks, excluding soil–structure interaction, may be derived using:

• the expression analytically derived by Haroun et al. in [219, 226] (approximating the fundamental mode shape by a cosine function $f(\zeta) = \cos\left(\frac{\pi}{2}\zeta\right)$) and by Rammerstorfer et al. in [166, 468, 470, 507] (Point A.3.3, EC8-4):

$$\omega'_{Vf} = \frac{1}{2R}\sqrt{\frac{2\pi Es(\zeta)}{\rho_l H(1-v^2)}\frac{I_1(\gamma_1)}{I_0(\gamma_1)}} \tag{5.55}$$

where $\zeta = \frac{z}{H} = \frac{1}{3}$ and s is the thickness of the wall; ρ_l (in kg/m^3) is the mass density of the liquid and $\gamma_1 = \frac{\pi R}{2H} = \frac{\pi}{2\gamma}$ where γ can be derived from Equation (2.6c); E (N/m^2) and v are Young's modulus and Poisson's ratio of the material of the tank wall respectively; I_0 and I_1 (Equation (2.7)) are the modified Bessel function of the first kind of order 0 and 1 respectively [561] (R, H, and s in metre);

• the same form of Equations (5.34) and (5.38) as suggested at Point C3.6 (Eq. (C3.27)), NZSEE-09 and Point 7.2, COVENIN 3623, 2000 Edition [320, 627]:

$$\omega'_{Vf} = \frac{2\pi}{T'_{Vf}} \qquad \text{where} \qquad T'_{Vf} = \frac{5.61\pi H}{K_v}\sqrt{\frac{\gamma_l}{Eg}} \tag{5.56}$$

---

3 The original expression has been modified according to ERRATA as of 3 December 2019, ACI 350.3-06, *Seismic Design of Liquid-Containing Concrete Structures and Commentary* (ACI 350.3-06), fifth to seventh editions Reported by ACI Committee 350. This ERRATA is also valid for Equation (5.54).

Table 5.4 Dimensionless coefficient K_v in order to evaluate the fundamental impulsive period of vibration T'_{Vf} for vertical mode excluding soil–structure interaction: steel and reinforced concrete tank [626, 635].

$\frac{H}{R}$	Steel tanks					Concrete tanks
	$\frac{S}{R} = 0.0005$	$\frac{S}{R} = 0.001$	$\frac{S}{R} = 0.002$	$\frac{S}{R} = 0.003$	$\frac{S}{R} = 0.005$	$\frac{S}{R} = 0.01$
0.2	–	0.0521	–	–	–	0.1135
0.3	0.0427	0.0611	0.0875	0.1078	0.1399	0.1250
0.4	0.0478	0.0682	0.0974	0.1197	0.1549	–
0.5	0.0518	0.0738	0.1053	0.1294	0.1675	0.1414
0.6	0.0550	0.0783	0.1116	0.1372	0.1776	–
0.7	0.0576	0.0819	0.1167	0.1434	0.1857	–
0.75	–	0.0834	–	–	–	0.1564
0.8	0.0596	0.0848	0.1207	0.1483	0.1922	–
0.9	0.0612	0.0870	0.1239	0.1522	0.1973	–
1.0	0.0625	0.0889	0.1263	0.1554	0.2014	0.1650
1.2	0.0644	0.0915	0.1301	0.1599	0.2073	–
1.4	0.0657	0.0933	0.1326	0.1630	0.2113	–
1.5	–	–	–	–	–	0.1726
1.6	0.0666	0.0946	0.1344	0.1651	0.2140	–
1.8	0.0673	0.0955	0.1356	0.1665	0.2158	–
2.0	0.0678	0.0961	0.1365	0.1676	0.2172	0.1756
2.2	0.0681	0.0966	0.1371	0.1684	0.2182	–
2.4	0.0684	0.0970	0.1376	0.1689	0.2189	–
2.6	0.0686	0.0973	0.1380	0.1694	0.2194	–
2.8	0.0688	0.0975	0.1383	0.1697	0.2198	–
3.0	0.0689	0.0977	0.1385	0.1700	0.2202	0.1770
5.0	–	0.0992	–	–	–	0.1775

for a steel tank (K_v in Table 5.4[4] [635] assuming a ratio between the density of the water ρ_w and the density of the wall ρ_s equal to $\frac{\rho_w}{\rho_s} = 0.127$ and a Poisson's ratio for the tank material equal to 0.3), or:

$$T'_{Vf} = \frac{3.16\pi H}{K_v}\sqrt{\frac{\gamma_l}{Eg}} \tag{5.57}$$

for a concrete tank (K_v in Table 5.4 [626] where the ratio between the density of the water ρ_w and the density of the wall ρ_s has been assumed to be equal to $\frac{\rho_w}{\rho_s} = 0.4$ and the Poisson's ratio for the tank material equal to 0.17).

---

4 Point 7.2, COVENIN 3623, 2000 Edition suggests explicitly in Figure 4 that if $\frac{S}{R} < 0.001$ to use $K_v = 0.07$.

- the relation suggested at Point 9.3.4, ACI 350.3-06 in the case of concrete circular tanks:

$$T'_{Vf} = 2\pi H \sqrt{\frac{R}{s} \frac{\gamma_l}{Eg}} \tag{5.58}$$

where E (in MPa) is the Young's modulus of the wall's material of the tank; the thickness of the wall s in mm; the liquid's height H and the radius R in metre; γ_l (the unit weight of the liquid) in kN/m^3.

The vertical T'_{Vf} impulsive period of vibration for flexible rectangular tanks excluding soil–structure interaction may be evaluated from relations (5.55)–(5.58), where R should be substituted by L.

The periods of vibration of the first impulsive tank-liquid horizontal and vertical mode including soil–structure interaction are respectively given by:

- the expression suggested at Point C3.6, (Equation (C3.29)), NZSEE-09 or at Point A.7.2.2, EC8-4 [625], for cylindrical and rectangular tanks [431]:

$$T_f^{\star} = T_f \sqrt{1 + \frac{K_f}{K_x} \left[1 + \frac{K_x h_f^2}{K_\theta} \right]} \tag{5.59}$$

where $K_f = \frac{4\pi^2 m_f}{T_f^2}$ (in N/m), K_x and K_θ may be derived from Equations (2.56) and (2.57) taking into account that in evaluating the dimensionless frequency parameter α in Equation (2.58), period T_0 must be substituted with $T_f^{\star}$; m_f (Equation (5.14) for a cylindrical tank and Equation (5.22a) for a rectangular tank) is the flexible impulsive mass placed at a height h_f from the base (Equation (5.16) for a cylindrical tank and Equation (5.22b) for a rectangular tank); the period T_f may be derived from expressions (5.34), (5.36), (5.39) or (5.40) (steel) and (5.38), (5.42) or (5.44) (concrete) for cylindrical tanks and from relation (5.49) or (5.52) for rectangular tanks;

- the expression suggested at Point C3.6, (Equation (C3.32)), NZSEE-09 or at Point A.7.2.2, in EC8-4 [625], for cylindrical and rectangular tanks:

$$T_{Vf} = T'_{Vf} \sqrt{1 + \frac{K_b}{K_V}} \tag{5.60}$$

where $K_b = \frac{4\pi^2 m}{T'^2_{Vf}}$ (in N/m), m is the total mass of the liquid and T'_{Vf} may be obtained from Equations (5.55), (5.56) (steel) or (5.57), (5.58) (concrete) for cylindrical tanks or (5.55)–(5.58), where R should be replaced by L, for rectangular tanks; K_V comes from Equation (2.60) taking into account that in evaluating the dimensionless frequency parameter α in Equation (2.58), period T_0 must be substituted with T_{Vf}.

5.5 Combination of Pressures

The hydrodynamic fluid pressure exerted on the wall of a flexible tank due to a ground motion $a(t)$ is given by the following contributions [158, 164, 224, 408, 419] (Point A.3, EC8-4):

1) The impulsive fluid pressure component $p_i(\xi, \zeta, \theta, t)$ (called the "rigid impulsive") which varies synchronously with the horizontal ground acceleration $a(t)$ and therefore remains unaffected by the

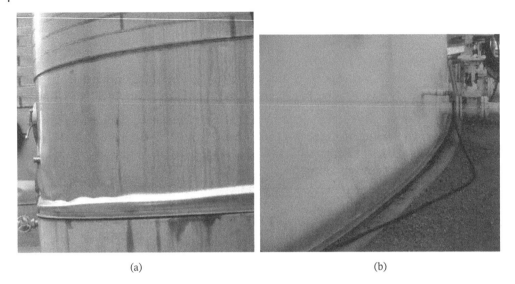

(a)   (b)

Figure 5.11 Elastic–plastic buckling or "elephant foot buckling": (a) Livermore earthquake, California (24 January 1980, Magnitude 5.9); (b) Imperial Valley earthquake, California (15 October 1979, Magnitude 7.0 (from Karl V. Steinbrugge Collection, University of California, Berkeley) [427, 428, 571].

rigid case. It is given by Equation (2.5) or (2.18) and (2.19) (for circular tanks) and Equation (2.125) (for rectangular tanks). According to Section 2.2.4, to the rigid impulsive value must be added the inertia force p_w of the tank itself given by expression (2.50) [626].

2) The long period component contributed by the convective fluid pressure $p_c(\xi, \zeta, \theta, t)$ driven by $a_{cn}(t)$ [215]. Because the convective components of the response are more or less insensitive to variations in wall flexibility, they may be considered to be the same as those obtained for rigid tanks from Equations (2.27) or (2.40) for cylindrical tanks and from Equations (2.128) or (2.129) in the case of rectangular tanks [635].

3) The short period fluid pressure $p_f(\zeta, \theta, t)$ caused by the wall deformation relative to the base circle due to the deformability of the tank wall (called the "flexible impulsive") [223, 624, 626, 646]. The difference in the dominant period of vibration of the convective and flexible impulsive response weakens the coupling between the two components and decreases the sensitivity of one component to changes in the characteristics of the other. Based on this reasoning, the coupling between the wall displacement and sloshing is not included in the usual design approaches (Point 4.3.3, HSE Research Report RR527). For cylindrical tanks the flexible impulsive component may be derived from expressions (5.5) and (5.8), while for rectangular tanks relation (5.20) may be used.

4) The fluid pressure $p_v(\zeta, t)$ (Equations (5.29) and (5.30)) due to the vertical motion of the tank. It is composed by the hydrodynamic rigid pressure $p_{vr}(\zeta, t)$ (Equation (2.46)) or $p_{vr}(z)$ (Equation (2.47)) and by the flexible wall component $p_{vf}(\zeta, t)$ (Equations (5.31) and (5.32)) induced by the vertical component of the ground acceleration $a_v(t)$.

In addition, the total hydrodynamic pressure referred to the above components is in excess of the hydrostatic pressure p_{hs}, and must be added to the latter to obtain the total pressure exerted on the wall.

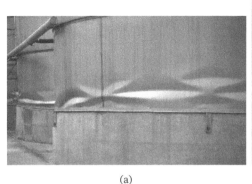

(a)   (b)

Figure 5.12 Elastic buckling or "diamond buckling" during the Livermore earthquake, California (24 January 1980, Magnitude 5.9) Source: From [428]/John Wiley & Sons, Inc. and [571]/U.S. Department of the Interior.

A conservative estimate of the maximum value of the horizontal hydrodynamic wall pressure p_h may be obtained by taking the sum of the numerical values of the maximum rigid impulsive pressure p_i, flexible impulsive p_f and of the maximum modal components of the convective pressure p_c (Point 4.8, HSE Research Report RR527) [214, 223]. This approach, in agreement with Point A.2.1.6, EC8-4 and Point 15.7.6.1, ASCE 7, 2010 Edition,[5] effectively uses the upper bound rule of adding the absolute values of the maxima, assuming that all of the $a_{cn}(t)$ occur simultaneously with the maximum ground acceleration and that maximum positive and maximum negative values of $a_{cn}(t)$ are equally likely [626]. In most cases it is adequate to consider only the contribution of the fundamental sloshing mode $a_{c1}(t)$. A preferable approach would be to add to the maximum value of the impulsive pressure (rigid and flexible) the square root of the sum of the squares of the maximum modal contributions of the convective pressure. According to Point 4.2, NZSEE-09, Chapter 4, ACI 350.3-06, Point E-5, FM Approvals Class Number 4020, Point E.6, (API 650, 2012 Edition), Point 13.5, AWWA D100-11, Point 4.3.1, AWWA D110-04 and D115-06, the total response resulting from the combined effects of impulsive (rigid and flexible) and convective actions will be found by combining the peak spectral responses of each action according to the square root of the sum of squares method (SRSS) as also suggested by Rammerstorfer, Fischer, and Scharf in [166, 469, 470, 506, 507].

The peak value of the hydrodynamic pressure on the tank wall due to horizontal p_h (rigid and flexible impulsive and convective) and vertical p_v seismic action, separately applied, may be obtained by applying Equation (2.101) and particularly the rule at Point 4.3.3.5.2 (4), UNI EN 1998-1:2004 (that derives from Points 3.2 (2) and 4.2, EC8-4) [5, 267, 344].

The most relevant types of tank damage [162, 166, 211, 244, 428, 481, 536], classified in Section 1.5, are elastic–plastic buckling [139, 377, 488] (usually called "elephant foot" buckling in Figures 5.11 and 1.14(a)) and elastic buckling [294, 325, 612] (or "diamond buckling" in Figures 5.12 and 1.14(b)) of the tank wall at the base (Figures 5.11 and 5.12), at the top (Figures 5.13(a)–5.13(c)) or along the wall

---

5 To be clear, Note e. in Point 15.7.6.1 clarifies that impulsive and convective seismic components are permitted to be combined also using the square root of the sum of the squares (SRSS) method in lieu of the direct sum method.

(a)

(b)

(c)

(d)

Figure 5.13 Buckling along and at the top of the tank's wall: (a)/(b) San Fernando earthquake, California (9 February 1971, Magnitudo 6.6) (Source: from Karl V. Steinbrugge Collection, University of California, Berkeley); (c) Izmit earthquake, Turkey (17 August 1999, Magnitude 7.4) (Source: from Izmit Collection, University of California, Berkeley); (d) Port-au-Prince earthquake, Haiti (12 January 2010, Magnitude 7) (Courtesy of Eduardo Fierro, BFP Engineers).

(Figure 5.13(d)). As shown by Scharf [506] and also by Rammerstorfer and Fischer in [166, 469, 470, 507], and according to Point 4.1.3, UNI ENV 1993-1-6 regarding the limit state of buckling LS3, different (compared to the rules proposed in Point 4.3.3.5.2 (4), UNI EN 1998-1:2004) superpositions of the individual pressure components, p_{hs}, p_h, and p_v, have to be considered with respect to different damage modes of the tank. Figure 5.14 shows three different possibilities of superposition of the contributions given by the axisymmetrically distributed hydrostatic pressure p_{hs}, the pressure due to the horizontal earthquake component p_h (obtained from p_i, p_f, and p_w using the sum of the numerical values of the maxima or the SRSS rule), and the axisymmetrically distributed pressure p_v due to the vertical earthquake component [166, 469, 470, 506, 507]:

$$\bar{p}_1 = p_{hs} + p_h + p_v \qquad \leftarrow \text{ Figure 5.14(a)} \qquad (5.61a)$$

$$\bar{p}_2 = p_{hs} + p_h - p_v \qquad \leftarrow \text{ Figure 5.14(b)} \qquad (5.61b)$$

$$\bar{p}_3 = p_{hs} - p_h - p_v \qquad \leftarrow \text{ Figure 5.14(c)} \qquad (5.61c)$$

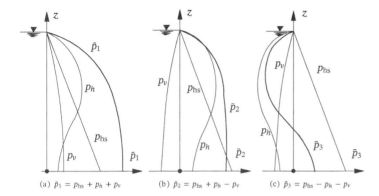

(a) $\bar{p}_1 = p_{\mathrm{hs}} + p_h + p_v$ (b) $\bar{p}_2 = p_{\mathrm{hs}} + p_h - p_v$ (c) $\bar{p}_3 = p_{\mathrm{hs}} - p_h - p_v$

Figure 5.14 Qualitative representation of three different modes of superposition, $(\bar{p}_1, \bar{p}_2, \bar{p}_3)$, of the contributions given by the hydrostatic pressure p_{hs}, the pressure due to the horizontal earthquake component p_h, and the pressure p_v due to the vertical earthquake component [166, 469, 470, 506, 507].

The combined pressure $\bar{p}_1$ causes the highest circumferential tensile stresses and leads, especially in the case of broad tanks, to elastic–plastic buckling [139, 377, 488] (Note 3, Point 5.3.2.2 and Point 5.3.2.4(7), UNI ENV 1993-4-1:2007; Point D.1.5.2(5), Annex D, UNI ENV 1993-1-6 where the pressurized meridional elastic–plastic buckling parameter explicitly requires the "largest design value of local internal pressure at the location of the point being assessed that can coexist with the meridional compression"). The "elephant foot" buckling results from the high circumferential tensile stress due to the internal pressure, in combination with the axial membrane stresses due to the overturning moment caused by the horizontal earthquake [469]. The resisting mechanism, developed by the axial compressive and hoop tensile stresses, cannot survive further vertical load increments since the annular strips at the base yield in tension [53]. The pressure $\bar{p}_2$ is most critical with respect to elastic buckling [294, 325, 612]. In this case the stabilizing effect of the internal pressure is reduced by subtracting the pressure p_v due to the vertical earthquake component [469] (Point D.1.5.2(3), Annex D UNI ENV 1993-1-6 where the pressurized meridional elastic buckling parameter explicitly requires the "smallest design value of local internal pressure at the location of the point being assessed, guaranteed to coexist with the meridional compression"). This damage pattern, less common than the previous one, occurs at small hoop stress levels and, hence, is particularly sensitive to internal pressure and imperfection amplitude: the buckling strength decreases, as the former reduces or as the latter increases [53]. The pressure $\bar{p}_3$ may lead to regions of low pressure mainly in the upper part of the tank's wall where the wall itself is rather thin [469], and buckling due to external pressure as well as cavitation may occur [199, 244, 420, 455, 536]; it may also happen at the same time as an elastic–plastic buckling, such as in Figure 5.13.

Which failure mode is relevant in a given case deeply depends on the characteristics of the earthquake, on the tank geometry and imperfections, on its liquid content and on the interaction or not with the soil. A common trend may be derived following the field observation by Niwa and Clough [428] during the Livermore earthquake in California on 24 January 1980 [571] and by Brunesi et al. in [53] observed during the Emilia earthquake in Italy, 20 and 29 of May 2012:

1) elastic–plastic buckling or "elephant foot" buckling (Figure 5.11 and 5.15(a)) is the most common damage in broad (squat/shallow/short) tanks, with a height to radius ratio $\frac{H}{R} < 2$;

(a)

(b)

(c)

(d)

(e)

(f)

Figure 5.15 Damaged tanks observed during the Emilia earthquake, Italy (20 and 29 of May 2012): (a) elastic–plastic buckling, (b) elastic buckling at the base and (c) "secondary" diamond-shaped buckling along the wall; (d) failed leg–supported, base–anchorage (e) and sliding of the tanks (f).

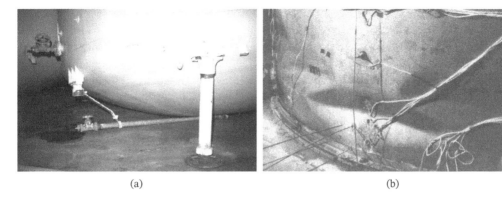

Figure 5.16 (a) earthquake event of Magnitude 5.9, 24 January 1980 in Livermore, California (from Karl V. Steinbrugge Collection, University of California, Berkeley); (b) "diamond"-shaped buckling spreading around the circumferences at the base obtained experimentally by Niwa and Clough [428]. Source: From [428]/John Wiley & Sons, Inc.

2) tall (slender) tanks with a ratio in the range $2 < \frac{H}{R} < 4$ suffered more "diamond"-shaped buckling spreading around the circumferences at the base, as shown in Figures 5.12 and 5.15(b) or along the wall in correspondence of the circumferential welds, where the amplitude of the imperfections is larger and their effects, concomitantly with the wall thickness reduction along the height, more visible (Figure 5.15(c));

3) fracture of piping at the connection of the tank (Figure 5.16(a)), leg–supported (Figure 5.15(d)) and base–anchorage failures in flat–bottomed systems (Figure 5.15(e)) or sliding of the tank (Figure 5.15(f)) in the case of very slender systems with $\frac{H}{R} > 4$.

The tank buckling mechanisms described above in Points 1. and 2. have been experimentally tested (Figure 5.16(b)) and verified by Niwa and Clough [428] comparing the experimental behaviour with the damage observed during the January 1980, Livermore, California, earthquake.

5.6 Tank Forces and Stresses

As clearly stated in Section 5.5, the time history of the total pressure exerted on the wall of a flexible tank is given by the time history contributions of the rigid impulsive pressure $p_i(\xi, \zeta, \theta, t)$, of the convective pressure $p_c(\xi, \zeta, \theta, t)$ and the flexible impulsive pressure $p_f(\zeta, \theta, t)$. With the temporal and spatial values of the hydrodynamic wall pressure determined, the corresponding tank forces and stresses may be computed. Similarly to the pressures, the time history of the base shear (and consequently of the bending moments), produced by these pressures, is governed by the following contributions [600] (Point A.3.2.1, EC8-4):

1) the impulsive rigid shear component $Q_i(t)$ which derives from $p_i(\xi, \zeta, \theta, t)$ and hence depends on the horizontal ground acceleration $a(t)$. It is given by Equation (2.9) or (2.23a) (for circular and rectangular tanks, taking care that masses must be obtained from Section 2.2 for cylindrical and Section 2.3 for rectangular tanks). According to Section 2.2.4, to the rigid impulsive value must be added the

instantaneous value of the base shear $Q_w(t)$ of the tank itself, given by expressions (2.51a) or (2.52a) (Equations (3.11) and (3.14), Point 3.5, NZSEE-09);

2) the long period component of the convective base shear $Q_c(t)$ contributed by the convective fluid pressure $p_c(\xi, \zeta, \theta, t)$ driven by the pseudo-acceleration $a_{cn}(t)$ induced by the prescribed free-field ground motion $a(t)$ [215]. It derives from Equation (2.31) or (2.42a) for cylindrical and rectangular [635] tanks, paying attention that masses should be obtained from Section 2.2.2 for cylindrical and Section 2.3.1 for rectangular tanks;

3) the resultant base horizontal shear $Q_f(t)$ from the short period fluid pressure $p_f(\zeta, \theta, t)$ caused by the wall deformation relative to the base circle due to the deformability of the tank wall (called the "flexible impulsive") [223, 624, 626, 646]. For cylindrical and rectangular tanks, the flexible impulsive shear component may be derived from expression (5.13) and the corresponding masses from Sections 5.2.1 and 5.2.2 respectively.

From a careful examination of Equations (2.5) and (5.8), it is clearly noted that the most significant distinction between the "flexible" and the "rigid" contribution to the hydrodynamic wall pressure involves the nature of the acceleration component employed: the impulsive "rigid" term is governed by the horizontal ground acceleration $a(t)$ while the impulsive "flexible" term is governed by the response acceleration $a_f(t)$, relative to the base, of a single degree of freedom linear oscillator induced by the prescribed free-field ground motion $a(t)$. According to Point A.3.2.1, EC8-4, the use of expressions (2.5) and (5.8) (or better (2.9) and (5.13)) poses the question of the combination of the peak of $a(t)$ with that of $a_f(t)$. Since the input and its response cannot be assumed as independent in the range of relatively high frequencies under consideration, the square root of the sum of the square rule is not sufficiently accurate (Point A.3.2.1, EC8-4) and, on the other hand, addition of the individual maxima could lead to overconservative estimations.

5.6.1 Vertical Cylindrical Tanks

Given the difficulties above, four approximate approaches for cylindrical tanks, are presented in detail below and are due to Veletsos and Yang [626, 646], Haroun, Housner, and Ellaithy [214, 215, 221, 223], Rammerstorfer, Fischer, and Scharf [166, 468, 506] and Malhotra [370] based on the research undertaken by Veletsos et al. in [591, 626]:

- the Veletsos and Yang procedure [626, 646] has been proposed only for tanks with values of $\frac{H}{R}$ in the range between 0.25 and 1.0 (Figure 5.17); for tanks with values of slenderness significantly in excess of unity, this procedure may be unduly conservative [626]. The convective effects (Q_c, M'_c, M_c) (for cylindrical and rectangular tanks) in this approach are computed by considering the tank's wall to be rigid according to Equations (2.31), (2.33) and (2.35) respectively. The resulting base shear $Q_i^{VY}(t)$ (where the superscript ^{VY} denotes the Veletsos and Yang procedure), the base moment due to the hydrodynamic wall pressure $M_i^{VY}(t)$ and the pressure exerted on the base $M_i^{\prime VY}(t)$ are expressed as:

$$Q_i^{VY}(t) = m_i \left[a(t) + a_f(t) \right] \tag{5.62a}$$

$$M_i^{VY}(t) = Q_i^{VY}(t) h_i \tag{5.62b}$$

$$M_i^{\prime VY}(t) = Q_i^{VY}(t) h_i' \tag{5.62c}$$

assuming the entire "rigid impulsive" mass m_i (Equations (2.10), (2.25a), (2.25b), and (2.131a) for cylindrical and rectangular tanks) to respond with the absolute response acceleration $a(t) + a_f(t)$; the

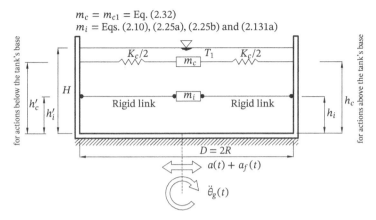

$$m_c = m_{c1} = \text{Eq. (2.32)}$$
$$m_i = \text{Eqs. (2.10), (2.25a), (2.25b) and (2.131a)}$$

Figure 5.17 Equivalent mechanical model, due to Veletsos and Yang [626, 646], for flexible cylindrical tanks with values of $\frac{H}{R}$ in the range between 0.25 and 1.0 under lateral translation $a(t)$ and base rocking motion $\ddot{\theta}_g(t)$.

Table 5.5 Effective non-dimensional mass coefficient ε_0 used to evaluate the effective structural mass $\varepsilon_0 m_w$.

H/R	0.5	1.0	3.0
ε_0	0.5	0.7	0.9

Source: From [626]/Praveen K. Malhotra.

maximum value of the impulsive shear Q_i^{VY} (and consequently of the moments M_i^{VY} and $M_i'^{VY}$) may then be determined by replacing $a(t) + a_f(t)$ by its spectral value S_e using the natural frequency (or period) of the first impulsive tank-liquid horizontal mode (Section 5.4 with or without soil–structure interaction) and damping value equal to 2% [162, 166, 223, 297] or 1% [161, 162]; h_i and h_i' may be derived from expressions (2.13), (2.15), (2.25c)–(2.26c) and (2.131c)–(2.131f) for cylindrical and rectangular tanks. The effects of tank inertia (wall and roof) are generally small and may be ignored particularly since the hydrodynamic effects are evaluated conservatively in the Veletsos and Yang procedure [626, 646]. However, the additional base shear $Q_w(t)$ and base moment $M_w(t)$ induced may be computed from expressions (2.51) by replacing m_w with the effective structural mass $\varepsilon_0 m_w$ [626]:

$$Q_w(t) = (\varepsilon_0 m_w + m_r)\left[a(t) + a_f(t)\right] \tag{5.63a}$$

$$M_w(t) = (\varepsilon_0 m_w h_w + m_r \overline{h}_r)\left[a(t) + a_f(t)\right] \tag{5.63b}$$

where m_w and m_r are the total masses of the tank wall and roof respectively and h_w and $\overline{h}_r$ are the distances from the base to the respective mass centres; the effective mass coefficient ε_0 is a function of the tank proportions, and may be determined from Table 5.5;

- the Haroun, Housner, and Ellaithy procedure [214, 215, 221, 223], suggested at Point C3.3.2, NZSEE-09, defines an equivalent mechanical model, depicted in Figure 5.18, for flexible cylindrical roofless[6] tanks

---

6 If the roof loading is sufficiently important to warrant its consideration, it may be incorporated in this formulation using the Dunkerley's expression given by Equation (5.46) as outlined by Veletsos in [626] (Point C3.6, NZSEE-09).

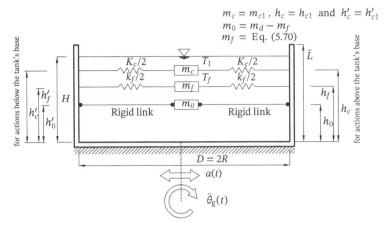

$$m_c = m_{c1}, \quad h_c = h_{c1} \quad \text{and} \quad h'_c = h'_{c1}$$
$$m_0 = m_d - m_f$$
$$m_f = \text{Eq. (5.70)}$$

Figure 5.18 Equivalent mechanical model, due to Haroun, Housner, and Ellaithy [214, 215, 221, 223], for flexible cylindrical roofless tanks of uniform wall thickness full of liquid under lateral translation $a(t)$ and base rocking motion $\ddot{\theta}_g(t)$.

of uniform wall thickness full of liquid, taking into consideration the effects of lateral translation $a(t)$ and base rocking motion $\ddot{\theta}_g(t)$ (Figure 2.1 and Section 5.7). The mechanical model (Figure 5.18) consists of two concentrated masses,[7] (m_0, m_f) with $m_0 = m_d - m_f$, in addition to the convective term $m_c = m_{c1}$ (or m_1; Section 2.2.2) computed by considering the tank's wall to be rigid: the rigidly attached mass $m_0 = m_d - m_f$, located at a distance $h_0 = \frac{m_d h_d - m_f h_f}{m_0}$ (or $h'_0 = \frac{m_d h'_d - m_f h'_f}{m_0}$) from the base, which simulates the hydrodynamic effects associated with the impulsive rigid-body component of wall motion, whereas the elastically supported mass m_f (with $m_f < m_d$),[8] located at a distance h_f (or h'_f) from the base, which considers the impulsive flexible component of the deformational wall using an equivalent spring constant equal to [218]:

$$k_f = \left(\frac{4\pi^5 E s m_f}{m} \right) \left[3.35 \cdot 10^{-4} \left(\frac{H}{R} \right)^3 - 2.10 \cdot 10^{-5} \left(\frac{H}{R} \right)^2 - 0.016361 \left(\frac{H}{R} \right) + 0.065598 \right]^2 \quad (5.64)$$

The resulting base shear $Q_i^{\mathrm{HHE}}(t)$ (where the superscript ^{HHE} denotes the Haroun, Housner, and Ellaithy procedure), the moment immediately above the base $M_i^{\mathrm{HHE}}(t)$ and the foundation moment due to the hydrodynamic wall pressure together with the pressure exerted on the base $M_i'^{\mathrm{HHE}}(t)$ are expressed as [223]:

$$Q_i^{\mathrm{HHE}}(t) = (m_d - m_f)a(t) + m_f \left[a(t) + a_f(t) \right] \quad (5.65a)$$

$$M_i^{\mathrm{HHE}}(t) = (m_d h_d - m_f h_f)a(t) + m_f h_f \left[a(t) + a_f(t) \right] \quad (5.65b)$$

$$M_i'^{\mathrm{HHE}}(t) = (m_d h'_d - m_f h'_f)a(t) + m_f h'_f \left[a(t) + a_f(t) \right] \quad (5.65c)$$

---

7 In the previous Veletsos and Yang procedure [626, 646], the two components were considered in a single step using mass m_i.

8 The mass m_f can be viewed as the portion of m_d which partecipates in the vibration of the flexible wall and, consequently, m_f is always smaller than m_d.

where $a(t) + a_f(t)$ is the same absolute response acceleration that appears in Equation (5.62a); the maximum value of the impulsive shear Q_i^{HHE} (and consequently of the moments M_i^{HHE} and $M_i'^{\mathrm{HHE}}$) may then be determined by replacing $a(t) + a_f(t)$ by its spectral value S_e using the natural frequency (or period) of the first impulsive tank-liquid horizontal mode (Section 5.4 with or without soil–structure interaction) and damping value equal to 2% [223, 297]; furthermore, by replacing $a(t)$ by its spectral value S_e using the impulsive mode period T_0 with soil–structure interaction (Equation (2.55)) or without soil–structure interaction, $S_e(T_0) = Sa_g$ where a_g is the peak value of the ground acceleration $a(t)$ and S the soil factor (Point 3.2.2.2, UNI EN 1998-1:2004). According to the nomenclature defined in Figure 2.18, the maximum radial base shear applied at the bottom of the tank walls and the resulting bending moments along the wall (above and below the base) of the tank and across sections normal to the tank axis may be determined using the SRSS rule (Point A.3.2.1, EC8-4) [223]:

$$Q_y^{\mathrm{HHE}} = \sqrt{\left[m_{c1}S_e(T_1)\right]^2 + \left[m_f S_e(T_f)\right]^2 + \left[(m_d - m_f)S_e(T_0)\right]^2} \tag{5.66a}$$

$$M_y^{\mathrm{HHE}} = \sqrt{\left[m_{c1}h_{c1}S_e(T_1)\right]^2 + \left[m_f h_f S_e(T_f)\right]^2 + \left[(m_d h_d - m_f h_f)S_e(T_0)\right]^2} \tag{5.66b}$$

$$M_i'^{\mathrm{HHE}} = \sqrt{\left[m_{c1}h_{c1}'S_e(T_1)\right]^2 + \left[m_f h_f' S_e(T_f)\right]^2 + \left[(m_d h_d' - m_f h_f')S_e(T_0)\right]^2} \tag{5.66c}$$

or using the rule suggested by Priestley et al. at Point C4.4.1, NZSEE-09:

$$Q_y^{\mathrm{HHE}} = \sqrt{\left[m_{c1}S_e(T_1)\right]^2 + \left[(m_f S_e(T_f)) + ((m_d - m_f)S_e(T_0))\right]^2} \tag{5.67a}$$

$$M_y^{\mathrm{HHE}} = \sqrt{\left[m_{c1}h_{c1}S_e(T_1)\right]^2 + \left[(m_f h_f S_e(T_f)) + ((m_d h_d - m_f h_f)S_e(T_0))\right]^2} \tag{5.67b}$$

$$M_i'^{\mathrm{HHE}} = \sqrt{\left[m_{c1}h_{c1}'S_e(T_1)\right]^2 + \left[\left(m_f h_f' S_e(T_f)\right) + \left((m_d h_d' - m_f h_f')S_e(T_0)\right)\right]^2} \tag{5.67c}$$

where, in both expressions (5.66) and (5.67), the hydrostatic bending moment M_{hs} should be added; convective mass m_{c1} and heights (h_{c1}, h_{c1}') may be derived respectively from Equations (2.32), (2.36) and (2.34) for cylindrical tanks and (2.131b), (2.131g), and (2.131h) for rectangular tanks; $S_e(\cdot)$ is the spectral acceleration obtained from a horizontal elastic response spectrum corresponding to the convective period of vibration of the 1st–sloshing mode of vibration T_1 of the liquid (Equation (2.54) for a cylindrical tank and (2.132) for a rectangular tank) and to the impulsive period with or without soil–structure interaction (T_f or $T_f^\star$ at Section 5.4) as well as to the impulsive period T_0 with soil–structure interaction (Equation (2.55)) or without soil–structure interaction, $S_e(T_0) = Sa_g$. The masses m_d and m_f, incorporating the effects of the inertia of the tank wall (thickness equal to s) via the density ρ_s (in kg/m^3), may be derived from the following Equations (5.68) and (5.70) as well as the heights h_d' and h_f', both including the effect of the inertia of the wall and the effect of the pressure variation on the base, from Equations (5.69) and (5.71), while the heights h_d and h_f, both including the effect of the inertia of the wall and just the effect of the pressure variation wall, are given in Figure 5.19 normalized with respect to the total height of the fluid and as a functions of both $\frac{H}{R}$ and $\frac{s}{R}$ [221]:

$$m_d = m\left[1 - \sum_{i=1}^{\infty}\left(\frac{2\tanh\bar{\lambda}_i}{\bar{\lambda}_i\lambda_i''}\right) + 2\frac{\rho_s\bar{L}s}{\rho_l HR}\right] \tag{5.68}$$

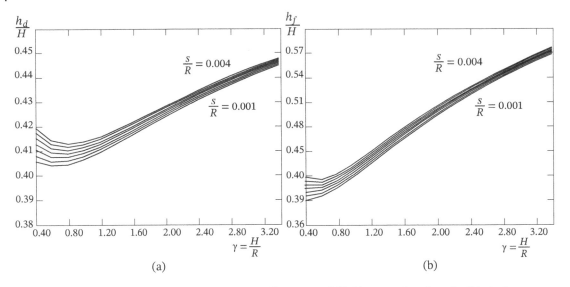

Figure 5.19 Equivalent heights h_d and h_f in Haroun, Housner, and Ellaithy procedure for a flexible tank [214, 215, 221, 223, 626].

$$h'_d = \frac{H\left[\frac{1}{2} - \sum_{i=1}^{\infty}\left(\frac{2\tanh\overline{\lambda}_i}{[\overline{\lambda}_i\lambda''_i]}\right) + \sum_{i=1}^{\infty}\left(\frac{4(\cosh\overline{\lambda}_i - 1)}{[\overline{\lambda}_i^2\lambda''_i\cosh\overline{\lambda}_i]}\right) + \frac{\rho_s\overline{L}^2 s}{\rho_l H^2 R}\right]}{\left[1 - \sum_{i=1}^{\infty}\left[\frac{2\tanh\overline{\lambda}_i}{[\overline{\lambda}_i\lambda''_i]}\right] + 2\frac{\rho_s\overline{L}s}{\rho_l HR}\right]} \tag{5.69}$$

$$m_f = \frac{m\left[\left(\frac{2}{\pi} - \sum_{i=1}^{\infty}\left[\frac{2\left(\overline{\lambda}_i\sinh\overline{\lambda}_i + \frac{\pi}{2}\right)}{\lambda''_i\cosh\overline{\lambda}_i\left[\overline{\lambda}_i^2 + \left(\frac{\pi}{2}\right)^2\right]}\right]\right)^2 + 2\frac{\rho_s s}{\rho_l R}\left(\frac{2}{\pi} + \frac{\overline{L}}{H} - 1\right)\right]}{\left[\frac{4}{\frac{R}{H}\pi^3}\sum_{i=1,3,5,...}^{\infty}\left[\left(\frac{I_1(\lambda'_i R)}{I'_1(\lambda'_i R)(2i-1)} + \frac{I_1(\lambda'_{i+1}R)}{I'_1(\lambda'_{i+1}R)(2i+1)}\right)\left(\frac{1}{i^2}\right)\right]\right] + 2\frac{\rho_s s}{\rho_l R}\left(\frac{\overline{L}}{H} - \frac{1}{2}\right)} \tag{5.70}$$

$$h'_f = \frac{H\left[\left(\frac{2}{\pi}\right)^2 - \sum_{i=1}^{\infty}\left[\frac{2\left(\overline{\lambda}_i\sinh\overline{\lambda}_i + \frac{\pi}{2}\right)}{\lambda''_i\cosh\overline{\lambda}_i\left[\overline{\lambda}_i^2 + \left(\frac{\pi}{2}\right)^2\right]}\right] - \sum_{i=1}^{\infty}\left[\frac{4\left(\overline{\lambda}_i - \frac{\pi}{2}\sinh\overline{\lambda}_i\right)}{\overline{\lambda}_i\lambda''_i\cosh\overline{\lambda}_i\left[\overline{\lambda}_i^2 + \left(\frac{\pi}{2}\right)^2\right]}\right]\right]}{\frac{2}{\pi} - \sum_{i=1}^{\infty}\left[\frac{2\left(\overline{\lambda}_i\sinh\overline{\lambda}_i + \frac{\pi}{2}\right)}{\lambda''_i\cosh\overline{\lambda}_i\left[\overline{\lambda}_i^2 + \left(\frac{\pi}{2}\right)^2\right]}\right] + 2\frac{\rho_s s}{\rho_l R}\left(\frac{2}{\pi} + \frac{\overline{L}}{H} - 1\right)}$$

$$+ \dfrac{2\dfrac{\rho_s s}{\rho_l R}\left(\dfrac{4}{\pi^2} + \dfrac{\overline{L}^2}{2H^2} - \dfrac{1}{2}\right)\Bigg]}{\dfrac{2}{\pi} - \sum_{i=1}^{\infty}\left[\dfrac{2\left(\overline{\lambda}_i \sinh \overline{\lambda}_i + \dfrac{\pi}{2}\right)}{\lambda_i'' \cosh \overline{\lambda}_i \left[\overline{\lambda}_i^2 + \left(\dfrac{\pi}{2}\right)^2\right]}\right] + 2\dfrac{\rho_s s}{\rho_l R}\left(\dfrac{2}{\pi} + \dfrac{\overline{L}}{H} - 1\right)} \tag{5.71}$$

where $\overline{\lambda}_i = \lambda_i \frac{H}{R}$, $\lambda_i'' = (\lambda_i^2 - 1)$ and $\lambda_1 = 1.8412$, $\lambda_2 = 5.3314$, $\lambda_3 = 8.5363$, $\lambda_4 = 11.7060$ and $\lambda_5 = 14.8636$ [221, 646] are the roots of the first derivative of J_1 which is the Bessel function of the first kind of order 1 [561]; $\lambda_i' = \frac{(2i-1)\pi}{2H}$; ρ_l (in kg/m^3) is the mass density of the liquid and R, H and $\overline{L}$ are geometrical properties of the tank depicted in Figure 5.18. In deriving expressions (5.68)–(5.71), the deflected shape of the tank has been approximated by assuming Equation (5.9a).

- The Rammerstorfer, Fischer, and Scharf procedure [166, 468, 506] suggests using the following expressions corresponding to the equivalent mechanical model depicted in Figure 5.4 and related variables in Section 5.2.1:

$$Q_i^{\mathrm{RFS}}(t) = m_i a(t) + m_f \left[a(t) + a_f(t)\right] \tag{5.72a}$$

$$M_i^{\mathrm{RFS}}(t) = m_i h_i a(t) + m_f h_f \left[a(t) + a_f(t)\right] \tag{5.72b}$$

$$M_i'^{\mathrm{RFS}}(t) = m_i h_i' a(t) + m_f h_f' \left[a(t) + a_f(t)\right] \tag{5.72c}$$

related to the consideration, deeply investigated by Scharf in [506], that in a wide frequency range for typical storage tanks, the maximum relative acceleration response is approximately equal to the maximum absolute acceleration response. In Equations (5.72) the superscript ^{RFS} denotes the Rammerstorfer, Fischer, and Scharf procedure; the "rigid impulsive" mass m_i may be derived from Equations (2.10), (2.25a), (2.25b), and (2.131a) for cylindrical and rectangular tanks, respectively; impulsive heights h_i and h_i' may be obtained from expressions (2.13), (2.15), (2.25c)–(2.26c) and (2.131c)–(2.131f) for cylindrical and rectangular tanks; m_f, h_f and h_f' come respectively from relations (5.14), (5.16) and (5.18) properly derived from Fischer in [158, 160]. The maximum value of the impulsive shear Q_i^{RFS} (and consequently of the moments M_i^{RFS} and $M_i'^{\mathrm{RFS}}$) may then be determined by replacing $a(t) + a_f(t)$ by its spectral value S_e using the natural frequency (or period) of the first impulsive tank-liquid horizontal mode (Section 5.4 with or without soil–structure interaction) and damping value equal to 2% [162, 166, 223, 297] or 1 % [161, 162]; furthermore $a(t)$ can be replaced by its spectral value S_e using the impulsive mode period T_0 with soil–structure interaction (Equation (2.55)) or without soil–structure interaction, $S_e(T_0) = Sa_g$ where a_g is the peak value of the ground acceleration $a(t)$ and S the soil factor (Point 3.2.2.2, UNI EN 1998-1:2004). According to the nomenclature defined in Figure 2.18, the maximum radial base shear applied at the bottom of the tank's wall and the resulting bending moments along the wall (above and below the base) of the tank and across sections normal to the tank axis may be determined using the SRSS rule (Point A.3.2.1, EC8-4) [166, 468, 506]:

$$Q_y^{\mathrm{RFS}} = \sqrt{\left[m_{c1}S_e(T_1)\right]^2 + \left[m_f S_e(T_f)\right]^2 + \left[m_i S_e(T_0)\right]^2} \tag{5.73a}$$

$$M_y^{\text{RFS}} = \sqrt{\left[m_{c1}h_{c1}S_e(T_1)\right]^2 + \left[m_f h_f S_e(T_f)\right]^2 + \left[m_i h_i S_e(T_0)\right]^2} \tag{5.73b}$$

$$M_y'^{\text{RFS}} = \sqrt{\left[m_{c1}h_{c1}'S_e(T_1)\right]^2 + \left[m_f h_f' S_e(T_f)\right]^2 + \left[m_i h_i' S_e(T_0)\right]^2} \tag{5.73c}$$

where the hydrostatic bending moment M_{hs} should be added; convective mass m_{c1} and heights (h_{c1}, h_{c1}') may be derived respectively from Equations (2.32), (2.36), and (2.34) for cylindrical tanks and (2.131b), (2.131g), and (2.131h) for rectangular tanks; $S_e(\cdot)$ is the spectral acceleration obtained from a horizontal elastic response spectrum corresponding to the convective period of vibration of the 1st–sloshing mode of vibration T_1 of the liquid (Equation (2.54) for a cylindrical tank and (2.132) for a rectangular tank) and to the impulsive period with or without soil–structure interaction (T_f or $T_f^\star$ at Section 5.4) as well as to the impulsive period T_0 with soil–structure interaction (Equation (2.55)) or without soil–structure interaction, $S_e(T_0) = Sa_g$.

- The Malhotra procedure [370], suggested in Point A.3.2.2, EC8-4, is based primarily on the work by Veletsos et al. in [591, 626] and is quite similar to the Veletsos and Yang procedure [626, 646] previously described. The roofless tank-liquid system is modelled by two single degree of freedom systems, as reported in Figure 5.20, one representing the first impulsive modal component m_f (and corresponding heights h_f and h_f') and the second one representing the first convective mode m_{c1} (and corresponding heights h_{c1}, h_{c1}'; Equations (2.32), (2.36), and (2.34) for cylindrical tanks and (2.131b), (2.131g), and (2.131h) for rectangular tanks). The maximum radial base shear applied at the bottom of the tank's wall and the resulting bending moments along the wall (above and below the base) of the tank and across sections normal to the tank axis may be determined using the direct sum method rather than the SRSS rule (Point A.3.2.2, EC8-4) as suggested by Malhotra in [370] (where the superscript ^{PKM} denotes the Praveen K. Malhotra procedure):

$$Q_y^{\text{PKM}} = (m_f + m_w + m_r)S_e(T_{\text{imp}}) + m_{c1}S_e(T_{\text{con}}) \tag{5.74a}$$

$$M_y^{\text{PKM}} = (m_f h_f + m_w h_w + m_r \overline{h}_r)S_e(T_{\text{imp}}) + m_{c1}h_{c1}S_e(T_{\text{con}}) \tag{5.74b}$$

$$M_y'^{\text{PKM}} = (m_f h_f' + m_w h_w + m_r \overline{h}_r)S_e(T_{\text{imp}}) + m_{c1}h_{c1}'S_e(T_{\text{con}}) \tag{5.74c}$$

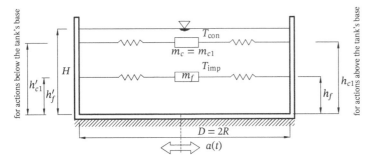

Figure 5.20 Equivalent mechanical model, due to Malhotra [370], for flexible cylindrical roofless tanks of uniform wall thickness full of liquid under lateral translation $a(t)$ as adopted by Point A.3.2.2, EC8-4. Source: Adapted from [370].

where m_w is the wall mass and m_r the roof mass; h_w and $\bar{h}_r$ are the distances from the base to the respective mass centres; the impulsive and convective masses and heights, $(m_f, m_{c1}, h_f, h_{c1}, h'_f, h'_{c1})$ are given respectively in columns 4–9 of Table 5.6 as a fraction of the total liquid mass m and total height H. In Equations (5.74), $S_e(T_{imp})$ is the spectral acceleration obtained from a horizontal elastic response spectrum corresponding to the impulsive period T_{imp} and to the equivalent viscous damping level[9] for horizontal mode consistent with the limit state (Point 2.3.3.1, EC8-4 and Point 4.1.3(1), EN 1998-2:2005), tank type (bolted and welded steel or reinforced and prestressed concrete; Point 4.1.3(1), EN 1998-2:2005); $S_e(T_{con})$ is the convective spectral acceleration obtained from a 0.5% damped elastic response spectrum (Point A.3.2.2.2, EC8-4 and also Point E.1, API 650, 2005 Edition). The natural periods of the impulsive and convective responses, in seconds, are [370]:

$$T_{imp} = C_i \frac{H\sqrt{\rho_l}}{\sqrt{s/R}\sqrt{E}} \tag{5.75a}$$

$$T_{con} = C_c \sqrt{R} \tag{5.75b}$$

where H (in metre) is the design liquid height (free surface), R (in metre) the tank's radius, ρ_l is the mass density of the liquid (in kg/m^3) and E is the Young's modulus of the tank material (in N/m^2); the coefficients C_i and C_c are derived respectively from columns 2 and 3 of Table 5.6; the coefficient C_i is dimensionless while C_c is in sec·m$^{-1/2}$; s (in metre) is an equivalent uniform thickness of the tank wall. For tanks with variable thickness, Malhotra in [370] suggests using weighted average over the wetted height H of the tank wall, assigning highest weight to the thickness near the base of the tank. Considering, as an example, in Figure 5.21, a tank with a wall made of four courses, the following weighted average thickness could be used, as suggested by Malhotra in [370]:

$$s = \frac{s_1 H_1 \overline{H}_1 + s_2 H_2 \overline{H}_2 + s_3 H_3 \overline{H}_3 + s_4 H_4 \overline{H}_4}{H_1 \overline{H}_1 + H_2 \overline{H}_2 + H_3 \overline{H}_3 + H_4 \overline{H}_4} \tag{5.76}$$

Table 5.6 Coefficients C_i and C_c, impulsive and convective masses and heights, $(m_f, m_{c1}, h_f, h_{c1}, h'_f, h'_{c1})$ used for the Malhotra procedure [370] adopted at Point A.3.2.2, EC8-4.

H/R	C_i	C_c	m_f/m	m_{c1}/m	h_f/H	h_{c1}/H	h'_f/H	h'_{c1}/H
0.3	9.28	2.09	0.176	0.824	0.400	0.521	2.640	3.414
0.5	7.74	1.74	0.300	0.700	0.400	0.543	1.460	1.517
0.7	6.97	1.60	0.414	0.586	0.401	0.571	1.009	1.011
1.0	6.36	1.52	0.548	0.452	0.419	0.616	0.721	0.785
1.5	6.06	1.48	0.686	0.314	0.439	0.690	0.555	0.734
2.0	6.21	1.48	0.763	0.237	0.448	0.751	0.500	0.764
2.5	6.56	1.48	0.810	0.190	0.452	0.794	0.480	0.796
3.0	7.03	1.48	0.842	0.158	0.453	0.825	0.472	0.825

9 Malhotra in [370] suggests using 2% damped elastic response spectrum for steel and prestressed concrete tanks and 5% damped elastic response spectrum for concrete tanks.

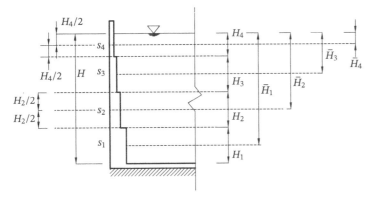

Figure 5.21 Evaluation of the weighted average thickness in the case of a tank with a wall made of four courses [370] (Point A.3.2.2.1, in EC8-4). Source: Adapted from [370].

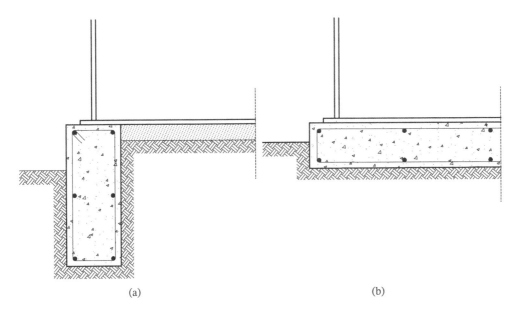

(a)   (b)

Figure 5.22 Foundations for stell cylindrical tanks: (a) circular concrete ring and (b) rigid concrete mat.

Furthermore, if the tank is supported on a ring foundation[10] [414] (Figure 5.22(a)), Malhotra in [370] suggests using the moment M_y^{PKM} (Equation (5.74b)) to design the tank wall, base anchors, and the foundation, whereas if the tank is supported on a mat foundation (Figure 5.22(b)), to use moment M_y^{PKM} (Equation (5.74b)) just to design the tank wall and base anchors, and to use moment $M_y'^{PKM}$ (Equation (5.74c)) to design the foundation.

10 Typical foundation will consist of a circular concrete ring wall if the foundations are soil-supported. The interior of the ring will be comprised of compacted backfill with a layer of compacted sand or asphalt to serve as a bearing surface for the tank bottom.

5.6.2 Rectangular Tanks

In the case of a dynamic analysis of fluid in flexible rectangular containers subjected to acceleration, the model suggested by Hashemi et al. in [229], depicted in Figure 5.5 and described in Section 5.2.2, suggests using the following expressions (where the superscript ^{HSK} denotes the Hashemi, Saadatpours and Kianoush procedure):

$$Q_y^{HSK} = \sqrt{[m_{c1}S_e(T_1)]^2 + [m_f S_e(T_f)]^2 + [(m_d - m_f)a_g]^2} \tag{5.77a}$$

$$M_y^{HSK} = \sqrt{[m_{c1}h_{c1}S_e(T_1)]^2 + [m_f h_f S_e(T_f)]^2 + [(m_d h_d - m_f h_f)a_g]^2} \tag{5.77b}$$

$$M_y'^{HSK} = \sqrt{[m_{c1}h_{c1}'S_e(T_1)]^2 + [m_f h_f' S_e(T_f)]^2 + [(m_d h_d' - m_f h_f')a_g]^2} \tag{5.77c}$$

according to the nomenclature defined in Figure 2.24; Q_y^{HSK} is the maximum base out of plane shear, M_y^{HSK} the maximum bending moment applied to the bottom of the wall and $M_y'^{HSK}$ the maximum foundation moment (or overturning moment); m_d, m_f and $m_{c1} = m_c$ may be derived respectively from expressions (5.21a), (5.22a) and (5.24a); h_d, h_f, h_{c1}, h_d', h_f' and h_{c1}' respectively from relations (5.21b), (5.22b), (5.24b), (5.21c), (5.22c) and (5.24c); the convective period of vibration of the first mode may be derived from Equation (2.132); the period T_f of vibration of the first impulsive rectangular tank-liquid horizontal mode excluding soil–structure interaction is given by Equation (5.49) and including soil–structure interaction ($T_f^\star$) by Equation (5.59); a_g is the peak value of the ground acceleration $a(t)$. Following the same approach adopted with relations (5.66) or (5.73), the following alternative to Equations (5.77) may be used:

$$Q_y^{HSK} = \sqrt{[m_{c1}S_e(T_1)]^2 + [m_f S_e(T_f)]^2 + [(m_d - m_f)S_e(T_0)]^2} \tag{5.78a}$$

$$M_y^{HSK} = \sqrt{[m_{c1}h_{c1}S_e(T_1)]^2 + [m_f h_f S_e(T_f)]^2 + [(m_d h_d - m_f h_f)S_e(T_0)]^2} \tag{5.78b}$$

$$M_y'^{HSK} = \sqrt{[m_{c1}h_{c1}'S_e(T_1)]^2 + [m_f h_f' S_e(T_f)]^2 + [(m_d h_d' - m_f h_f')S_e(T_0)]^2} \tag{5.78c}$$

where T_0 is the impulsive period with soil–structure interaction (Equation (2.55)) or without soil–structure interaction, $S_e(T_0) = Sa_g$ and S is the soil factor (Point 3.2.2.2, UNI EN 1998-1:2004).

5.7 Effects of Rocking Motion

As clearly explained above, the analysis of the seismic response of a tank to horizontal ground shaking is normally carried out on the assumption that the tank base moves horizontally without any rotation. In reality, because of the flexibility of the supporting soils [221, 638], the tank base experiences a rigid base rocking component of motion, $\theta_g(t)$ (corresponding to an angular acceleration of the base $\ddot{\theta}_g(t)$ in Figures 2.1 and 5.18 evaluated according to [204, 330, 337, 498, 593]), even under a purely translational free-field ground motion. The analytical mechanical model for flexible cylindrical tanks under a horizontal motion, proposed by Haroun, Housner, and Ellaithy [214, 215, 221, 223] and described in Section 5.6.1, has been extended to include the effect of a rigid base rocking motion by substituting expressions (5.1) and (5.3) to include the appropriate boundary conditions [221]:

$$\frac{\partial \Phi}{\partial z}(r, \theta, 0, t) = -r\dot{\theta}_g(t) \cos \theta \tag{5.79a}$$

$$\frac{\partial \Phi}{\partial r}(R, \theta, z, t) = \left[v(t) + \frac{\partial w}{\partial t}(\theta, z, t) + z\dot{\theta}_g(t) \right] \cos \theta \tag{5.79b}$$

According to the nomenclature defined in relations (5.65), the resulting base shear $Q_i^{\text{HHE}}(t)$ and the foundation moment due to the hydrodynamic wall pressure together with the pressure exerted on the base, $M_i'^{\text{HHE}}(t)$ are expressed as [221]:

$$Q_i^{\text{HHE}}(t) = (m_d - m_f)a(t) + m_f \left[a(t) + a_f(t) \right] + m_d h_d' \ddot{\theta}_g(t) \tag{5.80a}$$

$$M_i'^{\text{HHE}}(t) = (m_d h_d' - m_f h_f')a(t) + m_f h_f' \left[a(t) + a_f(t) \right] + (m_d h_d'^2 + I_d)\ddot{\theta}_g(t) \tag{5.80b}$$

where I_d is the central moment of inertia of the mass m_d and expression $(m_d h_d'^2 + I_d)$ can be expressed as [221]:

$$\begin{aligned} m_d h_d'^2 + I_d = mH^2 \Bigg[0.33 - \sum_{i=1}^{\infty} \left(\frac{2 \tanh \bar{\lambda}_i}{\bar{\lambda}_i \lambda_i''} \right) + \sum_{i=1}^{\infty} \left(\frac{8 \tanh \bar{\lambda}_i}{\bar{\lambda}_i^3 \lambda_i''} \right) - \\ - \sum_{i=1}^{\infty} \left(\frac{2(4 - \cosh \bar{\lambda}_i)}{\bar{\lambda}_i^2 \lambda_i'' \cosh \bar{\lambda}_i} \right) \Bigg] + mH \left(2\frac{\rho_s s \bar{L}^3}{3\rho_l R H^3} \right) \end{aligned} \tag{5.81}$$

whose variation is depicted in Figure 5.23 without the contribution of the shell mass $\frac{2\rho_s s \bar{L}^3}{3\rho_l R H^3}$; $\bar{\lambda}_i = \lambda_i \frac{H}{R}$, $\lambda_i'' = (\lambda_i^2 - 1)$ and $\lambda_1 = 1.8412$, $\lambda_2 = 5.3314$, $\lambda_3 = 8.5363$, $\lambda_4 = 11.7060$ and $\lambda_5 = 14.8636$ [221, 646] are the roots of the first derivative of J_1 which is the Bessel function of the first kind of order 1 [561]; ρ_l and ρ_s are the mass density of the liquid and of the tank wall's material.

For laterally excited tanks, a simple practical procedure has been proposed by Veletsos and Yang [626, 646] to evaluate the impulsive effects for flexible tanks from the corresponding solution for rigid tanks (Section 5.6 and Equations (5.62)). The procedure consists in replacing the input ground acceleration $a(t)$ in the relevant expressions of the rigid tank solution by $a(t) + a_f(t)$. The latter function refers to a similarly excited simple oscillator, the natural frequency of which is equal to the first impulsive tank-liquid horizontal mode (Section 5.4 with or without soil–structure interaction) and damping value equal to 2% [162, 166, 223, 297] or 1% [161, 162]. The maximum value may then be determined by replacing $a(t) + a_f(t)$ by its spectral value S_e. The same approximation has been proposed, in the case of a rocking response of a tank subjected to an angular acceleration of the base $\ddot{\theta}_g(t)$, and recommended for use in practice by Veletsos and Tang [638] defining the following impulsive expressions (derived from the corresponding solution for rigid tanks, Equations (2.116)):

$$p_i(z, \theta, t) = c_i^{\text{r}}(z)\rho_l R a_i^{\text{r}}(t) \cos \theta \tag{5.82a}$$

$$Q_i(t) = m_i^{\text{r}} a_i^{\text{r}}(t) = m_i \frac{h_i'}{H} a_i^{\text{r}}(t) \tag{5.82b}$$

$$M_i(t) = i_i mH a_i^{\text{r}}(t) \tag{5.82c}$$

$$M_i'(t) = i_i' mH a_i^{\text{r}}(t) \tag{5.82d}$$

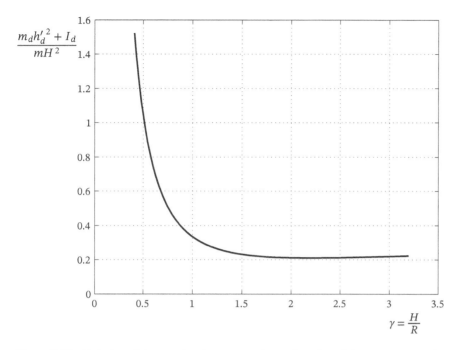

Figure 5.23 Variation of the rotational parameter I_d as a function of the slenderness of the tank.

and convective expressions [638] (identically re-written from the corresponding for rigid tanks, Equations (2.117)):

$$p_c(z, \theta, t) = \rho_l R \cos\theta \sum_{n=1}^{\infty} c_n^{\mathrm{r}}(z) a_{cn}^{\mathrm{r}}(t) \tag{5.83a}$$

$$Q_c(t) = \sum_{n=1}^{\infty} m_{cn}^{\mathrm{r}} a_{cn}^{\mathrm{r}}(t) = \sum_{n=1}^{\infty} m_{cn} \frac{h_{cn}'}{H} a_{cn}^{\mathrm{r}}(t) \tag{5.83b}$$

$$M_c(t) = \sum_{n=1}^{\infty} m_{cn}^{\mathrm{r}} h_{cn} a_{cn}^{\mathrm{r}}(t) \tag{5.83c}$$

$$M_c'(t) = \sum_{n=1}^{\infty} m_{cn}^{\mathrm{r}} h_{cn}' a_{cn}^{\mathrm{r}}(t) \tag{5.83d}$$

where the quantities involved have previously been described in Section 2.2.11. The tank-liquid system investigated is depicted in Figure 5.17 (and also Figure 2.1) and it is presumed to be of uniform thickness and clamped to a rigid base. The impulsive pressure exerted against the tank wall, Equation (5.82a), depends on the instantaneous value of the pseudo-acceleration $a_i^{\mathrm{r}}(t)$ induced by the prescribed horizontal acceleration $H\ddot{\theta}_g(t)$ in a single degree of freedom linear oscillator, with a prescribed damping as suggested by Makris et al. in [359, 360] (Equation (2.120)), and having the fundamental natural frequency of the tank-liquid system, ω_f, given in Section 5.4 for tanks excited laterally (with or without soil–structure interaction):

$$a_i^r(t) = \omega_f \int_0^t (H\ddot{\theta}_g(t)) \sin\left[\omega_f(t-\tau)\right] d\tau \tag{5.84}$$

The maximum value of the impulsive pressure p_i, shear Q_i and moments, M_i and M_i', may then be determined by replacing $a_i^r(t)$ by its rocking spectral value S_e^r [243, 359, 458] using the natural tank-liquid frequency ω_f, given in Section 5.4 for tanks excited laterally (with or without soil–structure interaction) and an equivalent viscous damping value according to the expressions (2.120)–(2.122).

6

Other Peculiar Principles

After reading this chapter you should be able to:

- Understand some peculiar principles in the analysis of tanks of any shape and material
- Evaluate the main characteristics in the case of soil–structure interaction, flow-dampening and base-isolation devices
- Evaluate the main characteristics in the case of underground rigid, horizontal, and conical tanks

6.1 Introduction

This chapter of the book represents a brief overview of some complex topics that have still not been fully investigated. This section summarizes some important considerations and specialist subjects involved in the analysis of liquid storage tanks for earthquakes:

- soil–structure interaction effects (Section 6.2);
- flow-dampening devices (Section 6.3);
- base-isolation devices (Section 6.4);
- underground rigid (Section 6.5), horizontal (Section 6.6), and conical tanks (Section 6.7).

6.2 Effects of Soil–Structure Interaction

It is widely recognized that, because of soil–structure interaction effects, for a specified free-field ground motion, a flexibly supported rigid or deformable tank may have an earthquake response significantly different from that of the same tank when rigidly supported [521] (Point 5.6, FEMA 450). Usually the flexibility of the support will change both the impulsive and convective effects, but the latter is consistently small and can be ignored (Point C3.6, NZSEE-09) [635]. Examination of the results presented by Veletsos et al. in [635, 640, 641] reveals that soil–structure interaction, for both the horizontal and rocking (rotation) modes of excitation, has a twofold effect: it increases the fundamental natural period of the

Seismic Design and Analysis of Tanks, First Edition. Gian Michele Calvi and Roberto Nascimbene.
© 2023 John Wiley & Sons, Inc. Published 2023 by John Wiley & Sons, Inc.

system and generally the corresponding damping factor (Figure 6.1(a)). As would be expected, these modifications increase with increasing flexibility of the soil, or decreasing values of the shear wave velocity v_s (Figure 6.1(b)). Furthermore [625], for a prescribed soil flexibility the increase in the period is more pronounced for tall (slender) tanks than for broad (squat/shallow/short) tanks because the contribution of the rocking foundation component is greater for tall tanks. By contrast, the reduction of the peak response is less pronounced for tall tanks because the radiational damping capacity for foundations in predominantly rocking motions is significantly less than for foundations in horizontal motion [640, 642]. In this approach [626, 635, 640, 641], soil–structure interaction effects for the fundamental mode of vibration are provided for by modifying the period and damping (Figure 6.1(a)) of an equivalent single degree of freedom system and evaluating its response to the prescribed free-field ground motion, considering the tank to be rigidly supported at the base. The effective damping factor ξ_f of the tank–foundation system for horizontal impulsive mode as a function of the shear wave velocity v_s (in m/sec) for several values of $\frac{H}{R}$ in the range between 0.3 and 3.0 is presented in Figures 6.2(a)–6.2(c) for steel tanks ($\frac{s}{R} = 0.0005, 0.001, 0.002$ where s is the uniform thickness of the wall) and in Figure 6.2(d) for a concrete tank ($\frac{s}{R} = 0.01$). In deriving Figures 6.2, it has been assumed a Poisson's ratios for steel and concrete equal to 0.3 and 0.17, respectively; a Poisson's ratio for the supporting soil $v_s = \frac{1}{3}$; in the system foundation damping ξ_s just the radiation damping has been included considering the hysteretic soil material damping to be negligible; a structural damping $\xi_m = 0.02$ in the case of flexible wall and rigid foundation is taken. Using a procedure similar to that described above for the horizontal mode damping values, Veletsos and Tang [636, 637, 639] obtained the vertical mode damping values depicted in Figures 6.3(a)–6.3(b) for steel tanks ($\frac{s}{R} = 0.0005, 0.001$) and in Figure 6.3(c) for a concrete tank ($\frac{s}{R} = 0.01$). Figures 6.2(d) and 6.3(c) are for circular concrete tanks and were derived from Veletsos et al. in [626, 635–637, 639–641]. The effective damping factor for rectangular concrete tanks can be derived from these figures by using an equivalent radius for a circular tank with the same plan area as the rectangular tank (see Section 4.3; Point CA2.1, NZS 3106 (2009)) [180].

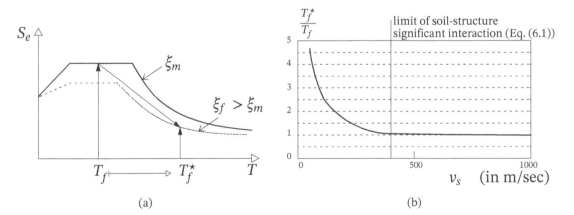

(a)   (b)

Figure 6.1 Soil–structure interaction effects: (a) schematic showing effects of period lengthening and effective damping factor on spectral acceleration [464, 569, 570]; (b) ratio of deformable ($T_f^\star$) and rigid foundation (T_f) first–mode natural period as a function of the shear wave velocity v_s (in m/sec). Source: From [327]/American Society of Civil Engineers.

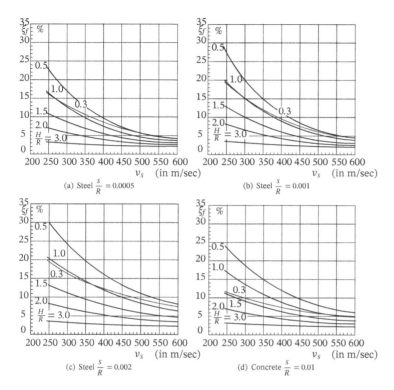

Figure 6.2 Effective damping factor ξ_f of the tank–foundation system for horizontal impulsive mode [635, 640]: (a)–(c) steel tank with $\frac{s}{R} = 0.0005, 0.001, 0.002$ and (d) concrete tank with $\frac{s}{R} = 0.01$. Source: Point 3.2, NZSEE-09.

Similar approaches based on an equivalent mechanical model capable of considering soil–tank interaction were introduced by Fischer and Seeber [167], Haroun and Abou-Izzeddine [218, 219] and more recently by Habenberger and Schwarz in [206, 208, 209].

An indication of the importance of including the soil–structure interaction may be seen from Equation (6.1) which has been derived by Veletsos [625] studying a range of buildings in order to determine those that will have significant foundation compliance by way of inertial interaction. When the following inequality is satisfied in Figure 6.1(b), it is expected there will be negligible interaction effects [327]:

$$\frac{\bar{L}}{v_s T_f} \sqrt[4]{\frac{\bar{L}}{R}} \leq \frac{1}{8} \tag{6.1}$$

where $\bar{L}$ is the length of the wall of the tank and T_f (i.e. Equations (5.34) and (5.49) in the case of cylindrical and rectangular tanks, respectively) is the rigid foundation first–mode natural period for a flexible tank. Figure 6.1(b) shows the limit of significant interaction derived from relation (6.1) and applied to a steel tank with $\frac{H}{R} = 1.5$ [327].

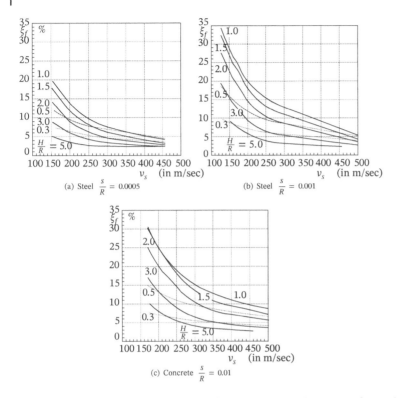

Figure 6.3 Effective damping factor ξ_f of the tank–foundation system for vertical impulsive mode [636, 637, 639]: (a)–(b) steel tank with $\frac{s}{R} = 0.0005, 0.001$ and (c) concrete tank with $\frac{s}{R} = 0.01$. Source: Point 3.2, NZSEE-09.

Using a simplified procedure, originally derived for buildings [625] (Point 5.6.2.1.2, FEMA 450), the effects of inertial soil–structure interaction[1] on the seismic response of flexible tanks can also be quantified, in the case of response spectrum–based demand analyses (Figure 6.1(a)), by the ratio of deformable ($T_f^\star$; Equation (5.59) in the case of cylindrical and rectangular tanks) and rigid foundation (T_f; i.e. Equations (5.34) and (5.49) in the case of cylindrical and rectangular tanks, respectively) first–mode natural period and by the effective damping ratio (ξ_f) of the tank–foundation system [156, 209, 343, 570] (Point A.7.2.3, EC8-4; Eq. (C3.34), Point C3.7, NZSEE-09; Eq. (5.6-9) Point 5.6.2.1.2, FEMA 450):

$$\xi_f = \xi_s + \frac{\xi_m}{\left(\dfrac{T_f^\star}{T_f}\right)^3} \tag{6.2}$$

where ξ_s is the system foundation damping including both the contribution of geometric or radiation and hysteretic soil material damping effects [111]; ξ_m is the structural damping in the case of a flexible wall and rigid foundation; usually $\xi_f > \xi_m$ [464] (Figure 6.1(a)). A closed–form expression of the damping

---

1 The focus of this paragraph is on inertial interaction, which can be a more important effect on foundations without large, rigid base slabs or deep embedding [569]. We do not explicitly consider kinematic interaction.

factor ξ_s including both radiation and hysteretic action is represented in Veletsos and Nair [630, 631] and Veletsos and Verbic [642]. Bielak [36] similarly expressed the effects of inertial interaction for embedded structures in terms of the period lengthening ratio $\frac{T_f^\star}{T_f}$ and the foundation damping factor ξ_s.

Approximate expressions for flexible tanks supported on non-dissipative perfectly elastic material, including just radiation damping [113, 593, 679], may be found in the work of Jennings and Bielak [37, 274] and Veletsos and Meek [392, 628, 629] in the case of horizontally excited tanks (Point A.7.2.3, EC8-4; Eq. (C3.35), Point C3.7, NZSEE-09):

$$\xi_s = \frac{2\pi^2 m_f \alpha}{K_x T_f^{\star 2}} \left(\frac{\beta_x}{\alpha_x} + \frac{K_x h_f^2}{K_\theta} \frac{\beta_\theta}{\alpha_\theta} \right) \tag{6.3}$$

where m_f at height h_f is the flexible tank mode impulsive mass (Equations (5.14) and (5.16)); the frequency parameter α is defined according to relation (2.58) in which T_0 must be substituted by $T_f^\star$; the horizontal translational and rocking stiffness of the foundation may be obtained from expressions (2.56) and (2.57); the dimensionless damping factors, β_x and β_θ [179, 352, 644], for translation and rocking motion are plotted in Figures 6.4(a) and 6.4(b) respectively and given by approximate closed–form expressions (6.6a) and (6.6b) [643]; these coefficients, like the stiffness coefficients α_x and α_θ (Figures 2.11(a) and 2.11(b) or expressions (2.61a) and (2.61b)), depend on the dimensionless frequency parameter α and the Poisson's ratio for the soil v_s.

The effective damping ratio for the case of a horizontally excited rigid tank may be obtained by Equations (6.2) and (6.3) putting $m_f = m_i$, $h_f = h_i$ and $T_f^\star = T_0$ (Point A.7.2.3, EC8-4):

$$\xi_f = \xi_s + \xi_m = \frac{2\pi^2 m_i \alpha}{K_x T_0^2} \left(\frac{\beta_x}{\alpha_x} + \frac{K_x h_i^2}{K_\theta} \frac{\beta_\theta}{\alpha_\theta} \right) + \xi_m \tag{6.4}$$

where m_i is the impulsive mass (Equation (2.10) or (2.25a)–(2.25b)), h_i is the distance from the tank base to the point of concentration of the impulsive liquid mass (Equation (2.15) or (2.25c)/(2.25d) and (2.26a)), and T_0 is the impulsive period of vibration of the tank–foundation system including the soil–structure interaction (Equation (2.55)).

Recommended levels of damping for the horizontal impulsive modes are given in Table 6.1, as derived by Whittaker and Jury in [666]: the total effective damping factor is made up of the tank–liquid system fixed-based damping plus the foundation radiation damping. Unless a more rigorous analysis is undertaken, Point C2.8 in the "Seismic Design of Storage Tanks", Recommendations of a Study Group of the New Zealand National Society for Earthquake Engineering, first edition, December 1986 by Professor M.J.N. Priestley, suggests using 7.5% for an effective damping factor, for the vertical impulsive mode, in the case of soft soil and 5% in the case of firm soil and rock.

The effective damping ratio for the case of a vertically excited rigid tank may be obtained by Point A.7.2.3, EC8-4:

$$\xi_f = \xi_s + \xi_m = \frac{2\pi^2 m_V \alpha}{K_V T_V^2} \frac{\beta_V}{\alpha_V} + \xi_m \tag{6.5}$$

where $m_V = m + m_b + m_w + m_r$, with m equal to the total mass of the liquid, m_b, m_w and m_r are the base, wall and roof mass, respectively; T_V is the vertical period of vibration of the tank foundation system including the soil–structure interaction (Equation (2.59)); α_V is the dimensionless factor (Equation (2.60)), required to convert the static stiffness value to a dynamic value and may be obtained

Table 6.1 Effective damping factor ξ_f for horizontal impulsive mode as a function of the liquid height H and radius R, tank wall thickness at base s and v_s foundation soil shear wave velocity averaged over a depth of $2R$ beneath the tank base.

Tank type	$\frac{H}{R}$	Effective damping factor (%)		
		$v_s = 1000$ m/s	$v_s = 500$ m/s	$v_s = 200$ m/s
Concrete and steel with $\frac{s}{R} = 0.002$ (rigid)	0.5	4	13	30
	1	4	10	20
	2	3	5	14
	3	2	3	7
Steel with $\frac{s}{R} = 0.001$	0.5	3	7	20
	1	3	6	15
	2	2	3	6
	3	2	2	3
Steel with $\frac{s}{R} = 0.0005$ (flexible)	0.5	2	4	12
	1	2	4	9
	2	2	3	4
	3	2	2	3

Source: Data from [666].

from Figure 2.11(c) or expression (2.61c) as a function of α given by relation (2.58) in which the period T_0 must be substituted by the vertical period T_V; the vertical stiffness of the foundation K_V may be derived from expression (2.60); the dimensionless damping factor for vertical motion β_V [179, 352, 644] is plotted in Figure 6.4(c) and given by an approximate closed–form expression (6.6c) [643]. In the case of a vertically excited flexible tank, the effects may be estimated [626] by application of the concepts described above for a rigid tank, mainly using expressions (5.60), (6.2), and (6.5).

Approximate closed–form expressions of the dimensionless factors, β_x, β_θ and β_V for translational, rocking and vertical damping, respectively, have been presented by Veletsos and Verbic in [643]:

$$\beta_x = \begin{matrix} 0.775 \\ 0.65 \\ 0.60 \end{matrix} \qquad \begin{matrix} v_s = 0 \\ v_s = 1/3 \\ v_s = 0.45 \text{ and } 0.5 \end{matrix} \tag{6.6a}$$

$$\beta_\theta = b_1 b_2 \frac{(b_2\alpha)^2}{1+(b_2\alpha)^2} \tag{6.6b}$$

$$\beta_V = b_4 + b_1 b_2 \frac{(b_2\alpha)^2}{1+(b_2\alpha)^2} \tag{6.6c}$$

where b_1, b_2, and b_4 are dimensionless functions of α (Equation (2.58)) and v_s as shown in Table 2.1.

In the case of elevated tanks (see Chapter 4), according to Point 5.6.2.1.2, FEMA 450 and evidence from Livaoğlu and Doğangün [343], Equation (6.2) could be used for the effective damping factor of the

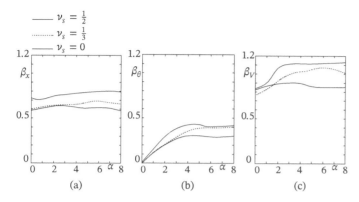

Figure 6.4 Dimensionless damping factors β_x, β_θ and β_v of rigid circular footings on homogeneous half-space [179, 352, 644, 662] used for the evaluation of system foundation damping in the case of soil–structure interaction.

elevated tank foundation system ξ_f, where ξ_s remains the system foundation damping, including both the contribution of geometric or radiation and hysteretic soil material damping effects, whereas ξ_m becomes the structural damping of the fixed base elevated tank (rigid foundation); T_f should be taken to be equal to the natural period of the fixed base elevated tank T_S (Equation (4.5) and Point C4.3.1.3, IITK-GSDMA 2007):

$$T_S = 2\pi \sqrt{\frac{m_i + \Delta m}{K_{\text{equiv}}}} \tag{6.7}$$

and $T_f^\star$ is the modified period of the structure $T_S^\star$ that verges on the flexibility of the supported system, and can be approximately estimated by relation (5.59):

$$T_S^\star = T_S \sqrt{1 + \frac{K_{\text{equiv}}}{K_x} \left[1 + \frac{K_x (h_i + \bar{H})^2}{K_\theta} \right]} \tag{6.8}$$

where $\bar{H}$ (Figure 4.7) is the height of the support staging, h_i the impulsive height (Equation (2.15) or (2.25c)–(2.25d), (2.26a) and (2.131c)–(2.131d)) and K_{equiv} is the equivalent stiffness of the impulsive system (first mode of vibration) and may be estimated with the following relations[2] [88, 342]:

$$K_{\text{equiv}} = \omega_2^2 M_{\text{equiv}}^\star \tag{6.9a}$$

$$M_{\text{equiv}}^\star = \frac{L_{\text{equiv}}^2}{M_{\text{equiv}}} \tag{6.9b}$$

$$L_{\text{equiv}} = (m_i + \Delta m)\phi_{11} + m_{c1}\phi_{21} \tag{6.9c}$$

$$M_{\text{equiv}} = (m_i + \Delta m)\phi_{11}^2 + m_{c1}\phi_{21}^2 \tag{6.9d}$$

---

2 For elevated tanks with moment-resisting type frame staging, the lateral stiffness can be evaluated by a computer analysis or by simple procedures [500, 501], or by the established structural analysis method.

where $\phi_{11} = \phi_1, \phi_{21} = 1$ and ω_2 comes from the two coupled masses model suggested in Section 4.4; Δm is the mass of the empty container plus one–third the mass of staging ($m_r + m_w + m_b$ in Figure 2.10 is the mass of the tank structure alone composed of the mass of the roof m_r, wall m_w, and base m_b); m_i is the impulsive mass of liquid which takes part in the lateral impulsive mode of vibration and may be derived from Equation (2.10) (or (2.25a)–(2.25b)) for cylindrical tanks and from expression (2.131a) for rectangular tanks.

6.3 Flow-Dampening Devices

The earthquake forces and moments in a tank are mainly caused by the structural mass of the tank itself and by the contents being accelerated by the horizontal and rocking ground motion [419]. As observed in Chapter 2, by solving tanks with increasing radius R, the hydrodynamic forces and moments, as well as wave heights (d_{max}), caused by the contents in fully filled tanks (or almost full) are much greater than the forces caused by the structure. If the liquid is allowed to slosh freely, this periodic motion may lead to structural failure, such as diamond-shaped buckling, "elephant foot" bulge, buckling at the top of the shell, connection and joint failures in many cases due to the impact of the wave on the roof (Section 2.2.8) determining loss of contents [251]. In an attempt to avoid this undesirable dynamic behaviour, it is important to introduce some means to suppress or to reduce the sloshing dynamic loads [250]. Hence, slosh-suppression devices should be used to damp the liquid motions and prevent this kind of damage to the tank [525]. Furthermore, these devices, or baffles, are used to reduce the structural forces induced by the sloshing of the liquid, to control the liquid position within a tank and to protect the roof, the tank bulkheads or other structures, against the liquid impact caused by the motion of the fluid. They act by changing the fluid's natural frequencies, by increasing the damping factor ξ (Equation (2.62)) by a damping ratio ξ_{rigid} (or ξ_{flex}), depending on their type (rigid or flexible), shape, size, and position [122, 123, 181, 250, 525]. Slosh-suppression devices damp the liquid by transforming the smooth flow along the tank's wall into a high speed turbulent flow around the devices. As clearly highlighted in Section 2.2.6, slosh damping, resulting only from the liquid viscosity in various tanks without baffles, could be sufficiently great that suppression devices may not be required or their design may be less critical [3]. Some common types of anti-slosh devices have been classified by Abramson in [3] and by Ibrahim in [250]:

1) Ring (or annular) horizontal baffles (Figure 6.5(a)) which can be rigid (Figures 6.6(a)–6.6(c)) [41, 42, 176, 177, 305, 326, 398, 529] or uniformly flexible cantilevered from the wall (Figure 6.6(d)) [42, 43, 120, 549, 568], full (Figures 6.6(a)–6.6(b)) or perforated section (Figure 6.6(c); usually called "orifices"), fixed and connected to the wall (Figures 6.5(a) and 6.6(a)–6.6(d)) or fixed at a certain height with a radial clearance measured from the wall (Figure 6.5(b)) [549], hinged-spring baffles (Figure 6.6(e); hinged to the wall and spring-loaded,) or slamming baffles (Figure 6.6(f); free to translate perpendicular to its plane between rigid stops) [518]. A system of ring baffles, fitted around the internal periphery of the tank, is preferred over a single submerged ring because, by using multiple rings, the minimum damping ratio can be controlled by properly spacing the baffles, and because if one of the baffles should fail, say, the one nearest the free surface, other submerged baffles will still remain to damp the liquid [326]. Flexible baffles, in general, seem to provide higher damping values than rigid baffles of the same

width and appear to be most efficient when the device's natural frequency equals the liquid's slosh frequency [3]. The damping effectiveness of perforated annular baffles is largely dependent upon the perforation hole size and the percentage area removed by perforation [172]. A suggested range for a perforated ring is around 20–25% of the area removed [3,4].

2) Conical baffles which could be placed upright (Figure 6.5(c)) or inverted (Figure 6.5(d)) [186, 549]. Usually the effectiveness of the conical inverted baffles is less than that of either the fixed ring or upright conical section of comparable size [549].

3) Complete 90°-, 60°- or 45°-sectored baffles (Figure 6.5(e)–6.5(g)) [661] or cruciform baffles as shown in Figure 6.5(h) [250]. If it is necessary to considerably shift the liquid's resonant frequencies into a more desirable range and, at the same time, to reduce the magnitudes of the sloshing masses, an attractive configuration is to compartmentalize the tank into sectors [3]. In order to reduce weight, the sectors can be perforated, allowing a portion of the liquid to flow between the various compartments [326]. Cruciform baffles are located physically in the same manner as stringers, so that for cylindrical tanks, there is the advantage of damping that is independent of the liquid height; however, cruciform baffles generally provide only a relatively small amount of damping [3].

4) A combination of ring and sectored baffles both described in previous Points 1. and 3. as well as in Figure 6.7 [326]; annular and sectored baffles could be also perforated [4].

5) Floating cans, floating lid devices, or floating porous mats [11, 91, 92, 121, 403, 528, 547, 661]. These movable devices or devices that act at the liquid's free surface (floating objects) have been shown to be either less effective than previously defined conventional baffles or to involve increased complexity.

Generally, it can be observed that, for the various configurations described, the location of the devices to substantially increase damping is limited to a narrow range near the liquid surface [549].

Many detailed experimental, and in a few cases also analytical, studies have been conducted to determine the damping due to various types of baffles present in tanks of various shapes [3, 44, 181, 230, 397, 519, 529, 549, 566–568] and many have been devoted to achieving the optimum design of an anti-slosh device [2, 55, 239, 326, 518, 528]. However, Miles [398], based on the work of Keulegan and Carpenter [298], derived a semi-empirical relations for damping ratio ξ_{rigid} (in %) in terms of tank and baffle dimensions and sloshing height amplitude for cylindrical tanks, with one annular rigid ring baffle (Point 1. and Figure 6.5(a), 6.6(a) and 6.8(a)), when the liquid height is considerably greater than the tank radius ($\frac{H}{R} > 1$):

$$\xi_{\mathrm{rigid}} = \left[2.83 e^{-4.60 \frac{d_s}{R}} a^{\frac{3}{2}} \sqrt{\frac{\eta_w}{R}} \right] \cdot 100 \tag{6.10}$$

where d_s (Figure 6.8(a)) is the distance between the baffle and the free level surface, R is the radius of the tank and η_w is the maximum amplitude of liquid wave height at the wall during one cycle ($\lambda_1 = 1.8412$, which corresponds to the first antisymmetrical mode in Figure 2.6(a); during the design stage it could be assumed to be equal to $0.1R$ [120] or $0.2R$ [119]; during the verification or validation stage Equation (2.87) can be used); a is the fraction of the tank area (πR^2) covered by the annular baffle ($\pi R^2 - \pi (R - w)^2$) (the fractional part of the cross-sectional area of the tank that is blocked by the baffle) [3, 549]:

$$a = \frac{R^2 - (R - w)^2}{R^2} \tag{6.11}$$

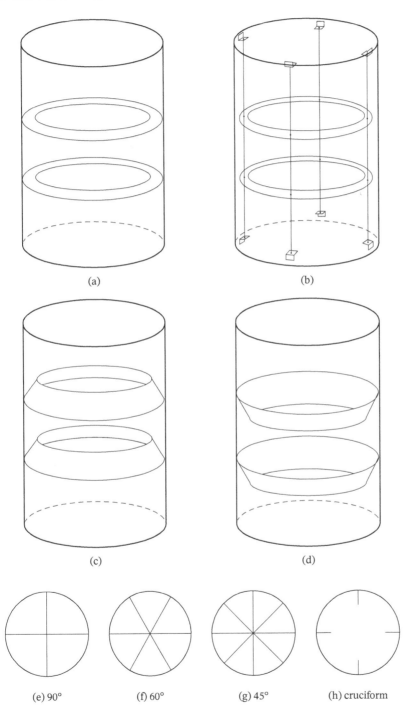

(a)

(b)

(c)

(d)

(e) 90°   (f) 60°   (g) 45°   (h) cruciform

Figure 6.5 Sloshing suppression devices: (a) ring horizontal rigid baffles fixed at the wall; (b) annular baffles fixed at a certain height with a radial clearance; (c) and (d) conical baffles and (e)–(h) sectored and cruciform devices [3, 250].

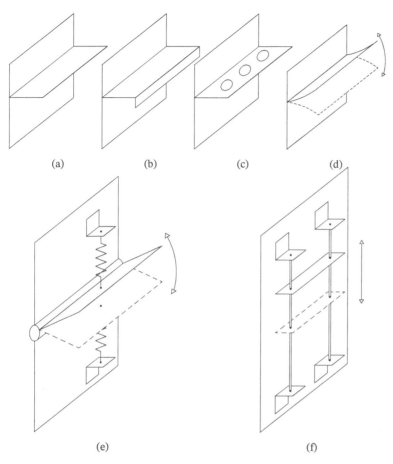

Figure 6.6 Rigid (a)–(c) and flexible baffles (d); (e) hinged-spring baffles and (f) slamming baffles free to translate perpendicular to its plane between rigid stops. Source: Adapted from [518].

where w is the plate baffle width (Figure 6.8(a)). The contribution given by Equation (6.10) should be added to the contribution given by ξ in Equation (2.62) and due to the viscosity of the liquid in the tank without anti-slosh devices [101, 103]. According to Abramson [3], the total damping provided by a series of annular j^{th} baffles may be obtained by a linear superposition of expression (6.10), although usually only those baffles near the liquid free surface (less than about $\frac{2}{3}$ of the tank radius below the liquid level; Figure 6.8(b)) contribute significantly to damping. The total damping contributions given by the jth baffles can be written as [55, 250, 326]:

$$\xi_{\text{rigid}} = \left[2.83 a^{\frac{3}{2}} \sqrt{\frac{d'_s}{2R}} \cdot 100 \cdot \sum_{j=1}^{N} e^{-4.60 \left[\frac{d_s + (j-1)d'_s}{R} \right]} \right] \tag{6.12}$$

where d'_s, depicted in Figure 6.8(b), is the vertical spacing between baffles; according to Langner [326], a design slosh height $\eta_w = \frac{d'_s}{2}$ has been assumed. Furthermore, if the spacing between the baffles becomes

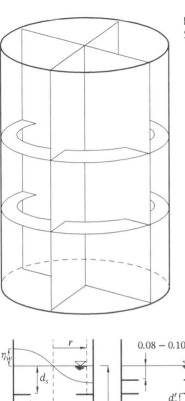

Figure 6.7 A combination of ring and sectored baffles. Source: Adapted from [326].

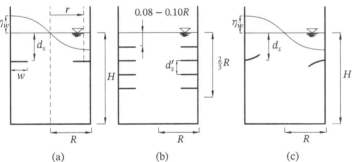

Figure 6.8 Tank and baffle system geometry: (a) cylindrical tank with a single rigid baffle, (b) a series of rigid baffles and (c) one single flexible baffle.

less than the baffle width w, the superposition rule does not accurately predict the damping [239], but as suggested by Hosseinzadeh et al. [239], a distance $\leq 0.2R$ can be considered a good design principle spacing of the ring baffles in the commonly used cylindrical tanks. To establish the proper position of the baffles, the following suggestions should be considered [3, 55, 119, 120, 250, 326, 549, 550]:

- uniform spacing d'_s between baffles [326] such that $\frac{d'_s}{R} \leq 0.2$ [3, 120];
- only those baffles positioned less than about $\frac{2}{3}$ of the tank radius below the liquid level (Figure 6.8(b)) can be considered effective [3];
- the first submerged baffle above the liquid surface is effective only when its distance from the mean liquid free surface is in the range 0.08–0.10R [3, 549] or 0.55d'_s [55] or 0.125R [4, 172].

In developing his damping Equation (6.10), Miles [398] ignored the effects of the Reynolds number, but he did incorporate the following assumption in his model [3, 56, 326, 549]:

- the fluid is oscillating in its fundamental mode, that is the first antisymmetrical 1st-sloshing mode (Figures 2.6(a) and 6.8(a)) [250];
- the slenderness ratio $\frac{H}{R} > 1$ [56, 250, 398, 550];
- the linearized potential theory accurately describes the flow except in a small region close to the ring and consequently the local flow in the region of the ring is unaffected by the free surface or the tank bottom. These assumptions require that $(H - d_s) > w$ and $(H - d_s) > \eta_w$ [56, 550] where w is the plate baffle width (Figure 6.8(a));
- $a \ll 1$ [398, 572].

Buchanan and Lott in [56] investigated the effect of the Reynolds number on damping; Isaacson and Premasiri [254] estimated the hydrodynamic damping of horizontal and vertical plate baffles in a rectangular tank; as opposed to Miles [398], who developed expression (6.10) for cylindrical tanks in which the liquid height is considerably greater than the tank radius, Maleki and Ziyaeifar in [362] suggest the following relation without any limitations on tank dimensions:

$$\xi_{\text{rigid}} = \left[4\left(\frac{w}{R}\right)^{\frac{3}{2}} \left(2 - \frac{w}{R}\right) \sqrt{\frac{\eta_w}{R}} \left(\frac{\sinh\left(1.841\frac{H-d_s}{R}\right)}{\sinh\left(1.841\frac{H}{R}\right)}\right)^{\frac{5}{2}} \tanh\left(1.841\frac{H}{R}\right) \right] \cdot 100 \tag{6.13}$$

For values $\frac{H}{R} > 1$ there is a conformity between expressions (6.10) and (6.13), but for $0 < \frac{H}{R} \leq 1$, Equation (6.13) shows quantities less than Miles' relation (6.10). A good comparison between values presented by Miles [398], Maleki and Ziyaeifar in [362] and a consistent experimental campaign on single baffled tank models, has been done by Hosseinzadeh et al. [239].

Assuming that only the 1st-sloshing mode (Figure 2.6(a)) of the fluid oscillation contributes to the pressure p on a rigid annular baffle, its value may be determined according to the expression derived from Buchanan in [55]:

$$p(r) = 1.7506 \, r^2 \frac{\rho_l g}{R} \left(\frac{\eta_w}{R}\right)^2 \tag{6.14}$$

where ρ_l is the mass density of the liquid and g the gravitational acceleration; r is the radius from the tank's centreline to a point on the baffle (Figure 6.8(a)). More refined expressions to evaluate pressure and forces on a rigid baffle and comparison with experimental tests available can be found in the work of Silverman and Abramson [550] and Garza [171].

Although a system of rigid ring baffles is effective in attenuating sloshing effects, they usually comprise a high percentage of the tank weight. Several experimental studies [3, 549, 568] demonstrated that flexible ring baffles give a consistent advantage in increasing damping effectiveness and reducing the baffles' weight. The damping factor ξ_{flex} associated with a flexible annular baffle (Figures 6.6(d) and 6.8(c)) [566, 568]) has been determined by Schwind et al. in [519]:

$$\xi_{\text{flex}} = \frac{1.841}{0.553} a \frac{\eta_w}{R} \left[\frac{\sinh\left(1.841\frac{H-d_s}{R}\right)}{\sinh\left(1.841\frac{H}{R}\right)}\right]^3 f(\mathsf{P}, \mathsf{F}) \cdot 100 \tag{6.15}$$

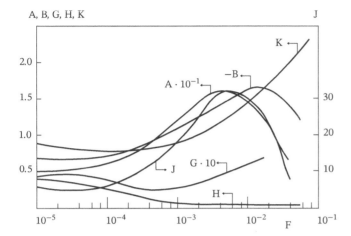

Figure 6.9 Flexibility functions A, B, G, H, K and J as a function of the flexibility parameter F. Source: From [519]/American Institute of Aeronautics and Astronautics.

where quantities d_s, η_w and R are depicted in Figure 6.8(c); $f(P, F)$ is a function with the following expression [519]:

$$f(P, F) = P^B A \int_0^{\frac{\pi}{2}} (\cos\theta)^{B+3} d\theta - J \int_0^{\frac{\pi}{2}} e^{-PK\cos\theta} \cos^3\theta d\theta + H \int_0^{\frac{\pi}{2}} e^{\frac{-1}{P^2 G^2 \cos^2\theta}} \cos^3\theta d\theta \qquad (6.16)$$

The flexibility functions A, B, G, H, K and J are shown in Figure 6.9 as a function of the flexibility parameter F; in the case of a rigid baffle $F = 10^{-5}$ and hence from Figure 6.9 we derive the constants $A = 5.2$, $B = -0.67$, $G = 0.04$, $H = 0.41$, $K = 0.92$ and $J = 7$. The damping of a rigid baffle, for a given R and η_w, thus depends only upon w and d_s while for a flexible baffle, the damping depends not only upon w and d_s, but also upon the thickness t and the elastic modulus E. It can be seen from Figure 6.10 [120, 568] that the flexible baffle damping ξ_{flex} is a multiple of the rigid baffle damping ξ_{rigid}, with baffles of the same width and under similar flow conditions, as a function of the period parameter P. In Equation (6.16), the non-dimensional period parameter[3] P and the flexibility parameter[4] F assume the values suggested by Dodge in [120] and by Stephens and Scholl in [568] (Poisson's coefficient $v = 0.3$):

$$P = 2\pi \frac{\eta_w}{w} e^{-1.841 \frac{d_s}{R}} \qquad (6.17a)$$

$$F = 0.0423 \frac{\rho_l g R}{E} \left(\frac{w}{R}\right)^5 \left(\frac{R}{t}\right)^3 \qquad (6.17b)$$

where E is the modulus of elasticity of baffle material (in N/m^2), ρ_l is the mass density of the liquid (in kg/m^3), t is the baffle's thickness (in metre) [102] and g is the acceleration due to gravity (in m/sec^2).

The average pressure across the width of a flexible baffle, derived from Dodge in [120], is:

$$p = 5.9 \rho g R \left(\frac{w}{R}\right)^{0.485} \left(\frac{\eta_w}{R}\right)^{1.515} e^{-2.79 \frac{d_s}{R}} \qquad (6.18)$$

---

3 It may be physically interpreted as the maximum distance travelled by a liquid particle at the antinode divided by the baffle width [568].

4 It may be defined as the deflection or flexibility characteristics of cantilever baffles; hence the parameter may be interpreted as the baffle deflection for a period parameter P of unity divided by the baffle width [568].

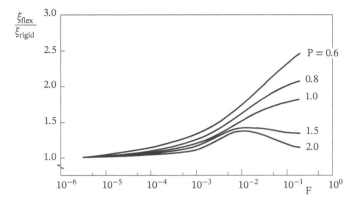

Figure 6.10 Damping ratio between flexible and rigid ring baffle $\frac{\xi_{flex}}{\xi_{rigid}}$ as a function of flexibility F and period P parameters. Source: Adapted from [120].

The ability of a flexible baffle to withstand the loads developed during the sloshing is derived primarily from the membrane stresses caused by the deflection of the baffle. Moving from a stress analysis for a circular ring flexible baffle, based on membrane theory, Dodge [120] predicted that the maximum stress should not exceed $\frac{f_y}{\gamma}$ where $\gamma > 1$ is a safety factor (UNI ENV 1993-1-1:2005) and that the corresponding minimum thickness should be evaluated according to the following expression:

$$t_{min} = 0.57R\sqrt{\frac{w}{R}}\left(\frac{\gamma}{f_y}\right)^{\frac{3}{2}}\sqrt{Ep} \tag{6.19}$$

where f_y is the nominal value of yield strength (Table 3.1, Point 3, UNI ENV 1993-1-1:2005) and E is the modulus of elasticity of the baffle material.

Flow-dampening devices, described above, can be successfully used in damping out the slosh amplitude of a liquid in a tank, but taking well in consideration that the sloshing frequency of a tank with internal anti-slosh devices is considerably higher than that of a tank without internals [355]. Hence the higher sloshing frequency reduces the sloshing wave height d_{max} on the free surface, but again increases the dynamic pressure in the fluid and consistently modifies the mass participation in the impulsive and convective mode [250, 355, 438]. Hence, the presence of devices can significantly alter the dynamic characteristics of the sloshing motion and in order to be able to investigate different solutions to mitigate liquid sloshing and to find the optimal selection of the variables involved (damping ratio, wave height, masses, frequencies, pressure and forces), an advanced numerical model can be built [251]:

- using "grid-based" methods such as the finite difference [13, 76, 77], the finite element method [43, 84, 115, 313, 358, 404, 525, 668, 674] often using the Arbitrary Lagrangian Eulerian (ALE) approach [33, 85, 234, 650], the volume of fluid method [235, 338, 339], and the boundary element method [153, 154, 182];
- adopting a meshfree approach such as the moving particle semi-implicit method [81, 248, 316, 592, 598, 681, 682, 685] or the smoothed particle hydrodynamics [340, 526, 527, 650];
- applying an analytically oriented approach, which provides accurate approximations of natural frequencies and modes, proposed by Gavrilyuk et al. [176, 177] in tanks with annular baffles;

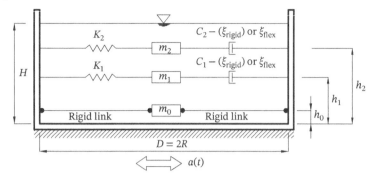

Figure 6.11 Equivalent mechanical model for response of contents of a tank including the effects of anti-slosh devices by inserting a mass–spring–dashpot system.

- creating an equivalent mechanical model such as the one proposed by Dodge [121] (Figure 6.11) or Ebrahimian et al. [138] where the only difference from the classical model in Figure 1.20 is the presence of a mass–spring–dashpot system.

6.4 Base-Isolation Devices

As clearly described in Sections 1.5, 4.1, and 5.5, liquid storage tanks have suffered substantial damage, including cracking at the corner of the bottom plate, compression buckling of the tank's wall, sliding of the base, anchorage failure, sloshing damage around the roof, failure of the piping system connected to the tanks, resulting in loss of functionality, fires, or environmental contamination due to the leakage of hazardous chemicals or flammable materials. Consequently, protection of storage tanks against severe seismic events is a crucial challenge [1]. Options available to the designers could be [16]:

- a thicker wall and anchorage at the base against lateral, vertical, and rocking movement, maintaining the selected tank geometry. Usually this is an excellent solution for water and oil storage tanks, however, it is often unfavourable for tanks storing liquids at cryogenic conditions or liquefied natural gas (LNG) tanks.
- a low aspect ratio modifying the tank geometry. This is often a costly alternative since it results in an uneconomical tank shape and uneconomical use of the site area.
- seismic isolation. This is an advanced technology that had consistently demonstrated its effectiveness in improving seismic performance of buildings, bridges and mission-critical facilities. Base-isolation devices have been recognized, simply by comparing the effects of a fully fixed leg-supported (Figure 5.15(d)) vs. a simply supported tank (Figure 5.15(f)), as one of the promising alternatives for protecting liquid storage tanks against severe earthquakes, alone or in combination with dampening devices. The main concept in isolation systems applied to tanks, curved surface sliding or elastomeric bearings, is to increase the fundamental period of structural vibration, to provide an additional means of energy dissipations, to reduce both the impulsive force transmitted to the walls and the sloshing wave heights to avoid an impact on the roof.

Hence, base-isolation systems are a cost-effective solution to protect very large storage tanks, such as LNG (liquefied natural gas) tanks that represent critical structures with stringent code-design requirements under accidental and earthquake conditions [299]. Herein, three main examples of worldwide LNG seismic isolated tanks are reported [16, 665]:

- the first project is located in Inchon on the west coast of South Korea where three storage tanks, each having a capacity of $100000\,m^3$, were built in 1997. The inner tank height and diameter are 30 m and 68 m, respectively. Due to poor soil conditions at the site, a pile foundation system is used. The isolation system consists of steel-laminated rubber bearings with a diameter of 600 mm and an overall height of 228 mm. The design isolation period is around 3 seconds. The design of a safe shutdown earthquake (SSE) has a maximum horizontal acceleration of 0.2g.
- the second example is located on Revithoussa Island in Greece where two tanks, each with a capacity of $65000\,m^3$, have been realized; the tanks are of the full containment type, with a 9% nickel steel unanchored inner tank of 65.7 m diameter and 22.5 m height [581, 582]; the isolation system consists of 212 isolators that are supported on a pedestal resting on the foundation slab (Figure 6.12); the isolators consist of Friction Pendulum System (FPS) bearings with a radius of curvature of 1880 mm and displacement capacity of 300 mm. The earthquake was defined in terms of elastic 5% damped spectra. Two levels were specified, the operating basis earthquake (OBE) and the safe shutdown earthquake (SSE). The SSE has a peak ground acceleration of 0.48g and spectral values of 0.61g and 0.29g at periods of 1.0 and 3.0 seconds respectively.
- the third LNG tank is located in Melchorita in Peru with a capacity of $130000\,m^3$ and supported using 256 isolators (Figure 6.13).

A comparative study of the performance of various isolation systems[5] for liquid storage tanks has been performed under real earthquake motions by Shrimali and Jangid in [539]. A summary of scientific research on seismic isolation of fluid storage tanks is presented in Table 6.2, as investigated by Panchal and Soni in [437]. A study on the sloshing of tanks subjected to horizontal seismic excitation considering simultaneously the effects of internal flow-dampening devices (Section 6.3) and external seismic isolations was done by Ma and Chang in [354], Maleki and Ziyaeifar in [361] and Shekari et al. in [530].

Based on the equivalent mechanical model proposed by Haroun, Housner, and Ellaithy [214, 215, 221, 223] (described in Figure 5.18 and Section 5.6.1), the hydrodynamic forces being exerted on the flexible isolated tank wall are composed, as consistently motivated in Section 1.7, of three components: (1) the convective component caused by the portion of the liquid sloshing in the tank (m_c); (2) the impulsive flexible component (m_f); and (3) the component associated with the rigid body or uniform component (m_d). The model considered for the base-isolated cylindrical liquid storage tanks is shown in Figure 6.14(b) in which a general isolation system is installed between the base and the foundation of the tank. The equations of motion of the base-isolated liquid tank system subjected to horizontal ground excitations can be written as follows [304, 436, 545, 546, 618]:

$$\begin{bmatrix} m_c & 0 & m_c \\ 0 & m_f & m_f \\ m_c & m_f & m_{\text{eff}} \end{bmatrix} \begin{bmatrix} \ddot{x}_c \\ \ddot{x}_f \\ \ddot{x}_b \end{bmatrix} + \begin{bmatrix} c_c & 0 & 0 \\ 0 & c_f & 0 \\ 0 & 0 & c_b \end{bmatrix} \begin{bmatrix} \dot{x}_c \\ \dot{x}_f \\ \dot{x}_b \end{bmatrix} + \begin{bmatrix} k_c & 0 & 0 \\ 0 & k_f & 0 \\ 0 & 0 & k_b \end{bmatrix} \begin{bmatrix} x_c \\ x_f \\ x_b \end{bmatrix}$$

$$= -\begin{bmatrix} m_c & 0 & m_c \\ 0 & m_f & m_f \\ m_c & m_f & m_{\text{eff}} \end{bmatrix} \begin{bmatrix} 0 \\ 0 \\ \ddot{u}_g \end{bmatrix} \tag{6.20}$$

---

5 The various isolation systems considered are: laminated-rubber bearings, lead-rubber bearings, a pure-friction system, the friction pendulum system, and a resilient-friction base isolator.

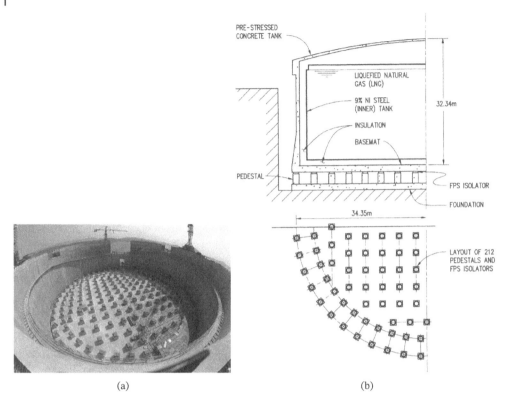

(a)   (b)

Figure 6.12 Overhead view of the isolation system and cross-section of the LNG tanks in Revithoussa, Greece. Source: From [16]/American Society of Civil Engineers, [108]/GeoScienceWorld, [581]/Eshop designed and developed by Pontemedia, [582], and [665].

(a)   (b)

Figure 6.13 Overview of the isolation bearings in Melchorita (Peru). Source: Courtesy of Earthquake Protection Systems, USA.

Table 6.2 Brief summary of scientific literature regarding seismic isolated tanks: Laminated Rubber Bearing (LRB), High Damping Rubber Bearing (HDRB), Lead-rubber bearing (NZ), Pure-Friction system (P-F), Friction Pendulum System (FPS), Variable Friction Pendulum System (VFPS), Electricité de France system (EDF), Vertical Isolation System (VIS), Sliding Concave Bearings (SCB) and Resilient-Friction Base Isolator (R-FBI).

Reference	Tank type	Isolation system	Motion
[48]	ground supported cylindrical steel	NZ	Unidirectional
[366, 367]	ground supported cylindrical steel	Ring of vertically soft rubber bearing and ring of horizontally flexible bearing	Uni-bidirectional
[531]	elevated cylindrical steel	LRB	Unidirectional
[441, 442]	pool-type rectangular concrete	HDRB	Unidirectional
[660]	ground supported cylindrical	FPS	Unidirectional
[303]	ground supported cylindrical	Horizontal $3.3 \cdot 10^5$ kg/m and vertical $3116.66 \cdot 10^5$ kg/m stiffness	Unidirectional
[539–546]	ground supported and elevated cylindrical steel	LRB, NZ, P-F, FPS, R-FBI, EDF	Uni-bidirectional
[83]	ground supported cylindrical steel	NZ	Unidirectional
[266]	ground supported cylindrical steel	LRB, NZ, FPS, R-FBI, EDF	Unidirectional
[191]	ground supported cylindrical steel and concrete	HDRB and NZ	–
[436]	ground supported cylindrical steel	FPS and VFPS	Unidirectional
[93]	ground supported cylindrical steel	FPS and NZ	Unidirectional
[228]	ground supported rectangular concrete	FPS and NZ	Uni-bidirectional
[472, 496]	ground supported cylindrical steel	FPS and NZ	Bidirectional
[282, 425]	ground supported cylindrical steel	FPS, NZ, VIS	Unidirectional and vertical
[107]	ground supported cylindrical steel	SCB	Unidirectional

or in matrix form:

$$\mathbf{M}\ddot{\mathbf{x}} + \mathbf{C}\dot{\mathbf{x}} + \mathbf{K}\mathbf{x} = -\mathbf{M}\mathbf{r}\ddot{u}_g \tag{6.21}$$

In Equation (6.21), $\mathbf{r} = \begin{bmatrix} 0 & 0 & 1 \end{bmatrix}^T$ is the unit influence coefficient vector and $\mathbf{x} = \begin{bmatrix} x_c & x_f & x_b \end{bmatrix}^T$ are the relative displacements as follows:

$$x_c = u_c - u_b \tag{6.22a}$$

$$x_f = u_f - u_b \tag{6.22b}$$

$$x_b = u_b - u_g \tag{6.22c}$$

where x_c is the displacement of the sloshing mass relative to bearing displacement, x_f is the displacement of the flexible impulsive mass relative to bearing displacement, and x_b is the displacement of the

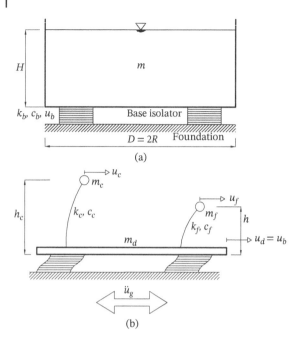

Figure 6.14 (a) Isolated model of liquid storage tank and (b) the corresponding equivalent mechanical; the base-isolated tank system has three degrees of freedom (u_c, u_f and $u_d = u_b$) under unidirectional earthquake excitation.

bearing relative to the ground; $\ddot{u}_g$ is the earthquake ground acceleration, $\dot{u}_g$ velocity and u_g displacement; ^{T} denotes the transpose and over-dots indicate derivative with respect to time. The base-isolated tank system in Figure 6.14(b) has three degrees of freedom under uni-directional earthquake excitation: these degrees of freedom are denoted by u_c, u_f and u_b which denote the absolute displacements of sloshing, flexible, and rigid masses respectively.

All the terms in expression (6.20) can be derived as follows:

- the total contained mass of the fluid $m = \rho_l \pi R^2 H$ where ρ_l is the mass density of the tank liquid (in kg/m^3);
- the impulsive mass m_d (Equation (5.68));
- the first mode convective mass $m_c = m_{c1}$ may be derived from Equation (2.32) for cylindrical tanks; the natural frequency ω_{c1} (Equation (2.30)) corresponds to the convective period of vibration of the 1st-sloshing mode of vibration T_1 of the liquid (Equation (2.54) for cylindrical tank); the equivalent stiffness of the spring k_c (Equation (1.37)); c_c the viscous damping coefficient of the sloshing mass;
- the elastically supported mass m_f (Equation (5.70)), which considers the impulsive flexible component of the deformational wall, is provided with an equivalent spring constant equal to expression (5.64); the circular frequency ω_f comes from Section 5.4 (starting from Equation (5.35)); c_f the viscous damping coefficient of the impulsive flexible mass;
- the total effective mass of the isolated liquid storage tank $m_{eff} = m_d + m_f + m_c > m$ [1]; furthermore, the self-weight of the tank is ignored, since it is very small in comparison to the effective mass of the tank liquid [545];

- the bearing is supposed to be modelled using a linear force-deformation behaviour having horizontal stiffness k_b, period T_b, isolation frequency ω_b and a damping ratio ξ_b:

$$T_b = 2\pi \sqrt{\frac{m_{\text{eff}}}{k_b}} \tag{6.23a}$$

$$\omega_b = \frac{2\pi}{T_b} \tag{6.23b}$$

$$\xi_b = \frac{c_b}{2m_{\text{eff}}\omega_b} \tag{6.23c}$$

where c_b is the viscous damping coefficient.

The relative displacement vector $\mathbf{x} = \begin{bmatrix} x_c & x_f & x_b \end{bmatrix}^{\mathsf{T}}$ can be approximated by a linear combination of three undamped modes by solving system (6.20) or (6.21) [546]:

$$\mathbf{x} = \Phi\mathbf{q} \tag{6.24}$$

where $\mathbf{q} = \begin{bmatrix} q_1 & q_2 & q_3 \end{bmatrix}^{\mathsf{T}}$ is the modal displacement vector (q_n is the nth-modal displacement) and $\Phi = \begin{bmatrix} \phi_1 & \phi_2 & \phi_3 \end{bmatrix}^{\mathsf{T}}$ is the undamped modal matrix (ϕ_n is the nth-modal vector). The solution of the following eigenvalue or characteristic value problem, provides the nth-modal vector:

$$\mathbf{K}\phi_n - \omega_n^2\mathbf{M}\phi_n = 0 \qquad n = 1, 2, 3 \tag{6.25}$$

where the quantities ω_n^2 are the eigenvalues or characteristic values indicating the square of the free-vibration frequencies. Hence, a nontrivial solution of Equation (6.25), ω_n with $n = 1, 2, 3$, is possible only when the determinant vanishes. In other words, finite amplitude free vibrations of the isolated tank are possible only when:

$$\left| \mathbf{K} - \omega_n^2\mathbf{M} \right| = 0 \tag{6.26}$$

Expanding the determinant gives the following algebraic expresion of the 6th-degree [296, 546]:

$$\begin{aligned} \omega_n^6 \left(\gamma_{mc} + \gamma_{mf} - 1 \right) + \omega_n^4 \left(\omega_c^2 + \omega_f^2 + \omega_b^2 - \gamma_{mc}\omega_f^2 - \gamma_{mf}\omega_c^2 \right) \\ - \omega_n^2 \left(\omega_c^2\omega_b^2 + \omega_f^2\omega_b^2 + \omega_f^2\omega_c^2 \right) + \omega_c^2\omega_f^2\omega_b^2 = 0 \end{aligned} \tag{6.27}$$

where:

$$\gamma_{mc} = \frac{m_c}{m_{\text{eff}}} \tag{6.28a}$$

$$\gamma_{mf} = \frac{m_f}{m_{\text{eff}}} \tag{6.28b}$$

Equation (6.27) may be simplified:

$$\omega_n^6 a + \omega_n^4 b + \omega_n^2 c + d = 0 \tag{6.29}$$

where coefficients a, b, c and d have the following form:

$$a = \gamma_{mc} + \gamma_{mf} - 1 \tag{6.30a}$$

$$b = \omega_c^2 + \omega_f^2 + \omega_b^2 - \gamma_{mc}\omega_f^2 - \gamma_{mf}\omega_c^2 \tag{6.30b}$$

$$c = -\omega_c^2\omega_b^2 - \omega_f^2\omega_b^2 - \omega_f^2\omega_c^2 \tag{6.30c}$$

$$d = \omega_c^2\omega_f^2\omega_b^2 \tag{6.30d}$$

Substituting variable $x = \omega_n^2$ into expression (6.29) and subdividing the same relation with the coefficient a, leads to the cubic equation:

$$x^3 + x^2\left(\frac{b}{a}\right) + x\left(\frac{c}{a}\right) + \left(\frac{d}{a}\right) = 0 \tag{6.31}$$

The three real roots of Equation (6.31) may be solved directly when $\bar{D} = \bar{Q}^3 + \bar{R}^2 < 0$, [561]:

$$x_1 = 2\sqrt{-\bar{Q}}\cos\left(\frac{\theta}{3} + 120°\right) - \frac{b}{3a} \tag{6.32a}$$

$$x_2 = 2\sqrt{-\bar{Q}}\cos\left(\frac{\theta}{3} + 240°\right) - \frac{b}{3a} \tag{6.32b}$$

$$x_3 = 2\sqrt{-\bar{Q}}\cos\left(\frac{\theta}{3}\right) - \frac{b}{3a} \tag{6.32c}$$

where θ (in degrees), $\bar{Q}$ and $\bar{R}$ are:

$$\theta = \arccos\left(\frac{\bar{R}}{\sqrt{-\bar{Q}^3}}\right) \tag{6.33a}$$

$$\bar{Q} = \frac{\frac{3c}{a} - \left(\frac{b}{a}\right)^2}{9} \tag{6.33b}$$

$$\bar{R} = \frac{\frac{9bc}{a^2} - \frac{27d}{a} - 2\left(\frac{b}{a}\right)^3}{54} \tag{6.33c}$$

Hence, the frequencies, $(\omega_1, \omega_2, \omega_3)$, of the isolated tank comes from $x = \omega_n^2$ using solutions (6.32); then using Equation (6.25) the corresponding eigenvectors or mode shapes (ϕ_1, ϕ_2, ϕ_3) can be derived.

6.5 Underground Rigid Tanks

The dynamically-induced lateral earth pressures on a flexible or rigid wall is a complicated soil–structure interaction problem [272]. The dynamic response depends on the mass and stiffness of the wall, the backfill and the underlying ground, the interaction among them and the nature of the input motions [446]. Usually the methods that are used to compute the dynamic earth pressure on the walls of a tank may be derived from the classical theories applied to the retaining walls [72, 446] and may be classified into three main groups [137]:

1) limit state (LS) analyses, in which a considerable relative movement occurs between the wall and soil to mobilize the shear strength of the soil [90, 289, 405, 417, 430, 477, 522]. These methods are classically applied to yielding flexible walls and have found widespread acceptance in practice (ATC-6, Point 4,

EN 1998-5, 2020 Edition and Point 2, EN 1997-1:2004). Limit states can occur either in the ground or in the structure or by combined failure in the structure and the ground (Point 1.2(3), EN 1997-1:2004).

2) elastic analyses, in which the relative movement in between the soil and wall is limited. Therefore, the soil behaves within its linear elastic range and this analysis is usually applied to non-yielding rigid walls [389, 520, 648, 649, 670].

3) numerical analyses, capable of accounting for non-linear, inelastic behaviour of the soil and of the interfaces between the soil and the wall [8, 190, 249, 548].

Earthquake-induced soil pressure can be found from flexible wall solutions [405, 430] or from rigid wall solutions [670, 671] following these two approaches, respectively, depending on how the wall and the soil are represented [412, 433, 647]:

1) the fist one is a pseudostatic method introduced by Okabe [430], Mononobe and Matsuo[6] [405] (Point 3.8.2, AIJ; Point E.4, UNI EN 1998-5:2004; Point 4.13.3, IITK-GSDMA 2007) following the great Kantō earthquake of 1923 [388]; as an alternative can be used; see Annex F, EN 1998-5, 2020 Edition "Evaluation of earth pressures on retaining structures". This is an extension of Coulomb's method for determining the earth pressures by considering the equilibrium of a triangular failure wedge [446]: the retained soil is considered dry cohesionless and perfectly plastic material, which fails along a planar surface, thereby exerting a limit thrust on the wall [415]. The increment in at-rest earth force, Δp_{0E} per unit length of the wall, due to the earthquake, is given by (Figure 6.15(a)):

$$\Delta p_{0E} = \frac{3}{8} \frac{S_e(0)}{g} \gamma_s H_e^2 \tag{6.34}$$

where $S_e(0)$ is the spectral acceleration obtained from a horizontal elastic response spectrum corresponding to a period $T = 0$ ($S_e(0) = Sa_g$ where a_g is the peak value of the ground acceleration $a(t)$ and S the soil factor as declared at Point 3.2.2.2, UNI EN 1998-1:2004); γ_s is the unit weight of soil in kN/m^3 and H_e is the embedded depth. Equation (6.34) is a simplified version of the Mononobe-Okabe expression, derived by Seed and Whitman [522] neglecting the vertical acceleration and applying the total force at $0.6H_e$;

2) the second one is the method described by Wood in [670, 671] considering the response of a rigid non-yielding wall (fixed at its base) retaining a homogeneous linear elastic soil and connected to a rigid base (Point C3.10, NZSEE-09 and Point A2.6, NZS 3106 (2009)). As depicted in Figure 6.15(b), the increment of pressure due to earthquake[7] is given by:

$$\Delta p_{0E} = \frac{S_e(0)}{g} \gamma_s H_e^2 \tag{6.35}$$

Equations (6.34) and (6.35) do not include pressures arising from gravity forces in the soil that must be evaluated separately. Furthermore, Equations (6.34) and (6.35) are intended to provide an evaluation of

---

6 As clearly demonstrated by Collins in [105, 281] the limit-equilibrium Mononobe-Okabe solution is based on kinematically admissible failure mechanisms in conjunction with a yield criterion and a flow rule for the soil material [212], both of which are enforced along pre-specified failure surfaces; on the other hand, Lancellotta in [323, 324] considers Coulomb's procedure as a limit equilibrium method consisting of finding the collapse load from the equilibrium of forces that act at the boundaries. Recently Mylonakis et al. in [415] suggested a closed-form solution of the stress type for assessing seismically-induced earth pressures, simpler if compared to the classical Coulomb and Mononobe-Okabe relations.

7 This pressure distribution will be conservative for most embedded tanks with horizontal backfill. and the pressures shown do not include the pressures arising from gravity forces in the soil (Point CA2.6, NZS 3106 (2009)).

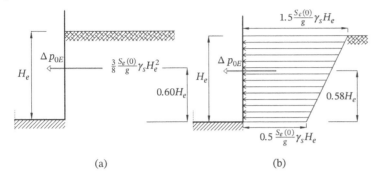

Figure 6.15 Typical dynamic pressure distributions proposed in seismic codes for seismic analysis of abutments [460]. Situations (a) and (b) correspond to the two extreme cases: (a) of yielding wall supporting elastoplastic soil in limit equilibrium (Mononobe-Okabe [405, 430] modified by Seed and Whitman [522]), and (b) of undeformable and non-yielding wall-supporting purely elastic soil (Point C3.10, NZSEE-09). Source: Adapted from [670].

the total force acting on a wall. The distribution of lateral pressure with depth H_e may be derived from Wood in [670] or from Veletsos and Younan in [647–649] considering a homogeneous linear viscoelastic material retained by a rigid [647, 648] or flexible wall [649]. The analytical solution of Veletsos and Younan [647–649], obtained by improving Scott's model [520], has been extended to account for soil inhomogeneity of the retained soil and translational flexibility of the wall foundation by Psarropoulos et al. in [460].

6.6 Horizontal Tanks

Horizontal pressure vessels are probably one of the most widespread types of equipment within the different industrial sectors [116]: steam boilers, horizontal above-ground and underground tanks, autoclaves, collectors, heat exchangers, gas and vapour storage. The structural design of most pressure vessels is done in accordance with the requirements contained in the "ASME Boiler and Pressure Vessel Code", Section VIII (ASME BPVC 2017). Vessels are usually obtained from the assemblage of different components: covers, heads, nozzle, saddle supports. Paragraph UG-22 of Division 1 specifies the loadings that must be considered to determine the minimum required thicknesses for the various vessel components. A simple procedure giving the seismic design of rigid horizontal circular cylindrical tanks is contained in Point A.5, EC8-4 and Point C3.3.1, NZSEE-09. Both codes suggest that horizontal tanks need to be analysed both along the longitudinal and the transverse axis and that approximate values for hydrodynamic pressures induced by horizontal excitation in either the longitudinal and transversal direction can be obtained from solutions for the rectangular tank of equal dimension at the liquid level and in the direction of motion, and of width required to give the equal liquid volume. Figure 6.16 shows an example of the equivalent rectangular tank configuration for longitudinal (x) and transversal (y) seismic action respectively. The approximation suggested by both codes is sufficiently accurate for design purposes over the range of $\frac{H}{R}$ between 0.5 and 1.6. When $\frac{H}{R}$ exceeds the value of 1.6, the tank should be assumed to behave as if it were full, with the total fluid mass rigidly connected to the tank. The total mass of the contents, linked rigidly to the tank's wall, acts as an impulsive mass at mid-height of the tank. Figure 6.17 shows the first

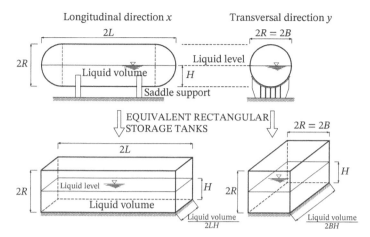

Figure 6.16 Nomenclature for horizontal axis circular cylindrical tank in longitudinal and transverse direction; equivalent rectangular storage tanks for both seismic actions. Source: Adapted from [116].

three ($n = 1, 2, 3$) sloshing modes dimensionless frequency (ω_{cn} in rad/sec) along the longitudinal and the transverse axis, respectively [2, 121]. The horizontal T_0 impulsive periods of vibration for the equivalent rigid rectangular tanks including soil–structure interaction may be evaluated from relation[8] (2.55). In the case of a rigid equivalent rectangular tank ignoring the soil–structure interaction, the impulsive period T_0 is theoretically zero, and $S_e(T_0) = Sa_g$ where a_g is the peak value of the ground acceleration $a(t)$ and S the soil factor (Point 3.2.2.2, UNI EN 1998-1:2004). Unless calculated on the basis of a more rigorous analysis, Point CA2.2.1, NZS 3106 (2009) suggests for a ground-supported rectangular (or vertical cylindrical) tank's horizontal mode to assume $T_0 = 0.1$ sec.

A more refined analysis, to derive the analytical distribution of the impulsive pressure for motion in the transversal direction, has been derived by Werner and Sundquist [663] and suggested in Point C3.3.1, NZSEE-09 and Point A.5, EC8-4:

$$p_i(\phi) = q_0(\phi)\frac{S_e(T_0)}{g}\gamma_l R \tag{6.36}$$

where ϕ is the angle in the polar coordinate system as depicted in Figure 6.18; γ_l is the unit weight of the liquid and $S_e(T_0)$ is the spectral acceleration obtained from a horizontal elastic response spectrum corresponding to the impulsive period with or without soil–structure interaction. For the case of the tank hall full, $H = R$, the dimensionless pressure coefficient $q_0(\phi)$ is given by the following expression and plotted in Figure 6.18:

$$q_0(\phi) = \frac{4}{\pi}\sum_{n=1}^{\infty}\frac{(-1)^{n-1}}{(2n)^2 - 1}\sin(2n\phi) \tag{6.37}$$

where the distribution of the impulsive pressure is in the radial direction and it is antisymmetric with respect to the vertical centreline of the tank. The impulsive m_i and first convective mode m_{c1} masses, as a

---

8 R_b is the radius of an equivalent circular foundation, or alternatively Equation (2.133) can be used.

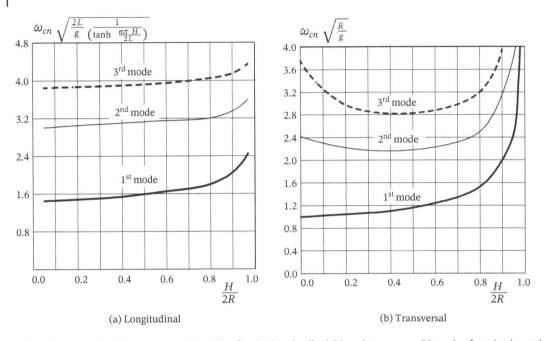

(a) Longitudinal   (b) Transversal

Figure 6.17 Natural frequency ω_{cn} in rad/sec for the longitudinal (a) and transverse (b) modes for a horizontal cylinder tank [2, 121].

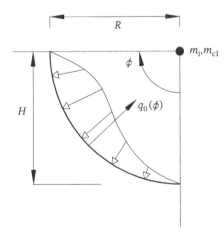

Figure 6.18 Radial impulsive pressures $p_i(\phi)$ on half-full horizontal cylinder tank with $H = R$ under transverse seismic action; both the impulsive and the convective masses are assumed to be at the centre of the circular section. Source: Adapted from [663].

portion of the total contained mass of the fluid m, may be derived by integrating the pressure distribution (6.36) (half-full tank $H = R$):

$$m_i = 0.4m \tag{6.38a}$$

$$m_{c1} = 0.6m \tag{6.38b}$$

Equations (6.38a and b) are expected to be reasonable approximations for liquid levels given by $\frac{H}{R}$ ranging from 0.8 to 1.2 (Point C3.3.1, NZSEE-09 and Point A.5, EC8-4). Because the pressures $p_i(\phi)$ are in the radial direction, the forces acting on the horizontal tank pass through the centre of the transverse circular section, and both the impulsive and convective masses should be assumed to act at this point.

The seismic design force Q_y, in the case of a rigid horizontal cylindrical tank under transverse excitation, can be calculated through the classical square root of the sum of the squares (SRSS) combination of the impulsive ($m_i^{tot}S_e(T_0)$) and convective ($m_{c1}S_e(T_1)$) values, as suggested by Fiore et al. in [157]:

$$Q_y = \sqrt{\left[m_{c1}S_e(T_1)\right]^2 + \left[m_i^{tot}S_e(T_0)\right]^2} \tag{6.39}$$

where $m_i^{tot} = m_i + m_w$ is the total moving mass, m_w being the wall's mass of the empty horizontal container; the impulsive and sloshing mass ratios are presented in Table 6.3 [291] where it is clear that when the cylindrical rigid tank is nearly full, the entire mass responds impulsively. Furthermore, when the container is nearly empty, the impulsive mass is almost zero and sloshing motion dominates the liquid response. For the cases where the liquid height has an intermediate value ($H = R$ and $\frac{H}{R} - 1 = 0.0$), values in Table 6.3 are in full agreement with expressions (6.38a and b). Considering only the first sloshing mode, the corresponding equivalent mechanical model, as proposed by Karamanos et al. in [291, 439, 440, 447], is shown in Figure 6.19 (a damping ratio $\xi_{c1} = 1\%$ is suggested in [291]). In Figure 6.19 y_1 represents the motion of the liquid mass associated with the first mode sloshing term, while y_2 the impulsive motion.

The seismic design force Q_y, in the case of a rigid horizontal cylindrical tank under longitudinal excitation, can be calculated through the classical square root of the sum of the squares (SRSS) combination of the impulsive ($m_i^{tot}S_e(T_0)$) and convective ($m_{c1}S_e(T_1)$) values, as suggested by Karamanos et al. in [291]:

$$Q_y = \sqrt{\left[m_{c1}S_e(T_1)\right]^2 + \left[m_i^{tot}S_e(T_0)\right]^2} \tag{6.40}$$

where an equivalent rectangle with liquid height H_{eq} can easily be computed in terms of the horizontal rigid cylinder diameter ($D = 2R$) and the liquid height H ($0 \leq H \leq 2R$) [290, 291]:

$$H_{eq} = \frac{1}{2}\left(H - \frac{D}{2}\right) + \frac{D^2}{8}\left[\frac{\sin^{-1}\left(\frac{2H-D}{D}\right) + \frac{\pi}{2}}{\sqrt{H(D-H)}}\right] \tag{6.41}$$

The following expressions can be used to calculate the sloshing frequencies (or periods) and the corresponding convective masses:

$$\frac{\omega_{cn}^2 R}{g} = \frac{(2n-1)\pi R}{2L}\tanh\left[\frac{(2n-1)\pi H_{eq}}{2L}\right] \tag{6.42a}$$

Figure 6.19 Equivalent mechanical model representing the impulsive and first sloshing mode in a rigid cylindrical tank under transverse excitation [157, 291, 439, 440, 447].

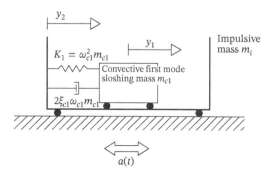

Table 6.3 Impulsive (m_i) and convective (m_{c1}) masses with respect to the liquid height ($0 \leq H \leq 2R$) and to the total contained mass of the fluid m in a rigid horizontal cylindrical tank [291, 439, 440, 447] under transverse excitation.

$\frac{H}{R} - 1$	$\frac{m_i}{m}$	$\frac{m_{c1}}{m}$
−1.00	0.0	1.0
−0.95	0.01991	0.97994
−0.90	0.03968	0.95974
−0.80	0.07895	0.91895
−0.60	0.15736	0.83566
−0.40	0.23690	0.74989
−0.20	0.31890	0.66125
0.0	0.40543	0.56916
0.20	0.49593	0.47276
0.40	0.59247	0.37077
0.60	0.69835	0.26115
0.80	0.81900	0.14032
0.90	0.90955	0.07361
0.95	0.94060	0.03793
1.00	1.0	0.0

$$\frac{m_{cn}}{m} = \frac{8 \tanh\left[\frac{(2n-1)\pi H_{eq}}{2L}\right]}{\frac{(2n-1)^2 \pi^3 H_{eq}}{2L}} \qquad (6.42b)$$

using $n = 1, 2, 3 \ldots$ A long cylindrical container with large values of the aspect ratio $\frac{2L}{R}$ (or $\frac{H_{eq}}{2L} < 0.1$) can be considered a shallow equivalent rectangular tank and the following simplified expressions can be used [291]:

$$\frac{\omega_{cn}^2(2L)}{g} = \frac{(2n-1)^2 \pi^2 H_{eq}}{2L} \qquad (6.43a)$$

$$\frac{m_{cn}}{m} = \frac{8}{(2n-1)\pi^2} \qquad (6.43b)$$

The total impulsive moving mass in Equation (6.40) comes from $m_i^{tot} = m + m_w - m_{c1}$, m_w being the wall's mass of the empty horizontal container and m the total contained mass of the fluid.

6.7 Conical Tanks

The containment vessels in the oil industry are commonly built in cylindrical shapes. Anyway, conical tanks (on ground or elevated) offer a working solution for applications that require 100% drainage and mainly are required for efficient processing of water-filled tanks. These conical-shaped tanks are also referred to as sloped bottom tanks, truncated conical tanks, full drain tanks, fermenter tanks, brewing tanks, and mixing tanks. They are frequently used in the industrial, agricultural, commercial, manufacturing, and water treatment sectors. One of the main reasons is that the conical bottoms enable quick and complete drainage. Furthermore, elevated conical tanks are considered one of the most popular containment vessels since they provide greater liquid retaining capacity for the same base radius of their cylindrical counterpart. These conical tanks require also a lower height of water for the same containing volume of the cylindrical shape. Consequently, the hydrostatic pressure acting on the vessel base is minimized, leading to an increase in its structural efficiency. Moreover, a large containing volume can be achieved without having the base over-hanging and cantilevered from the supporting tower as in the case of elevated cylindrical tanks [21]. Point A.6, EC8-4 and Point 4.2.3, IITK-GSDMA 2007 suggest that the slenderness value of $\frac{H}{2R} = \frac{H}{D}$ should correspond to that of an equivalent circular tank of the same volume and diameter $D = 2R$ equal to the diameter of the tank at the top level of the liquid, and impulsive, convective masses, stiffnesses and heights of an equivalent circular tank should be used.

The refined behaviour of conical tanks under seismic ground excitations has been widely investigated by El Damatty, Sweedan, and coworkers in many publications [143–147, 277, 575–579]. The first numerical and analytical studies, able to predict the vibration of a liquid–filled conical tank, were developed in [145, 146] based on a boundary element added mass formulation. Then, experimental shake table tests

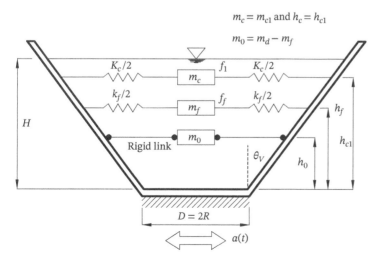

Figure 6.20 Equivalent mechanical model, due to El Damatty and Sweedan [576, 577], for flexible conical roofless tanks of uniform wall thickness filled with liquid under lateral translation $a(t)$.

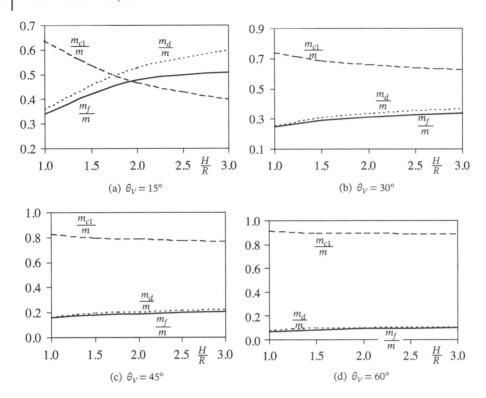

Figure 6.21 Equivalent mechanical quantities [576, 577]: three concentrated masses (m_{c1}, m_f, m_d) as a function of the slenderness $\frac{H}{R}$ and four values of the angle θ_V.

used to validate the above numerical model, were conducted by El Damatty and Sweedan in [576, 577]. The equivalent mechanical model proposed by El Damatty and Sweedan in [147], and mainly based on the original idea for cylindrical flexible tanks proposed by Haroun, Housner, and Ellaithy [214, 215, 221, 223] (described in Figure 5.18 and Section 5.6.1), is illustrated in Figure 6.20, based on eight parameters, three masses and corresponding heights, and two natural frequencies. The mechanical model (Figure 6.20) consists of two concentrated masses, (m_0, m_f) with $m_0 = m_d - m_f$, in addition to the convective term $m_c = m_{c1}$ (or m_1; Section 2.2.2) computed by considering the tank's conical wall to be rigid: the rigidly attached mass $m_0 = m_d - m_f$, located at a distance h_0 from the base, simulates the hydrodynamic effects associated with the impulsive rigid-body component of the wall's motion, whereas the elastically supported mass m_f, located at a distance h_f from the base, is considered with regard to the impulsive flexible component of the deformational wall associated with the $\cos\theta$-type first mode ($m = 1$ in Figure 5.1(b)). According to the nomenclature defined in Figure 2.18, the maximum radial base shear applied

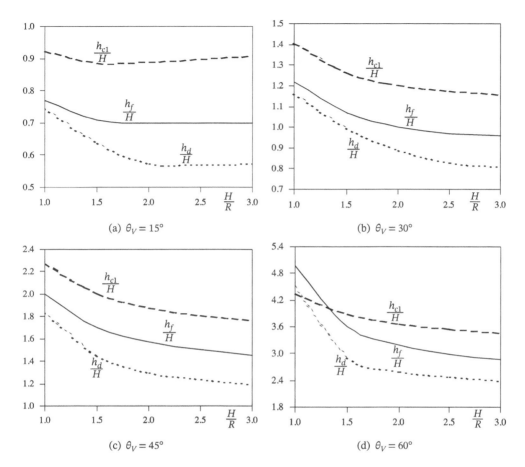

Figure 6.22 Equivalent mechanical quantities [576, 577]: three heights (h_{c1}, h_f, h_d) as a function of the slenderness $\frac{H}{R}$ and four values of the angle θ_V.

at the bottom of the tank walls and the resulting bending moments along the wall (above the base) of the tank may be determined using the SRSS rule (Point A.3.2.1, EC8-4):

$$Q_y^{EDS} = \sqrt{\left[m_{c1}S_e(T_1)\right]^2 + \left[m_f S_e(T_f)\right]^2 + \left[(m_d - m_f)S_e(T_0)\right]^2} \tag{6.44a}$$

$$M_y^{EDS} = \sqrt{\left[m_{c1}h_{c1}S_e(T_1)\right]^2 + \left[m_f h_f S_e(T_f)\right]^2 + \left[(m_d h_d - m_f h_f)S_e(T_0)\right]^2} \tag{6.44b}$$

where masses (m_{c1}, m_f, m_d) and corresponding heights (h_{c1}, h_f, h_d) may be derived from Figures 6.21 and 6.22 as a function of the slenderness $\frac{H}{R}$ and four values of the angle θ_V between the wall's tank

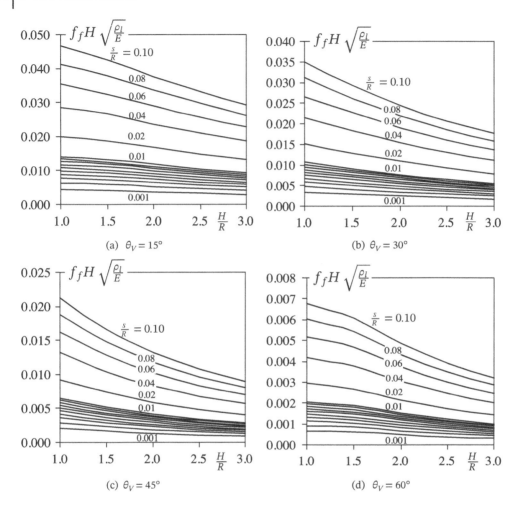

Figure 6.23 The first impulsive flexible component f_f of the deformational wall as a function of the slenderness $\frac{H}{R}$ for four values of the angle θ_V.

generator and the vertical direction (Figure 6.20; $\theta_V = 15°, 30°, 45°$ and $60°$); $S_e(\cdot)$ is the spectral acceleration obtained from a horizontal elastic response spectrum corresponding to the convective period of the vibration of the 1st-sloshing mode of vibration $T_1 = \frac{1}{f_1}$ of the liquid and to the impulsive flexible period $(T_f = \frac{1}{f_f})$ as well as to the impulsive period T_0 without soil–structure interaction, $S_e(T_0) = Sa_g$. The first impulsive flexible component f_f of the deformational wall associated with the $\cos\theta$-type mode can be derived from Figure 6.23 where, for steel tanks, the ratio between the thickness and the tank's bottom wall radius $(\frac{s}{R})$ is assumed to vary between 0.001 and 0.01, while, for concrete tanks, this ratio is extended up to 0.10 (ρ_l is the mass density of the fluid and E the Young's modulus of the material of the tank wall). The fundamental sloshing frequency f_1 is presented in Figure 6.24 for different values of the angle θ_V.

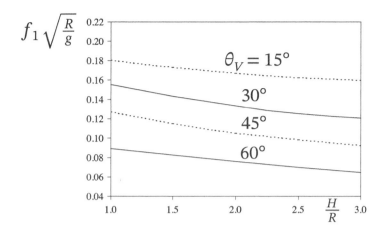

$$f_1 \sqrt{\frac{R}{g}}$$

Figure 6.24 The fundamental sloshing frequency f_1 [567, 577] (rigid wall configuration) as a function of the slenderness $\frac{H}{R}$ for four values of the angle θ_v.

7

General Design Principles

After reading this chapter you should be able to:

- Understand how to design steel tanks
- Understand how to design reinforced concrete tanks
- Identify the critical components in steel and reinforced concrete tank's design

7.1 Introduction

The tank design basis includes the tank capacity, service, design code, materials, design conditions, external loads (wind, snow, or other loads), physical properties for the range of liquids stored, appurtenances (such as openings, nozzles, access platform, roof or shell manhole, open or pressure vacuum vent, ladder, support legs), and connections; geotechnical reports. A compilation of all these relevant data should be made prior to a detailed tank design [414].

The storage of large volume of fluids, sometimes highly flammable, is a topic that was able to attract many codes, regulations, standardization from many interested private and public parties [344]. In the following we will mainly refer to US and European documents focusing on steel (Section 7.2) and concrete (Section 7.3) containment structures.

7.2 Requirements for Steel Tanks

There are five structural elements which are key to ensuring a storage tanks achieve their maximum lifespan:

- **Base plate**: in the analysis, tanks are usually modelled as fixed to the ground, so that it is not a problem to know exactly the shape of the bottom. The classical way to design tank bottoms is to assume they are membranes for liquid containment with no real structural requirements [414]. Anyway, a tank bottom may be broadly classified as a flat bottom or conical (cone up or cone down) and attention should be paid to the high local stresses that can occur in the bottom-to-sidewall connection.

Seismic Design and Analysis of Tanks, First Edition. Gian Michele Calvi and Roberto Nascimbene.

- **Sidewall and openings**: most steel tanks have a cylindrical body, which is used as storage volume. The walls may have a constant thickness or a tapered wall with different values of the thickness at different elevations. The cylinder itself is formed by curved plates that are welded or bolted. Furthermore, the distribution of stresses and strains near structural and manufacturing openings, inclusions, and junctions, may result in damage to the structural elements, and cannot be ignored.
- **Roof**: it can be shaped in many different ways as a dome, conical, geodesic, umbrella, fixed or floating (external open-top or internal); it can be self-supported or column-supported.
- **Foundation**: it supports the tank and prevents it from settling or sinking into the ground. Classical information required is site and soil conditions, amounts and rate of settlement, pressures (earth and pore water).
- **Stiffeners**: theoretically there are two ways to provide adequate stiffness to the wall: increase the thickness or provide additional stiffness. In most cases it is uneconomical to make the shell thick enough to provide all the necessary stiffness. Therefore additional external systems to resist shell deformations should be provided, by welding circumferential rings or vertical stringers around the outside of the tank's wall.

7.2.1 Base Plate

When the foundation is ready for the erection of the tank, the bottom plate will start to be laid on top of the foundation itself and welded in the right sequence in order to avoid out-of-plane weld distortion. Two typical bottom layouts of tanks are depicted in Figures 7.1(a) and 7.1(b): for (small) tanks up to and including 12.5 m diameter, a typical floor arrangement usually called rect-and-sketch layout is reported in Figure 7.1(a) with a classical connection (Figure 7.1(c)); for tanks over 12.5 m diameter a ring of peripherical plates are butt welded together using backing strips (Figures 7.1(b) and 7.1(d)) [344]. According to Point 5.4.1, API 650 all bottom plates should have a corroded thickness[1] of not less than 6 mm (0.236 in). Furthermore, bottom plates should be ordered such that, when trimmed, al least 50 mm (2 in) width will project beyond the tank's wall.

Point 5.5, API 650 introduces annular bottom plates (Figures 7.1(b) and 7.1(d)) which are usually rings used on larger tanks and placed on the outside of the base plate under the shell. The purpose of this annular plate is to support the weight of the tank's courses, improving the seismic and wind design. Point 5.5.3, API 650 gives Table 7.1 specifying the minimum annular plate thickness according to the first maximum shell stress and corroded thickness, again of the first shell course connected to the annular plate. It is clear that API 650 has a very straightforward requirement both on the bottom and annular plate thickness and width requirements.

Table 7.2 from Point 11.4, UNI ENV 1993-4-2 (in agreement with Point 8.2.3, UNI EN 14015:2006), gives the minimum thickness of the bottom plate excluding corrosion allowance (as in API 650, it is a corroded thickness), depending on material and weld type (lap or butt welded). Mainly in the case of seismic design, larger values should be used if required to resist uplift due to the internal negative pressure (impulsive or convective), unless a minimum guaranteed residual liquid level is used to assist in resisting this uplift (for more details, see Chapter 3 and Section 3.2.4). In determining the thickness,

---

1 The corroded thickness is the design nominal thickness less any specified corrosion allowance. If we select a corrosion allowance equal to 1.5 mm, we obtain a minimum thickness of $6 + 1.5 = 7.5$ mm, which means a probable nominal bottom thickness of 8 mm.

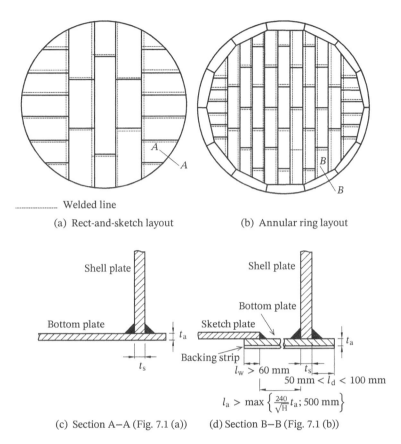

Welded line

(a) Rect-and-sketch layout   (b) Annular ring layout

Shell plate

Bottom plate

t_a

t_s

(c) Section A–A (Fig. 7.1 (a))

Shell plate

Bottom plate

Sketch plate

Backing strip

$l_w > 60$ mm t_s

50 mm $< l_d < 100$ mm

t_a

$l_a > \max \left\{ \frac{240}{\sqrt{H}} t_a ; 500 \text{ mm} \right\}$

(d) Section B–B (Fig. 7.1 (b))

Figure 7.1 Typical floor arrangements with (a) bottom and (b) annular plates at the perimeter and corresponding connections (Point 8.3, UNI EN 14015:2006). Source: From [344]/with permission from John Wiley & Sons.

Table 7.1 Minimum annular plate thickness in mm according to Table 5-1a in API 650 and based on the condition that the foundation is able to provide a uniform support under the full width of the annular plate itself.

Corroded thickness in mm	Stress in MPa			
	≤ 190	≤ 210	≤ 220	≤ 250
≤ 19	6	6	7	9
> 19 and ≤ 25	6	7	10	11
> 25 and ≤ 32	6	9	12	14
> 32 and ≤ 40	8	11	14	17
> 40 and ≤ 45	9	13	16	19

Table 7.2 Minimum bottom plate thickness in mm according to Table 11.1 in Point 11.4, UNI ENV 1993-4-2.

Material type	Lap-welded bottom plates	Butt-welded bottom plates
Carbon steels	6	5
Stainless steels	5	3

Table 7.3 Minimum thickness in mm (wall and bottom) for ground-supported or other tanks as a function of the nominal diameter D and nominal height H, according to Point 3.10.3, AWWA D100-11.

Diameter in m	Height in m	Ground-supported tank	Other tank
$D \leq 6.1$	All	4.76	6.35
$6.1 < D \leq 15.2$	$H \leq 14.6$	4.76	6.35
$6.1 < D \leq 15.2$	$H > 14.6$	6.35	6.35
$15.2 < D \leq 36.6$	All	6.35	6.35
$36.6 < D \leq 61.0$	All	7.94	7.94
$D \geq 61.0$	All	9.52	9.52

the effects of corrosion should be taken into account: it depends upon the stored liquid, the type of steel, the heat treatment, and the measures taken to protect the construction against corrosion. Furthermore, Point 11.4(5), UNI ENV 1993-4-2 (also Point 8.3.1, UNI EN 14015:2006) suggests that bottoms for tanks greater than 12.5 m diameter should have an annular bottom plate (similar to the one defined in API 650) with a minimum corroded thickness equal to $t_a = \max\left(\frac{t_s}{3} + 3 \text{ mm}; 6 \text{ mm}\right)$ where t_s is the thickness of the attached first shell course (as explicitly declared in Section 3.2.4, this minimum thickness of bottom plate may lead to the formation of a plastic hinge in the annular ring plate, avoiding alternating plasticity in the weld detail at the bottom of the shell wall or the formation of a plastic hinge in the first shell course).

According to Point 3.10.1/2, AWWA D100-11, all parts of the tanks (wall and bottom) in contact or not with the content should have a minimum thickness, respectively, of 6.35 mm (0.25 in) and 4.76 mm (3/16 in). The minimum thickness of double butt-welded knuckles (in contact with water) for ground-supported flat-bottom tanks with shell height less than 14.6 m (48 ft) and nominal diameter[2] not greater than 15.2 m (50 ft) may be 4.76 mm (3/16 in). Anyway, according to Point 3.10.3, AWWA D100-11, all cylindrical shell plates in contact with the fluid should have minimum thicknesses as indicated in Table 7.3.

---

2 The nominal diameter is the mid-thickness diameter (or centreline diameter of the bottom shell course, according to Note 1, Point 5.6.1.1, API 650) of the bottom ring-wall unless the owner specifies otherwise.

7.2.2 Sidewall

One of the main purposes of a steel tank is to provide storage of liquids to meet the demands of the area it will service. Hence, the design of the vertical wall of a tank is a crucial point in controlling, limiting, and avoiding leaks and spills that can have substantial environmental impact. Table 7.4 from Point 5.6.1.1, API 650 and Point 9.1.5, UNI EN 14015:2006 give the minimum nominal shell thickness, including any corrosion allowance, required for the tank's sidewall to be acceptable. However, the full set of criteria that affects shell design for small and large tanks is completely different. This is the main reason why Point 5.6, API 650 suggests three methods to determine the required plate thickness of the shell:

1) The first method is the "1 foot method" (one-foot-method) which is based on the hoop stress obtained from the classical "membrane theory" [22]. Point 5.6.3.1, API 650 limits the applicability of this method to tanks up to 61 m (200 ft) in diameter. The wall thickness for each shell course is calculated using the circumferential hoop stress at a point 0.3 m (1 foot) above the lower horizontal weld seam of the shell course due to hydrostatic pressure of the stored liquid. The reasoning behind this assumption is that the tank bottom base plates provide restraint to reduce circumferential stress in

Table 7.4 Minimum thickness of the wall in mm for ground-supported or other steel tanks (carbon and stainless) as a function of the nominal diameter D (in m), according to Point 5.6.1.1, API 650 and Point 9.1.5, UNI EN 14015:2006.

API 650	
Nominal diameter in m	**Nominal thickness in mm**
< 15	5
15 to < 36	6
36 to 60	8
> 60	10

UNI EN 14015:2006		
Nominal diameter in m	**Carbon steels**	**Stainless steels**
< 4	5	2
4 to < 10	5	3
10 to < 15	5	4
15 to < 30	6	5
30 to < 45	8	6
45 to < 60	8	Agreed between purchaser and manufacturer
60 to < 90	10	Agreed between purchaser and manufacturer
≥ 90	12	Agreed between purchaser and manufacturer

the lowest shell course, compared to the membrane solution, due to hydrostatic pressure [344, 414]. The required minimum thickness is the larger of the two values computed by the following formulas:

$$t_d = \frac{4.9D(H - 0.3)G}{S_d} + CA \tag{7.1a}$$

$$t_t = \frac{4.9D(H - 0.3)}{S_t} \tag{7.1b}$$

where t_d is the design shell thickness in mm resulting from hydrostatic pressure of the liquid to be stored and desired corrosion allowance CA (in mm) for the design life of the storage tank; t_t is the hydrostatic shell thickness in mm resulting from hydrostatic test pressure of the test liquid; H (in m) is the distance from the maximum liquid level to the bottom of the shell course under consideration (it is called the design liquid level); G is the design-specific gravity of the stored liquid and is classically defined as a dimensionless unit (in the case of water it is equal to 1.0) which is the ratio of density of a fluid to the density of water; S_d and S_t (in MPa) are the allowable stresses for design and hydrostatic test condition, respectively, and are defined according to Points 5.6.2.1/2, API 650:

- S_d should be the minimum between $\frac{2}{3}$ the yield strength and $\frac{2}{5}$ the tensile strength;
- S_t should be the minimum between $\frac{3}{4}$ the yield strength and $\frac{3}{7}$ the tensile strength.

The one-foot-method has been used successfully for many years for the majority of tanks. However, the designs based on the one-foot-method may become slightly conservative [22, 344, 414] and cost-prohibitive for larger diameter tanks. To include this better understanding, Point 5.6.4, API 650 includes the option to use the "variable design point method" to reduce the wall thickness.

2) The second method to calculate the required wall thickness is the "variable design point method" that is also based on the "membrane theory" [22]. The variable design point method takes into consideration the restraint provided by the tank bottom plates to the first shell course and the restraint provided by each lower shell course to the upper shell course [22]. The variable design point method uses a variable distance instead of fixed distance, as in the one-foot-method, as a function of the shell plate thickness above and below the seam for each shell course. The method can be used for tanks more than 61 m (200 ft) in diameter when the following inequalities are true:

$$\frac{L}{H} \leq \frac{1000}{6} \tag{7.2}$$

where $L = (500Dt)^{0.5}$ in mm; D is the tank diameter in m, t the bottom course corroded shell thickness in mm, H (in m) is the maximum design liquid level (as previously defined in Equation (7.1)). For the first shell course from the bottom, the greater of the two thicknesses should be used as the design shell thickness (Point 5.6.4.4, API 650):

$$t_{1d} = \left(1.06 - \frac{0.0696D}{H}\sqrt{\frac{HG}{S_d}}\right)\left(\frac{4.9HDG}{S_d}\right) + CA \tag{7.2a}$$

$$t_{1t} = \left(1.06 - \frac{0.0696D}{H}\sqrt{\frac{H}{S_t}}\right)\left(\frac{4.9HD}{S_t}\right) \tag{7.2b}$$

Note that if the calculated first shell course thickness using Equations (7.3) is more than the thickness obtained using the traditional one-foot-method using expressions (7.1), the thickness calculated from one-foot-method may be used as the first shell course thickness. Furthermore, the shell thickness should not be less than the shell thickness required by Table 7.4. Calculation of the shell thickness for the second and upper shell courses requires an iterative process, as defined in Point 5.6.4.5-8, API 650. The thickness of the second course t_2 is dependent upon the height of the bottom first course h_1 (in mm) and the radius times thickness value $r \cdot t_1$ of the bottom course, where r is the nominal tank radius and t_1[3] is the calculated corroded thickness of the bottom shell course (both variables in mm). The three governing conditions, for calculating the minimum design second shell course thickness t_2 (in mm), come from Point 5.6.4.5, API 650 and are the following:

$$\text{if} \quad \frac{h_1}{(rt_1)^{0.5}} \leq 1.375 \qquad\qquad \Rightarrow t_2 = t_1 \tag{7.4}$$

$$\text{if} \quad \frac{h_1}{(rt_1)^{0.5}} \geq 2.625 \qquad\qquad \Rightarrow t_2 = t_{2a} \tag{7.5}$$

$$\text{if} \quad 1.375 < \frac{h_1}{(rt_1)^{0.5}} < 2.625 \quad \Rightarrow t_2 = t_{2a} + (t_1 - t_{2a}) \left[2.1 - \frac{h_1}{1.25(rt_1)^{0.5}} \right] \tag{7.6}$$

where t_{2a} in mm is the corroded thickness of the second shell course and is calculated iteratively as follows. As a first step, to calculate the upper course (corroded) thicknesses for both the design and the hydrostatic test condition, a preliminary starting value t_u for the upper course thickness should be evaluated using expressions (7.1) excluding corrosion. Then the location (the variable design point) x of the maximum hoop stress from the bottom of the course is found using the lowest of the values:

$$x_1 = 0.61(rt_u)^{0.5} + 320CH \tag{7.7a}$$

$$x_2 = 1000CH \tag{7.7b}$$

$$x_3 = 1.22(rt_u)^{0.5} \tag{7.7c}$$

where $C = \frac{[K^{0.5}(K-1)]}{(1+K^{1.5})}$ and $K = \frac{t_L}{t_u}$; t_L (in mm) is the corroded thickness of the lower course and H (in m) is the maximum design liquid level (as previously defined in Equations (7.1) and (7.2)). The minimum thickness t_u of the upper course is then found using the minimum value of the location x from expressions (7.7) substituted in the following relations both for the design and the hydrostatic test condition:

$$t_{dx} = \frac{4.9D \left(H - \frac{x}{1000} \right) G}{S_d} + CA \tag{7.8a}$$

$$t_{tx} = \frac{4.9D \left(H - \frac{x}{1000} \right)}{S_t} \tag{7.8b}$$

The steps just described should be repeated (two to three times is normally sufficient) until there is a small difference between the calculated values of $t_{dx} = t_u$ or $t_{tx} = t_u$ in succession.

---

3 To be clear, for the design condition the calculated thickness $t_1 = t_{1d} - CA$ from Equation (7.3a); while for the hydrostatic test condition $t_1 = t_{1t}$ from relation (7.3b).

Most of the time the variable design point method provides a reduction in wall thicknesses [344] and total material weight. Anyway for some tank geometries the variable design point method may become unconservative. Therefore, API 650 limits the applicability of this method for tanks with $\frac{L}{H} \leq \frac{1000}{6}$.

3) The third method given in Point 5.6.5, API 650 for the evaluation of the shell thickness is based on a linear analysis. This method should be used when $L/H > 1000/6$. API 650 does not describe a specific linear analysis method, anyway Azzuni and Guzey in [22] developed a new method using thin shell theory to perform a linear analysis for the shell thickness calculation.

After the shell thickness is determined, API 650 requires that the shell be checked for stability against buckling and overturning due to wind or seismic loads. In the case of instability of tank shells under wind loads, stiffeners or wind girders may be designed [118]. In the case of a seismic event, from the moment calculated, the tank is checked for overturning stability, and consequently the meridional axial forces on the wall are examined to prevent buckling.

According to Point 5.1.2(1), UNI ENV 1993-4-2, the cylindrical shell wall of the tank should be checked for the following phenomena under the limit states defined in Point 1.3.2, UNI ENV 1993-1-6:

- global stability and static equilibrium;
- LS1: the plastic limit is the ultimate limit state where the structure develops zones of yielding in a pattern such that its ability to resist increased loading is deemed to be exhausted (Points 4.1.1 and 6, UNI ENV 1993-1-6);
- LS2: the cyclic plasticity is the ultimate limit state where repeated yielding is caused by cycles of loading and unloading, leading to a low cycle fatigue failure where the energy absorption capacity of the material is exhausted (Points 4.1.2 and 7, UNI ENV 1993-1-6);
- LS3: the wall buckling is the ultimate limit state where the structure suddenly loses its stability under membrane compression and/or shear (Points 4.1.3 and 8, UNI ENV 1993-1-6);
- LS4: the fatigue limit is the ultimate limit state where many cycles of loading cause cracks to develop in the shell plate that in further load cycles may lead to rupture (Points 4.1.4 and 9, UNI ENV 1993-1-6).

The serviceability limit states for shell walls should be taken as deformations, deflections, and vibrations that adversely affect the effective use of the structure or cause damage to non-structural elements. The specific limiting values, appropriate to the operational use, should be agreed between the designer, the client, and the relevant authority, taking account of the intended use and the nature of the liquids to be stored. In a shell thickness calculation, according to Point 11.3.1, UNI ENV 1993-4-2, the circumferential normal stress due to liquid loads and internal pressure should be verified in each shell course using:

$$\left(\gamma_F \rho g H_{\text{red}} + p_d\right)\big|_{j^{\text{th}}-\text{course}} \left(\frac{r}{t}\right) \leq f_{yd} \tag{7.9}$$

where ρ is the density of the fluid (in kg/mm^3), g (in m/sec^2) the acceleration due to gravity, and γ_F the partial factor for actions as given by Point 2.9.2.1, UNI ENV 1993-4-2; r is the radius of the middle surface of the cylindrical wall of the tank and t its thickness (both in mm); f_{yd} is the design yield strength of steel (in MPa); p_d in MPa is the design value of the pressure above the liquid level (i.e. the characteristic value, according to Annex A of UNI ENV 1993-4-2, multiplied by the partial factor for actions). In Equation (7.9) the value of H_{red} (in mm) for the jth course, denoted by H_{redj}, is determined according to its relationship with the value for the course below it, which is the $(j-1)^{th}$ course:

$$H_{redj} = H_j - \Delta H \quad \text{if} \quad \frac{H_{red(j-1)}}{f_{yd,(j-1)}} \geq \frac{H_{redj}}{f_{yd,j}} \tag{7.10a}$$

$$H_{redj} = H_j \quad \text{if} \quad \frac{H_{red(j-1)}}{f_{yd,(j-1)}} < \frac{H_{redj+1}}{f_{yd,j+1}} \tag{7.10b}$$

where $\Delta H = 0.3$ m and H_j is the vertical distance from the bottom of the jth course to the liquid level. As a starting point, Equation (7.9) and Table 7.2 can be also used to determine a minimum corroded thickness t in mm, whichever is the greater. Furthermore, an expression similar to relation (7.9) may be found in Point 9.2.2, UNI EN 14015:2006, except for the units used and the presence of the corrosion allowance in mm:

$$e_c = \frac{D}{20S} \left[98W(H_c - 0.3) + p \right] + c \tag{7.11}$$

where e_c (in mm) is the design nominal shell thickness; D in m is the tank diameter and S is the allowable design stress (MPa); W is the density of the contained liquid in kg/l; H_c (in m) is the distance[4] from the maximum liquid level (the top capacity level) to the bottom of the shell course under consideration; p in mbar is the design pressure.

The thickness t in mm of cylindrical shell plates stressed by pressure of the tank contents should be calculated by the following formula (Point 3.7, AWWA D100-11) which is quite similar to expressions (7.1):

$$t = \frac{4.9h_p DG}{sE} \tag{7.12}$$

where h_p in m is the height of the liquid from the top capacity level[5] to the bottom of the shell course being designed, D is the nominal tank diameter (in m), s is the allowable design stress of the steel course in MPa, E is a joint efficiency parameter depending on the type of welding[6] and G is the design-specific gravity of the stored liquid and is classically defined as a dimensionless unit (in the case of water, it is equal to 1.0) which is the ratio of density of a fluid to the density of water.

7.2.3 Openings

In many industrial applications, shells are equipped with openings of various shape, size, and location within their lateral surface. Even if openings are invariably necessary in shell structures for a large variety of functional requirements, some work has been done in the literature on this subject. Early contributions include the research of Tennyson [607], Almroth and Holmes [10], Starnes [562], Toda [614], Knödel and Schulz [311]. In the last two decades, the most significant contributions to improve the understanding of the effects of cutouts on the critical buckling load of thin cylindrical shells were provided by Samuelson and Eggwertz [502], Jullien and Limam [279], Hilburger et al. in [232] and Brunesi [52]. Furthermore,

---

4 As in Equations (7.1) and (7.10a) the hoop stress in each course should be computed at 0.3 m above the centreline of the horizontal joint in question (Point 9.2.3, UNI EN 14015:2006).
5 The water level defined by the lip of the overflow (Point 1.2, Definition 10.3, AWWA D100-11).
6 The joint efficiency parameter may be derived from Table 15, Point 3.7, AWWA D100-11 in the case of single/double-groove butt joints with complete or partial penetration or a transverse lap joint.

field observations following past earthquakes have revealed the seismic vulnerability of storage steel tanks typical of the past design practice, highlighting structural deficiencies mostly related to the lack of structural seismic design and detailing, lack of redundancy, and inadequate anchorage design and execution. Figure 7.2(a) shows an example of the buckling mechanism that occurred in correspondence to an elliptical opening in the lateral surface of the shell. Even though reinforcement was provided in this case, the damage pattern at failure reveals the significant influence of the cutout in the evolution of this collapse mode (Figure 7.2(b)), characterized by a large outward bulging, which began to develop at mid-height of the hole. Stress/strain concentrations suddenly spread throughout the shell, causing more prominent deformation to occur between the opening and the base of the tank wall. In light of this scenario, practitioners mainly face the need for operative guidelines or codes for the evaluation of the buckling strength of cylindrical steel thin shells, when structural openings (nozzle or manhole) are included in the lateral surface of the structure or in the roof.

Manholes are used to guarantee access inside fixed roof tanks or shell walls for maintenance and inspection and cleaning purposes. Manholes through the roof have the advantage that they are always accessible, even when the tank is full, while access through the shell wall is more convenient for cleaning out. Nozzles are required through the shell roof, wall, and bottom for inlet, outlet, and drainage pipes, and for vents. They are normally made by welding a cylindrical section of plate into a circular hole in the structural plate. Stress concentrations cause high local stresses at openings cut into the shell of tanks. The idea behind codes, of reinforcing the opening, is to "replace" the amount of material removed adjacent to the opening through which the stresses may flow [414]. Furthermore, the addition of material (reinforcement) must be placed within definite limits around the openings, as it is ineffective beyond these limits. Point 5.7.2.1, API 650 suggests that the minimum cross-sectional area[7] of the required reinforcement should not be less than the product of the vertical diameter of the hole cut in the shell and the nominal plate thickness. The "area replacement method" is suggested also in Point 13.1.4, UNI EN 14015:2006

(a)   (b)

Figure 7.2 Buckling mechanism in correspondence to an opening during the Emilia earthquake, Italy (20 and 29 May 2012, Magnitudes 6.1 and 5.9 respectively). Source: From [52, 53]/with permission from SAGE Publications.

---

7 The areas are measured vertically, along the diameter of the opening, and must be within a distance above or below the horizontal centreline equals to the vertical dimension of the hole.

where the cross-sectional area of reinforcement provided, measured in the vertical plane containing the axis of the mounting, should be not less than (in agreement with Point 5.4.6.3(3), UNI ENV 1993-4-2):

$$0.75 \cdot d \cdot e_1 \tag{7.13}$$

where d is the diameter of the hole cut in the shell plate (in mm) and e_1 (in mm) is equal to e_c in expression (7.11), or the minimum nominal shell thickness, including any corrosion allowance in accordance with Table 7.4. It is clear, as highlighted by Myers in [414], that when the standard nozzle and manhole details of API 650 are used, the amount of reinforcing actually available is usually conservative. Point 13.2, UNI EN 14015:2006 specifies that no additional reinforcement is required for nozzles less than 80 mm in outside diameter, which is in agreement with the request coming from Point 5.4.6.2(1), UNI ENV 1993-4-2. Conversely, Point 3.13, AWWA D100-11 requires reinforcement for openings greater than 102 mm (4 in) in diameter in the tank shell, suspended bottom, riser plating, and other locations that are subject to membrane tension stress caused by fluid pressure. In a way similar to API 650 and UNI EN 14015:2006, Point 3.13.1, AWWA D100-11 suggests that the minimum cross-sectional area of the reinforcement should not be less than the product of the maximum dimension of the hole cut in the tank plating perpendicular to the direction of the maximum stress (approximately the the vertical diameter of the hole cut in the shell) and the required shell plate thickness.

Point 5.4.6.4(1), UNI ENV 1993-4-2 ignores the effect of openings on the stability of the shell wall, provided that the dimensionless opening size satisfies the following inequalities:

$$\frac{r_0}{\sqrt{rt}} < 0.6 \tag{7.14}$$

where r is the radius of the cylindrical shell near the opening, t is the thickness of the unstiffened shell wall near the opening and r_0 is the radius of the opening equal to $(a + b)/4$ where the opening is rectangular (a and b are the horizontal and vertical side lengths of the opening, respectively). According to Point 5.4.6, UNI ENV 1993-4-2, where an opening in the cylindrical shell wall reduces the load, carrying capacity or endangers the stability of the shell, the opening should be reinforced:

- increasing the thickness of the shell as depicted in Figure 7.3(a) by inserting a variable (or uniform) transition plate. In Appendix P of API 650 [332], a procedure has been described for evaluating the allowable loads on tank shell openings. This procedure is a practical solution to a very complex problem of stress localization, especially when low-type nozzles are close to the plate bottom and thus are affected by the bottom-to-shell junction. As mentioned by Billimoria and Hagstrom in [38], this procedure is conservative, but, even though Appendix P in API 650 is not mandatory, many designers use this method for lack of any other guidance.
- adding a (circular) reinforcing plate (Figure 7.3(b)), the limit of reinforcement being such that $1.5d < d_r < 2d$, where d_r is the effective diameter of reinforcement (in mm). A non-circular reinforcing plate may be used provided the minimum requirements are met.
- inserting a thickened nozzle body as in Figure 7.3(c). The portion of the body nozzle which may be considered as reinforcement is that lying within the shell plate thickness e and within a distance of four times the body thickness $4e_n$ from the shell plate surface, where e_n is the thickness of the body nozzle.

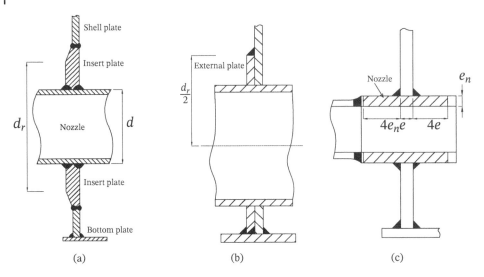

Figure 7.3 Reinforcing details: addition of a thickened shell insert plate, linear (a) internal and (b) external; or (c) addition of a thickened nozzle shell reinforcement or manhole body.

7.2.4 Roof

Tank roofs are basic components in a steel containment structures: they possess the basic function of keeping the external air pollution out of the fluid/material stored product and keeping the inside vapours out of the atmosphere. Fixed roof and floating roof are the two main types of tank roof and are available in a number of different forms. Fixed roofs for cylindrical tanks may be basically subdivided into two categories: self-supported or column-supported dome or cone roofs. Roof design starts with the definition of the roof dead and live loads and, consequently, the most severe load combination. Then, according to Point 5.10.2.2, API 650, roof plates should have a nominal thickness (excluding any corrosion allowance) of not less than 5 mm, while in Point 10.3.3, UNI EN 14015:2006 (fully in agreement with Points 11.2.2(1) and 11.2.3(1), UNI ENV 1993-4-2) the specified thickness of all roof plating should be not less than 5 mm for carbon and carbon manganese steels, and 3 mm for stainless steels (both of them excluding any corrosion allowance). Roof plating without supporting structure (membrane roof) should be designed to resist buckling (Point 7.1.2, UNI ENV 1993-4-2 and Point 10.4.2, UNI EN 14015:2006). Unfortunately, the problem of the instability of steel spherical or conical domes is not explicitly/theoretically treated in UNI ENV 1993-1-6 (or in the US guidelines). This is the reason why in the following, reference is made to the classic formulations reported by Tymoshenko in [612], Nazarov in [421], Flügge in [169] or Donnell in [131], Ramm in [466], and Gioncu in [185]. We will present the main results, then referring to detailed texts, like the one written by Calvi and Nascimbene [66], in order to obtain the analytical derivation.

The most important applications of the dome in a tank's construction are that of shallow shells, the rise of which is consistently small compared to the span. Vlasov in [654] suggests the ratio $\frac{f_{max}}{d_{min}} < \frac{1}{5}$ as limit of applicability of the relations of shallow shells (Figure 7.4) where f_{max} is the maximum rise and d_{min} is the minimum dimension of the shell contour. The complete solution of the problem of the stability of a shallow spherical dome subject to a uniform pressure normal to the middle surface [66], may be obtained

Figure 7.4 Spherical shallow shell roof subjected to external uniform pressure: geometry and buckling load derivation.

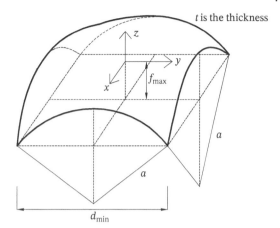

t is the thickness

starting from the formulation given by Marguerre [385] and Vlasov [653, 654], which gives the same recommendations highlighted in [493]. In the derivation of their equations, Marguerre and Vlasov, in addition to the general assumptions used for plates and shallow surface (originally written by Kirchhoff),[8] introduce supplemental hypotheses following the Donnell-Mushtari-Vlasov[9] theory [188, 456]:

1) the effects of the transverse shear forces in the in-plane equilibrium equations are negligible;
2) the influence of the normal displacement will predominate over the influence of the in-plane displacements in the bending response of the shell.

A spherical cap without imperfections, like the one in Figure 7.4, preserves its state of equilibrium produced by a uniform radial pressure $-p$ by compression axial stresses, uniformly distributed over the whole surface ($N_{xy} = 0$):

$$N_x = N_y = \frac{ap}{2} \tag{7.15}$$

---

8 Kirchhoff [306] based his reasoning on several assumptions (used by Aron in [14]), analogous to those used by de Saint Venant in his theory of beams:

1) material is homogeneous, isotropic, and linearly elastic;
2) thickness is small compared to the radius and deflection is small compared to the radius (which means that slopes are small compared to unity) and to the thickness;
3) straight fibres which are perpendicular to the middle surface before deformation remain so after deformation and do not change their length (plane sections remain plane and shear deformations can be ignored);
4) transverse normal stress is negligibly small.

A few inaccuracies were noticed and corrected by Love in [347] and this is the main reason why these assumptions are usually called *Kirchhoff-Love hypotheses*.

9 Donnell [130], Mushtari [413] and Vlasov [653, 654] ("DMV theory" [188, 456]; independently derived the expressions in the USA (Donnell) and in the Soviet Union (Mushtari), then Vlasov provided explicit solutions. Their equations give the same information in UNI ENV 1993-1-6 [29]. DMV theory is usually chosen because it is believed to be the simplest possible form of equations for buckling analysis and hence to be especially well suited for professional purposes. This theory is a clear simplification of the more general curvilinear formulation derived in 1888 by Love [347, 348] and then independently improved by Sanders [504], Morley [407], and Koiter [310]. Love's formulation may be considered an extension to the shell of the Kirchhoff's theory (1876) [615] originally derived for plates [658].

Moving from this state of stress, it was Zoelly in [689] and Schwerin in [517] who first derived the elastic upper critical loading p_{cr} of the complete spherical shell, assuming an axisymmetric buckling pattern[10]:

$$p_{cr} = \frac{2}{\sqrt{3(1-v^2)}} E\left(\frac{t}{a}\right)^2 \tag{7.16}$$

where t and a are the uniform thickness and radius of the shell middle surface, respectively. Many experimental tests [314] have shown substantially lower values than the linear critical load in expression (7.16). The reason for such a difference, has been explained for the first time by von Karman and Tsien in [293] where it has been proved that states of equilibrium exist with surfaces where a *snap-through* occurred under loads smaller than the upper critical one (7.16). To be clear, if we consider a clamped spherical arc in Figure 7.5(a), the linear theory (Equation (7.16)) gives for the critical load of this arc a festoon curve, extended to the whole surface, with a horizontal lower tangent (Figure 7.5(b)). However, the loss of stability can appear also by a *snap-through* in a local portion (Figure 7.5(c)) or by non-symmetrical deformations extended to the whole surface (Figure 7.5(d)). These results obtained for an arc can also be extended to domes, as reported in Figure 7.6, where a local *snap-through* buckling phenomena appears evident.

In order to account for these phenomena, Wunderlich and Albertin in [676, 677] (and mostly Wunderlich's recommendations given in Chapter 15 [493, 675]), based on extensive numerical (and experimental) analyses, including geometrically perfect and imperfect spherical shells of constant thick-

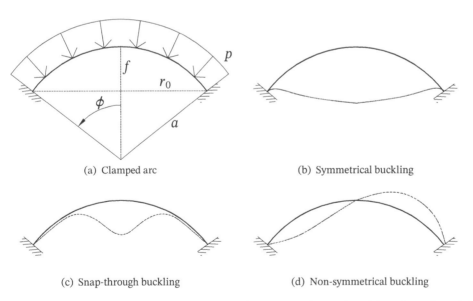

(a) Clamped arc

(b) Symmetrical buckling

(c) Snap-through buckling

(d) Non-symmetrical buckling

Figure 7.5 Buckling of (a) a spherical arc and different kinds of loss of stability: (b) buckling of the whole shell; (c) local snap-through and (d) non-symmetrical mode (ϕ is the semi-angle and r_0 the radius of the base circle of the spherical cap). Source: From [555]/with permission from John Wiley & Sons.

10 Later Arie van der Neut in [422] showed that the same value is obtained also assuming an asymmetric buckling pattern.

Figure 7.6 Buckled locally surface (*snap-through*) in domes on a square boundary. Source: From [185]/with Courtesy of Professor Gioncu.

ness, different boundary conditions and made of elastic-plastic material, subjected to external uniform pressure, modified expression (7.16):

$$p_{Rcr} = \frac{2}{\sqrt{3(1-v^2)}} C_c E \left(\frac{t}{a}\right)^2 \tag{7.17}$$

The rules, derived from Equation (7.17), are applicable only within these ranges $\frac{a}{t} \leq 3000$, $\phi \leq 135°$ and $\phi = 180°$; furthermore, they are derived in case the steel shell segments are connected by welded butt-joints or by bolted symmetrical double-lap-joints. The non-dimensional coefficient C_c describes the buckling resistance reduction as a function of the boundary conditions as depicted in Figure 7.7 where:

1) **BC2** (Figure 7.7(a)) corresponds to a spherical cap with clamped edges (fixed).
2) **BC3** (Figure 7.7(b)) corresponds to a spherical cap with radially and meridionally restrained, hinged edges.
3) **BC4** (Figure 7.7(c)) corresponds to a spherical cap with radially and meridionally restrained edges, but radially free edges normal to the shell midsurface (radially restrained but pinned).
4) **BC5** (Figure 7.7(d)) corresponds to a spherical cap with radially free edges in the plane of the base circle (support free to slide radially).

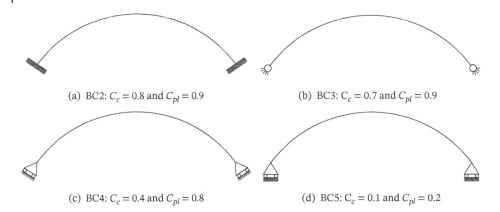

(a) BC2: $C_c = 0.8$ and $C_{pl} = 0.9$

(b) BC3: $C_c = 0.7$ and $C_{pl} = 0.9$

(c) BC4: $C_c = 0.4$ and $C_{pl} = 0.8$

(d) BC5: $C_c = 0.1$ and $C_{pl} = 0.2$

Figure 7.7 Factors C_c used for the evaluation of the elastic critical buckling pressure p_{Rcr} and C_{pl} used for the evaluation of the plastic reference resistance p_{Rpl} in expression (7.19); both functions of the different boundary conditions. Source: Adapted from [675].

5) Previous constraints are very simple support conditions used at the dome edge and usually only three (BC2/4 and 5) can potentially be applied to the practical design of a tank; real dome shells are always constructed with an eaves ring, edge ring, curb angle, or wind girder. More insight into this practical question can be found in a very recent paper from Rotter et al. in [492].

Following the notations at Point 8 (LS3), UNI ENV 1993-1-6:2007 and the recommendations given by Wunderlich in [675], the buckling strength verification should be (Point 8.6.3 (1), UNI ENV 1993-1-6:2007):

$$p_{Ed} \leq p_{Rd} = \frac{p_{Rk}}{\gamma_M} \tag{7.18}$$

where p_{Ed} is the design value of the pressure, p_{Rd} the design buckling resistance, p_{Rk} the characteristic buckling pressure and $\gamma_M = 1.1$ the safety factor taken from Note to Point 8.5.2(2), UNI ENV 1993-1-6:2007. The characteristic buckling pressure should be determined as a function of the plastic reference resistance p_{Rpl} following explicitly the buckling reduction factor approach proposed in Point 8, UNI ENV 1993-1-6:2007:

$$p_{Rk} = \chi p_{Rpl} = \chi f_{yk} C_{pl} \frac{2t}{a} \tag{7.19}$$

where the coefficient C_{pl} covers the yield load reduction caused by the different boundary conditions of the elastic-plastic spherical caps: it is equal to one just for a perfect shell and smaller than one as reported in Figure 7.7 [676, 677] because the effects of boundary conditions cause earlier plastification. The buckling reduction factor $\chi = f(\lambda)$ for elastic-plastic buckling comes from Equations (7.93) where:

$$\lambda = \sqrt{\frac{p_{Rpl}}{p_{Rcr}}} = \sqrt{\frac{f_{yk} C_{pl} \frac{2t}{a}}{p_{Rcr}}} \tag{7.20a}$$

$$\lambda_0 = 0.20 \tag{7.20b}$$

$$\eta = 1.0 \tag{7.20c}$$

$$\beta = 0.70 \tag{7.20d}$$

$$\lambda_p = \sqrt{\frac{\alpha}{1 - \beta}} \tag{7.20e}$$

The elastic imperfection factor α in Equation (7.20e) should be derived in this way [677]:

$$\alpha = \frac{0.70}{1 + 1.90\left(\frac{\Delta w_k}{t}\right)^{0.75}} \tag{7.21}$$

where the characteristic imperfection amplitude $\Delta w_k = \frac{\sqrt{at}}{Q}$ is a function of the non-dimensional fabrication quality parameter Q depending on the specified fabrication tolerance quality: Class A (excellent) $Q = 40$, Class B (high) $Q = 25$, Class C (normal) $Q = 16$ (Appendix D.1.2.2, UNI ENV 1993-1-6:2007).

Expressions (7.17)–(7.21), derived by Wunderlich in [675], are applicable within the ranges given by $\frac{t}{a} \geq \frac{1}{3000}$ and $\phi \leq 135°$ (or $\phi = 180°$ that correspond to a complete sphere); furthermore, it is possible to avoid checking the resistance to buckling if the following conditions are satisfied:

$$\frac{a}{t} \leq \frac{EC_c}{20f_{yk}} \tag{7.22}$$

or, for very flat spherical shells, when (Point 7.2, DIN 18800-4):

$$\frac{r_0}{a} \leq \frac{1.1}{\sqrt{\frac{a}{t}}} \tag{7.23}$$

where $r_0 = a \sin \phi$ as depicted in Figure 7.5(a).

Like the dome roof seen above, the thickness of a self-supported cone roof is based on the elastic stability of a cone under external pressure, as analytically and explicitly derived in [344, 414]. According to Point 5.10.5.1, API 650, the nominal thickness of a self-supported cone roof with a slope ϕ within the range of 9.5° to 37° should not be less than the greater of:

$$\frac{D}{4.8 \sin \phi} \sqrt{\frac{T}{2.2}} + CA \qquad \frac{D}{5.5 \sin \phi} \sqrt{\frac{U}{2.2}} + CA \quad \text{and} \quad 5 \text{ mm} \tag{7.24}$$

where D is the nominal diameter of the tank in m and ϕ in degrees is the angle of the cone roof to the horizontal and the desired corrosion allowance is CA; the corroded thickness should not be more than 13 mm; T and U in kPa may be derived by the greater between the two following load combination using balanced snow load S_b (determination of T) and unbalanced snow load S_u (determination of U) respectively (Appendix R, API 650 and Point 7, ASCE 7):

$$D_L + (S_b \text{ or } S_u) + 0.4P_e \tag{7.25a}$$

$$D_L + 0.4(S_b \text{ or } S_u) + P_e \tag{7.25b}$$

where D_L is the dead load and P_e the external pressure which should be considered equal 0 kPa in the case of tanks with circulation vents meeting Appendix H requirements in API 650. Expressions similar to (7.24) can also be used for a self-supported dome roof, according to Point 5.10.6.1, API 650:

$$\frac{a}{2.4} \sqrt{\frac{T}{2.2}} + CA \qquad \frac{a}{2.7} \sqrt{\frac{U}{2.2}} + CA \quad \text{and} \quad 5 \text{ mm} \tag{7.26}$$

where, as previously defined, a is the spherical radius of the dome in m in the range 0.8D to 1.2D.

7.2.5 Foundation

It is really important that uniformly supported flat-bottomed tanks are provided with suitable foundations, although tank foundations in many aspects are similar to other foundations. In the early stages of foundation design, there are many considerations to be made, such as [20, 344, 414]:

- shape of the foundation that is mainly governed by the soil characteristics, the design loadings, the presence or absence of facilities around the tank eventually for future expansion;
- site and soil conditions, including soil-bearing capacity and soil characteristics and stability (soil report which creates an important advantage in the design process of the tank). Once the suitability of the site has been established, also based on a visual reconnaissance, the geotechnical composition of the soil should be investigated, also defining boring locations (minimum three [20]), load tests, sampling, and laboratory testing. If the soil is found to be inadequate and the tank cannot be relocated, soil improvements may be used (stabilization by grout injection, sub-soil drainage, removal of inadequate material and replacement by suitable compacted fill, vibrocompaction or dynamic compaction);
  - applicable guidelines, codes and standards;
  - acceptable amounts and rate of settlement (and eventual costs associated with ground contamination).

Many possible types of tank foundation are common for specific applications, generally with different order of costs:

1) compacted soil without a ringwall: it is classically used when soil quality and bearing capacity are good. The material in the foundation should serve to raise the tank above the surrounding area and provide a competent founding material for the tank (Point I.4.2, UNI EN 14015:2006 and Figure 7.8(a)). Furthermore, one main advantage is the relatively low cost, but the underside tank bottom is more susceptible to accelerated corrosion. This foundation should be avoided when a tank is designed for uplift because anchorage cannot be provided easily [414]. Generally, the top 6 to 15 cm are removed and replaced by bitumen sand mix, usually called sand pad foundation which is laid directly on the earth.
2) earth foundation with a concrete ringwall: it is classically used in large diameter tanks, with good to medium properties of soil, where it is the most cost-effective solution. The ringwall is of reinforced concrete and it provides good anchorage restraint for the tank's wall. The ring beam should be designed to withstand horizontal pressures from the contained earth mound, including all surface effects from the tank and contents (Point I.4.3, UNI EN 14015:2006 and Figure 7.8(b)). It usually minimizes differential shell settlement and corrosion problems. The membrane vapour barrier at the base should be selected considering the temperature and stresses to which it may be subject in service and under exceptional conditions. The load acting on the ringwall is the sum of the tank's wall weight, the portion of the roof supported by it, the liquid content projecting vertically on the ringwall and the weight of the ringwall itself.
3) earth foundation with a crushed stone or gravel ringwall: this solution has the huge advantage of costing less than the reinforced concrete ringwall, however, still providing uniform support and application on any size of tanks.
4) concrete slab foundation: it is usually limited to tanks with relatively small diameter and provides a good plane base for tank construction minimizing the entry of water from the base and thus corrosion problems. The edge of the slab can be thickened to provide for anchorage as in the case of a single ringwall (Figure 7.8(c)). In the design of the slab, provision should be made to accommodate the

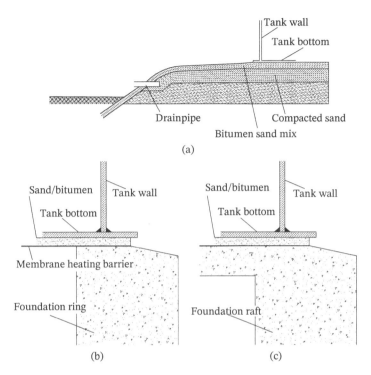

Figure 7.8 Typical tank foundations: (a) pad, (b) ring beam, and (c) raft foundation (UNI EN 14015:2006).

effects of local differential settlements, drying shrinkage, creep, and thermal strains during service or under upset conditions (Point I.4.4, UNI EN 14015:2006; ACI 209R-92). Such provisions may involve additional reinforcement, prestress or post-tensioned cables.

5) concrete slab foundation with supporting piles: this solution is used if the soil-bearing pressure is very low and it can be applied to any size of tanks. Unfortunately this is a very costly solution that requires a lot of geotechnical soil information.

In the case of tanks that require holding-down anchors against uplift resulting from wind, internal pressures or seismic overturning moment (Point 12.1, UNI EN 14015:2006), the foundation will usually be a reinforced concrete ringwall or a slab.

Overturning loads can be mainly induced by either of two possibilities that have to be studied separately: seismic and wind action. In high seismicity zones, usually wind action will not govern either the foundations or the anchors design requirements. Instead, the design conditions will be established by the overturning due to the hydrodynamic components of the movement due to the earthquake. According to Point E.6.2.1.1, API 650, the tank can be assumed to be self-anchored providing that the following five conditions are met:

1) the anchorage ratio, defined using Equation (3.9), must be ≤ 1.54;
2) the maximum width of the annulus for determining the resisting force must be 3.5% of the tank diameter (for more details see Equation (3.13));
3) the compression in the shell must satisfy Equations (3.11) and (3.12);

4) the thickness of the bottom annular plate ring, s_b, must not exceed the thickness of the bottom shell course, s_t;
5) piping flexibility requirements are satisfied.

In the case that the self-anchored requirements are not met for the tank's configuration, the tank must be anchored with mechanical devices. The anchor bolts or straps should be sized to provide for at least the following minimum anchorage resistance (Point E.6.2.1.2, API 650 and Point G.5.1, UNI EN 14015:2006):

$$w_{AB} = \frac{1.273M_y}{(2R)^2} - w_t(1 - 0.4S_{vd}(T_V))$$ (7.27)

plus the uplift w_{int}, in N/m, of the shell circumference, due to design internal pressure (see Equation (3.9) for an explanations regarding the symbols involved in expression (7.27)). Hence, the anchor seismic design load P_{AB} is:

$$P_{AB} = w_{AB}\left(\frac{2\pi R}{n_A}\right)$$ (7.28)

where n_A is the number of equally spaced anchors around the tank circumference. In Point 3.5.2.3 (1)P, EC8-4 suggests that anchoring systems should generally be designed to remain elastic in the seismic design situation. However, they should also be provided with sufficient ductility, so as to avoid brittle failures.

Overturning moments induced by wind action on a large height to radius ratio tank can be very significant. According to Point 5.11.2, API 650, unanchored tanks must satisfy the following two uplift criteria:

$$0.6M_w + M_{Pi} < \frac{M_{DL}}{1.5} + M_{DLR}$$ (7.29a)

$$M_w + F_P \cdot M_{Pi} < \frac{M_{DL} + M_F}{2} + M_{DLR}$$ (7.29b)

where, as clearly depicted in Figure 7.9, M_w is the overturning moment about joint A in Figure 7.9 from the horizontal plus the vertical wind pressure; M_{Pi} is the moment about joint A from the design internal pressure P_{int}; M_{DL} is the moment about joint A from the nominal weight of the shell wall and roof structure supported by the shell that is not attached to the roof plates (this means the moment about the shell-to-bottom joint from the nominal weight of the shell plus attachment weight); M_{DLR} is the moment about joint A from the nominal weight of the roof plate plus any attached structure (simply the moment about the shell-to-bottom joint from the nominal weight of the roof plate plus any attached structure); M_F is the moment about joint A from the liquid weight. Equation (7.29a), according to Myers in [414], says that to maintain a factor of 1.5, for unanchored tanks, the overturning moment, more or less, should not exceed two-thirds of the dead load resisting moment. When the requirements of (7.29) cannot be satisfied, anchorage must be provided to resist each of the net uplift load (U) cases listed in Point 5.12.2, API 650 and specifically: design, test and failure pressure, wind and seismic load, design pressure plus wind and design pressure plus seismic, frangibility pressure. The load per anchor is:

$$L_{anchor} = \frac{U}{N}$$ (7.30)

where U (in Newton) is the net uplift load according to the previously defined cases and N the number of anchors where a minimum of four is required. Point 3.8.9.1, AWWA D100-11 in a fashion similar to

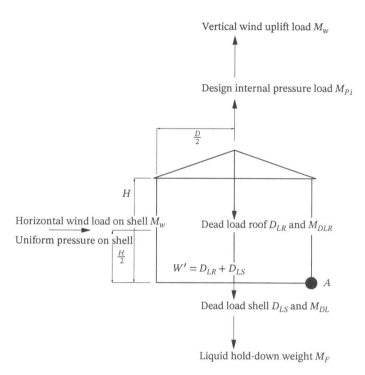

Vertical wind uplift load M_w

Design internal pressure load M_{Pi}

$\frac{D}{2}$

H

Horizontal wind load on shell M_w

Uniform pressure on shell

$\frac{H}{2}$

Dead load roof D_{LR} and M_{DLR}

$W' = D_{LR} + D_{LS}$

A

Dead load shell D_{LS} and M_{DL}

Liquid hold-down weight M_F

Figure 7.9 Overturning check in case of unanchored tanks: wind actions (Point 5.11.2, API 650).

Point 5.12.2, API 650, suggests that anchors should be designed for the maximum effect of the design uplift forces, including the wind uplift force P_W and the seismic uplift force P_S (both in N). Then, the design uplift forces P_W and P_S represent the net uplift force L_{anchor} to be resisted by the anchor after consideration for any reductions resulting from the dead weight of the structure. When the calculated net uplift force for both P_W and P_S results in a negative value, no uplift anchorage is required. Expression (7.30) in Point 5.12.2, API 650, has been transformed in Point 3.8.9.1, AWWA D100-11 into the following two expressions:

$$P_W = \frac{4M_w}{Nd_{anc}} - \frac{9.81W'}{N} \tag{7.31a}$$

$$P_S = \frac{4M_y}{Nd_{anc}} - \frac{9.81W'}{N} \tag{7.31b}$$

where d_{anc} (in m) is the diameter of the anchor circle; M_w and M_y are in Nm; W' (in Kg) is the shell weight plus roof dead load reaction on the shell (this means the weight of the tank shell and portion or roof supported by it as in Figure 7.9).

The four most commons anchors used (Point 3.8.3, AWWA D100-11 and Annex M in UNI EN 14015:2006) are holding-down strap (Figure 7.10(a)), holding-down bolt with individual chair (Figures 7.10(b) and 7.10(c)) or with continuous support ring (Figures 7.10(b) and 7.10(d)), and a combination using strap and bolt anchorage [344]. Regardless of the type of anchorage system, it is important that it fails (in ductility) before the tank's shell, the anchor chair or the foundation attachment,

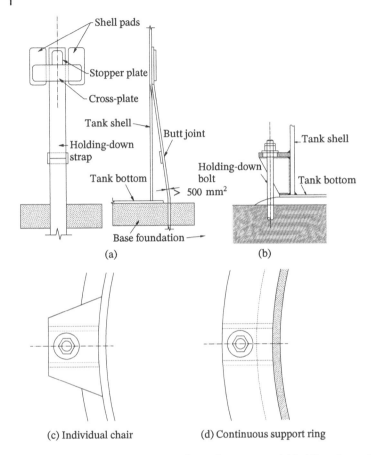

Figure 7.10 Typical arrangement for anchor systems: (a) holding-down strap; (b) and (c) holding-down bolt with an individual chair; (b) and (d) holding-down bolt with a continuous support ring (Annex M, UNI EN 14015:2006).

in order to prevent the loss of containment contents. Furthermore, Point 12.3.2, UNI EN 14015:2006 states that each holding-down bolt or strap, without any initial tension applied, should have a minimum cross-sectional area of 500 mm² (according to Point 3.8.5.1, AWWA D100-11 the minimum anchor-bolt diameter should be 1 in (equal to 25.4 mm)); it is also recommended that anchor points are spaced at a maximum of 3 m intervals and should, as far as possible, be spaced evenly around the circumference. Point 3.8, AWWA D100-11 for ground-supported flat-bottom reservoirs, establishes that the maximum anchor spacing in mechanical anchorage should not exceed 3 m (10 ft – same in Point E.7.1.2, API 650) and that the minimum number of anchor bolts should be 6.

Tank and foundation may experience movements due to earthquake loading. Hence, piping systems connected to the wall of the tank have to consider potential movement of the connection points and provide enough flexibility to avoid loss of product. Unless otherwise numerically evaluated, Point E.7.3, API 650 provides for a minimum design displacement (Table 7.5) intended as a compromise of practical design consideration, economics, and probability that the piping connection will be at the point of

Table 7.5 Design displacement (Point E.7.3, API 650) in the case of piping flexibility.

Type of anchorage at the base	Displacement (mm)
Mechanically-anchored tanks	
Upward vertical displacement relative to support or foundation	25
Downward vertical displacement relative to support or foundation	13
Radial/tangential displacement relative to support or foundation	13
Self-anchored tanks	
Upward vertical displacement relative to support or foundation	
Anchorage ratio (Equation (3.9)) ≤ 0.785	25
Anchorage ratio (Equation (3.9)) > 0.785	100
Downward vertical displacement relative to support or foundation	
Tanks with a ringwall/slab or mat foundation	13
Tanks without a ringwall or with berm foundation	25
Radial/tangential displacement relative to support or foundation	50

maximum uplift. As clearly underlined by Point EC.7.3, API 650 mechanically anchoring the tank to reduce piping flexibility demands should be a "last resort". The cost of anchoring a tank that otherwise need not be anchored will often be greater than altering the piping configuration and, as a consequence, the cost of the anchors, the foundation type and the attachments should be weighted against piping flexibility devices and configuration changes.

7.2.6 Stiffeners

In the case of fixed roof tanks, wind load or, more in general, external pressure act only on the outer surface of the container, whereas in open top or floating roof tanks pressure also acts on the inner surface which can cause the effect of a vacuum load (due also to tank venting problems or extreme events to open vents and pressure relief valves). A fixed roof can assist the wall of a tank in keeping the shell stiff, but generally many tanks do not have the benefit of this shell rigidity. Hence circumferential ring stiffeners in the case of external pressure and longitudinal stiffeners (usually called stringers) in the case of meridional compression are provided at or near the top (primary girder) or along the height of the shell (secondary girder). The theory behind the design expression for buckling of rings (Section 7.2.7) and stringers (Section 7.2.8) goes back to many analytical, numerical and experimental activities carried out in several places, as clearly explained by Schmidt, Greiner, and Samuelson in [509, 512].

7.2.7 Rings

Rings are widely used in conjunction with shell structures such as pressure vessels, tanks, silos, and roofs over large enclosed spaces. Such ring-stiffened shell junctions may fail due to buckling if compression forces are dominant [502]. This section deals with the buckling of isolated rings of arbitrary cross-section and rings at steel tank transition junctions, as defined by Teng and Zhao in [605]:

1) for an isolated ring of arbitrary cross-section, buckling can occur:
   - in plane, as depicted in Figures 7.13 and 7.14, following the classical stability solution proposed by Timoshenko in [612] involving only periodical flexural deformations;
   - out of plane using a closed-form solution for the elastic buckling obtained by Timoshenko in [612] (Figure 7.15); more insight into flexural-torsional buckling of monosymmetric arches can be derived in the research of Trahair and Papangelis mainly in [616];
2) for a ring at steel silo or tank transition junctions, rules change for "heavy ring" (stiff[11]) or "light ring" (flexible). The following checks should be performed for light ring stiffeners, as proposed by Schmidt and Greiner in [509], Teng and Zhao in [605] and Greiner in [195]:
   - verify the ratio of ring bending stiffness to the plate-shell-wall stiffness, as described by Greiner in [192];
   - verify the flexural global buckling (wall-ring system) of the ring stiffeners in their circular plane which is crucial because it controls the global buckling strength of the stiffened shell [563]. This is to avoid the global collapse buckling represented in Figure 7.11(a).
   - verify shell wall buckling adjacent to the ring if the ring is stocky but the shell segment is thin [605]; this mode of buckling is reported in Figures 7.11(b) and 7.11(c);
   - verify the distorsional (Figure 7.12(a)) and local (Figure 7.12(b)) buckling mode of the ring stiffeners;
   - verify the out-of-plane torsional buckling of the ring stiffeners as depicted in Figure 7.12(c);
3) the heavy ring stiffeners (including end rings) should be checked according to the verification proposed by Schmidt and Greiner in [509].

7.2.7.1 In-Plane Buckling of Isolated Ring

Let us consider a circular ring of radius r and uniform bending stiffness EJ, under uniform external pressure p per unit length of the centreline of the ring (Figure 7.13(a)). The compressive force in the undeformed configuration is:

$$T = pr \tag{7.32}$$

and the bending moment $M = 0$. The critical value p_{cr} of the external uniform pressure is the minimum value which is necessary to keep the ring in equilibrium in the deformed two waves ($n = 2$) circumferential buckled mode depicted in Figure 7.13(b) (full line); the dotted line can be considered as a funicular curve for the uniform pressure. The bending moment at any cross-section of the buckled ring, considering only inextensional deformation [390], is $M = Tw = prw$ and the deflection curve can be written as [30, 32, 612]:

$$\frac{d^2w}{d\theta^2} + w + \frac{pr^3}{EJ}w = 0 \tag{7.33}$$

or using a different notation:

$$\frac{d^2w}{d\theta^2} + n^2w = 0 \qquad \text{where} \qquad n^2 = 1 + \frac{pr^3}{EJ} \tag{7.34}$$

---

11 The heavy ring is supposed, according to Greiner in [192], to maintain the circularity of the shell.

Figure 7.11 Basic examples of collapse/buckling modes for a ring-stiffened cylindrical shell due to hydrostatic pressure: (a) global collapse buckling; (b) and (c) local buckling asymmetric and symmetric modes. Source: From [319]/with permission of the American Institute of Aeronautics and Astronautics.

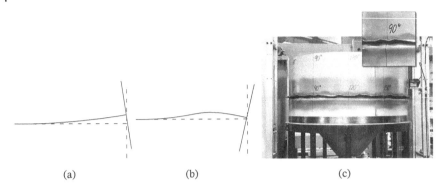

(a) (b) (c)

Figure 7.12 Two buckling modes of a clamped T-section ring: (a) distorsional buckling and (b) local buckling; (c) torsional buckling. Source: From [605, 606]/Elsevier.

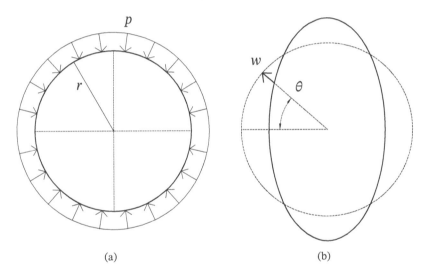

(a) (b)

Figure 7.13 (a) Full circular ring under uniform external pressure in its initial undeformed configuration and (b) a two-waves circumferential buckled mode. The dotted lines indicate the initial undeformed circular shape of the ring and the full line represents the deflected ring under a uniformly distributed pressure.

The solution of the homogeneous second-order Equation (7.34) is:

$$w = C_1 \sin n\theta + C_2 \cos n\theta \tag{7.35}$$

From the symmetry of the buckled ring, let us impose two conditions at the cross-section $\theta = 0$ and $\theta = \frac{\pi}{2}$ (Figure 7.13(b)), where the rotation $\frac{dw}{d\theta}$ must be equal to zero in order to find coefficients C_1 and C_2:

$$\left(\frac{dw}{d\theta}\right)_{\theta=0} = 0 \quad \Rightarrow \quad C_1 = 0 \tag{7.36a}$$

$$\left(\frac{dw}{d\theta}\right)_{\theta=\frac{\pi}{2}} = 0 \quad \Rightarrow \quad n = 2 \tag{7.36b}$$

The smallest root (7.36b) ($n = 2$), corresponding to the buckled shape in Figure 7.14(a), can be substituted into expression (7.34) to obtain the value of the critical pressure:

$$p_{cr} = \frac{3EJ}{r^3} \tag{7.37}$$

The corresponding critical compressive force can be derived substituting expression (7.37) in (7.32) [612]:

$$T_{cr} = p_{cr}r = \frac{3EJ}{r^2} \tag{7.38}$$

By taking $n = 3$ we obtain a critical pressure $p_{cr} = \frac{8EJ}{r^3}$ (Figure 7.14(b)), while using $n = 4$ we obtain $p_{cr} = \frac{15EJ}{r^3}$ (Figure 7.14(c)), both corresponding to higher-order buckling forms.

7.2.7.2 Out-of-Plane Buckling of Isolated Ring

In the case of rings at a steel tank/silo transition joint, out-of-plane buckling (Figure 7.15), involving twisting deformations, is generally the critical failure mode. The same ring depicted in Figure 7.13(a) has a critical out-of-plane compressive force equal to [32, 612]:

$$p_{cr} = \frac{EJ_y}{r^3} \frac{9}{4 + \frac{EJ_y}{C}} \tag{7.39}$$

where $C = \frac{GJ_p}{q}$; GJ_p is the torsional rigidity and G is the shearing modulus of elasticity; J_p is the (primary) polar moment of inertia of the cross-section (torsional constant); q is a numerical non-dimensional factor depending upon the shape of the section [610, 611]: $q = 1$ in the case of a circular section and greater than one for all the other types of section (for an infinite long rectangular section with $\frac{b}{d} \to \infty$, we have $\frac{J_p}{q} = \frac{1}{3}bd^3$, while for a square section with $b = d$, the expression is $\frac{J_p}{q} = \frac{1}{7.114}b^4$)[12]; in Equation (7.39) the

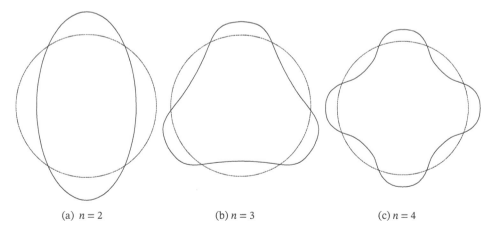

(a) $n = 2$ (b) $n = 3$ (c) $n = 4$

Figure 7.14 Series of the possible shape of buckled circular rings under uniform external pressure corresponding to three circumferential waves number: $n = 2$, 3, and 4.

---

12 More data on the twist of various rectangular sections can be found in Volume 1, Section 60 page. 270, of *Strength of Materials* by Timoshenko [610].

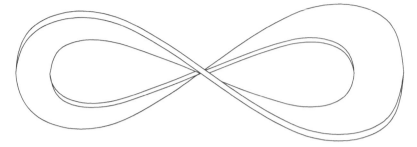

Figure 7.15 A complete ring buckled (out of plane) under the action of a radial pressure in four half-waves. Source: Adapted from [32].

coordinates are designated such that y is measured radially inwards in the plane of the ring while z is measured normal to the ring plane, and x, the circumferential coordinate, satisfying the right-hand-screw rule [612]. In deriving Equation (7.39) it has been assumed that the ring is free to rotate with respect to the principal axes (y, z) but unable to rotate with respect to the tangents to the centreline of the ring. For these kinds of rings, out-of-plane buckling is uncoupled from in-plane buckling, so out-of-plane buckling can be studied independently. The theory behind these assumptions considers only rings of a monosymmetric section with the axis of symmetry lying in the plane of the ring, and with the load and/or restraint located on this symmetry axis.

7.2.7.3 Buckling of a Light Ring-Stiffened Cylindrical Wall: Limiting Stiffness

Let us consider a circular cylindrical shell (Figure 7.16) of constant wall thickness t and radius r of the middle surface. The shell is stiffened by equidistant flexible (light) rings with cross-sectional area A_D and second moment of area J_{Dy} about the meridional axis y, and by heavy ring with cross-sectional area A_R and second moment of area J_{Ry}. The ring stiffeners may be attached externally or internally to the shell wall; l_R is the distance between the heavy stiffeners, l the distance between a heavy ring and end (or base) closure while l_D the length of unstiffened shell between light stiffeners. Two type of loads are applied to the tank as depicted in Figure 7.16:

- a uniform horizontal external pressure p_{Ed} along the vertical wall and a vertical uniform pressure on end closures if relevant;
- a vertical (meridional) external axial line load P_{Ed} acting along the top end closures of the tank.

The appropriate boundary conditions, according to Point 5.2.2, UNI ENV 1993-1-6, are summarized in Table 7.6 corresponding to the meridional displacement u, displacement normal to the shell surface w and the meridional rotation β_ϕ; BC1r is a clamped condition, while BC2f and BC3 are pinned and free edge conditions, respectively. The rules herein used, and derived from Schmidt and Greiner in [509], are applicable only to rings with radially (w) restrained edges providing boundary conditions BC1 or BC2. Possible cross-sectional forms of ring stiffeners that can be externally or internally attached to the wall of a tank are shown in Figures 7.17(a) to 7.17(c) [509] with the main geometrical properties defined (G represents the centroid/barycentre of the section). As depicted in Figure 7.17(d), the effective section[13]

---

13 The effective (or equivalent) section is made of the section of the ring plus an appropriate part of the wall where the ring is attached.

Figure 7.16 Steel-stiffened cylindrical tank under uniform external pressure p_{Ed} and axial load in the form of a meridional line force P_{Ed}. Source: [509]/European Convention for Constructional Steelwork.

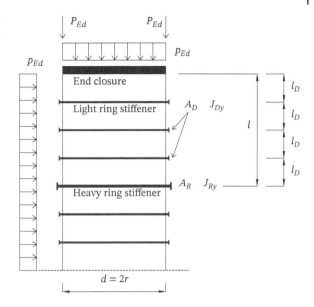

Table 7.6 Boundary conditions for shells, according to Point 5.2.2, UNI ENV 1993-1-6.

Codice	w	u	β_ϕ
BC1r	0	0	0
BC1f	0	0	$\neq 0$
BC2r	0	$\neq 0$	0
BC2f	0	$\neq 0$	$\neq 0$
BC3	$\neq 0$	$\neq 0$	$\neq 0$

properties of a ring stiffener should be determined prior to the buckling verification (G_e represents the centroid/barycentre of the effective section):

• cross-sectional area of the effective light stiffener:

$$A_{De} = A_D + b_{eN}t \tag{7.40}$$

where b_{eN} is the effective width of the cylinder wall for circumferential force:

$$b_{eN} = b_0 + 1.56\eta_l \sqrt{rt} \tag{7.41a}$$

$$\eta_l = 0.64\frac{l_D - b_0}{\sqrt{rt}} \leq 1 \tag{7.41b}$$

• second moment of area of the effective light stiffener:

$$J_{De} = J_{De}^\star + \frac{b_{eM}t^3}{12} \tag{7.42}$$

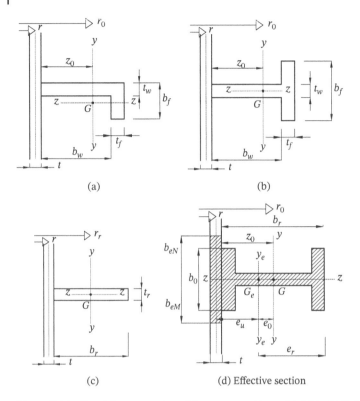

(a)

(b)

(c)

(d) Effective section

Figure 7.17 Possible geometries of ring cross-section with (a) L, (b) T, (c) I (or flange) and (d) H shape; furthermore, (d) is the effective section for buckling verification.

where:

$$J_{De}^{\star} = J_{Dy} + A_D e_0^2 + b_{eM} t e_u^2 \tag{7.43}$$

The distance (eccentricity) e_0 in a stiffener cross-section (G) relative to the axis of the effective cross-section (G_e) in Figure 7.17(d) is:

$$e_0 = \left(\frac{b_r}{2} + \frac{t}{2} \right) \frac{b_{eN} t}{A_{De}} \tag{7.44}$$

from which we can derive $e_u = \left(\frac{b_r}{2} + \frac{t}{2} \right) - e_0$. The effective width of the cylinder wall for circumferential bending b_{eM} in expressions (7.42) and (7.43), should be determined as given in Table 7.7 where:

$$\lambda_r = 1.285 \frac{l_D}{\sqrt{rt}} \tag{7.45a}$$

$$C_1 = 2 - (0.17m - 0.31) \left(\frac{290 - \frac{r}{t}}{260} \right)^{3.9} \tag{7.45b}$$

and $m = 2, \ldots, 6, \ldots$ is the waves number (Figure 7.14) corresponding to the possible shape of the buckled circular ring;

Table 7.7 The effective width of the cylinder wall b_{eM}, according to Schmidt and Greiner in [509], as a function of λ_r and C_1 (Equations (7.45)).

$\dfrac{r}{t}$	λ_r	$\dfrac{b_{eM}}{l_D}$
$30 \le \dfrac{r}{t} < 290$	$\lambda_r < 2$	$0.5 \cdot C_1 - 0.25(\lambda_r - 2) \le 1$
$30 \le \dfrac{r}{t} < 290$	$\lambda_r \ge 2$	$\dfrac{C_1}{\lambda_r}$
$\dfrac{r}{t} \ge 290$	$\forall \lambda_r$	$\left[2 + 0.1 \left(\dfrac{\frac{r}{t} - 290}{620} \right) \right] \dfrac{1}{\lambda_r} \le 1$

Source: From [509]/European Convention for Constructional Steelwork.

- section modulus of the effective light stiffener:

$$W_{De} = \frac{J_{De}}{e_r} \tag{7.46}$$

The first common verification to be used in the case of ring stiffened cylinders is the use of the "limiting ring stiffness" principle which shows that the load-carrying capacity of the unstiffened shell between the rings is equivalently provided by the stiffeners [195]. It should be verified that:

$$\gamma_r \ge \gamma_r^{\star} \tag{7.47}$$

where the non-dimensional stiffness parameter [493]:

$$\gamma_r = \frac{12(1 - v^2)J_{De}^{\star}}{lt^3} \tag{7.48}$$

while the minimum value for $\gamma_r^{\star}$ can be derived from Figure 7.18 originally derived by Greiner in [192] as a function of $\frac{l}{r}\sqrt{\frac{r}{t}}$ and the numbers of equidistant stiffener n_D (extrapolation up to $n_D = 10$ can be done with values in the range $5 < n_D < 10$). Similar values have been derived by Blackler in [45] resulting in a formulation in good accordance with the data in Figure 7.18.

Figure 7.18 Limiting stiffness $\gamma_r^{\star}$ of ring-stiffened cylinder with one to five equidistant rings [192, 195].

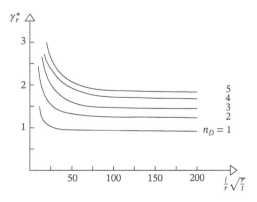

7.2.7.4 Buckling of a Light Ring-Stiffened Cylindrical Wall: Flexural Buckling of the Ring Stiffeners

The general instability (Figure 7.11(a)) load of a ring-stiffened circular cylindrical shell has been determined by Bodner in [49] based on a linear buckling theory, assuming a smeared ring stiffeners approach. The bifurcation pressure $p^\star_{Rcr}$ (Equations (7.95)) is written as a function of the assumed number m of circumferential buckling waves (Figure 7.14) [509]:

$$p^\star_{Rcr} = \frac{1}{6(1-v^2)} E \left(\frac{t}{r}\right)^3 \overline{p} \tag{7.49}$$

where:

$$\overline{p} = \frac{1}{\overline{\lambda}_r^2} + \frac{m^4}{\left(\frac{1}{\overline{\lambda}_r^2} + 2m^2\right)} \left[1 + \gamma_r \left(\frac{l}{l_D}\right)\right] + \frac{12(1-v^2)\left(\frac{r}{t}\right)^2}{\left(1 + m^2\overline{\lambda}_r^{-2}\right)^2 \left(\frac{1}{\overline{\lambda}_r^2} + 2m^2\right)} \tag{7.50}$$

in which $\overline{\lambda}_r = \frac{l}{\pi r}$. The critical buckling load pressure p_{cr} may be obtained by minimizing $p^\star_{Rcr}$ in Equation (7.49) with respect to m. The minimization procedure may be done by substituting $m = 2, \ldots, 6, \ldots$ in expression (7.50) and then matching p_{cr} to the lower $p^\star_{Rcr}$ derived from (7.49). The number m which minimizes $p^\star_{Rcr}$ is the critical buckling wave number:

$$m_{cr} = \frac{r}{l} 10^{\overline{m}} \qquad\qquad \text{if}\quad \beta \le 10^4 \tag{7.51a}$$

$$m_{cr} = \left[\frac{3200}{1 + \gamma_r\left(\frac{l}{l_D}\right)}\right]^{0.125} \left(\frac{r^3}{l^2 t}\right)^{0.25} \qquad \text{if}\quad \beta > 10^4 \tag{7.51b}$$

where:

$$\beta = \frac{0.95}{\sqrt{1 + \gamma_r\left(\frac{l}{l_D}\right)}} \left(\frac{l^2}{rt}\right) \tag{7.52a}$$

$$\overline{m} = 0.443 + 0.248 \log\beta - 7 \cdot 10^{-4}(4 - \log\beta)^{4.51} \tag{7.52b}$$

The flexural global buckling check should be performed as follows:

$$\sigma_{\theta,Rcr} = \left(\frac{N_D}{A_{De}}\right) + \left(\frac{M_D}{W_{De}}\right) \le \frac{f_{yk}}{\gamma_M} \tag{7.53}$$

where $\gamma_M \ge 1.1$ is the partial factor for resistance (Table 2.2, UNI ENV 1993-4-1) and:

$$N_D = p_{Ed} r b_{eN} \tag{7.54a}$$

$$M_D = \left[\frac{EJ_{De}}{r^2}(m^2 - 1)a_{eq}\frac{p_{Ed}}{p^\star_{Rcr}} + \left(P_{Ed}\frac{2a_{eq}}{l_D}\frac{r^2}{m^2 - 1}\right)\right]\frac{1}{1 - \frac{p_{Ed}}{p^\star_{Rcr}}} \tag{7.54b}$$

The first term in the square brackets of Equation (7.54b) represents the effect of the external pressure p_{Ed} on the bending moment of the imperfect stiffener (Figure 7.16) [509], whereas the second term represents the effect of the radial deviation force from the axial load P_{Ed} (Point 5.3.2/5.3.3, UNI ENV

1993-1-1:2005). This second term is based on the assumption that, of the three neighbouring rings stiffeners, each at a distance l_D, the middle one has the full imperfection amplitude a_{eq} and the two others have zero imperfections [509]; furthermore, in the same second term of Equation (7.54b), related to the contribution of P_{Ed}, the ratio $\frac{a_{eq}}{l_D}$ should be limited $\frac{a_{eq}}{l_D} \leq \frac{1}{200}$. The amplitude imperfection value $a_{eq} = a \cdot \eta_2$ where $\eta_2 = 1.5$ for cold-formed ring stiffeners and $\eta_2 = 1.3$ for warm-formed or welded ring stiffeners; a is the out-of-roundness tolerance which is the maximum absolute radial deviation from the true mean circle [509]:

$$a \leq U_{ra}r \qquad \text{if} \quad m \leq 6 \tag{7.52}$$

$$a \leq U_{rb}\frac{r}{m} \qquad \text{if} \quad m \geq 6 \tag{7.53}$$

where U_{ra} and U_{rb} are taken as function, of the fabrication tolerance quality class: Class A (excellent) $U_{ra} = 0.0035$ and $U_{rb} = 0.020$, Class B (high) $U_{ra} = 0.0050$ and $U_{rb} = 0.030$, Class C (normal) $U_{ra} = 0.0075$ and $U_{rb} = 0.045$.

The number m of circumferential buckling waves and the bifurcation pressure $p_{Rcr}^{\star}$ to be introduced in Equation (7.54b) depend on the critical wave number m_{cr} according to the rules given for the general GMNIA approach in UNI ENV 1993-1-6:2007:

- if $m_{cr} > 6$ we assume $m = m_{cr}$ and $p_{Rcr}^{\star} = p_{cr}$;
- if $m_{cr} \leq 6$, then m and $p_{Rcr}^{\star}$ should be taken as those values which maximize the circumferential stress $\sigma_{\theta,Rcr}$ in Equation (7.53).

Based on a Ritz (energy) formulation, Tian et al. in [608] derived useful governing eigenvalue equations to solve the overall buckling pressure of ring-stiffened shells under general lateral pressure distribution, with different combinations of end conditions, and ring-stiffeners of general cross-sectional shape and distribution along the length of the cylindrical shell.

7.2.7.5 Buckling of Light Ring-Stiffened Cylindrical Wall: Shell Wall Local Buckling

The unstiffened cylindrical wall between light ring stiffeners, with boundary conditions BC2-BC2, should be designed against local buckling according to Point 8, UNI ENV 1993-1-6 described at Section 7.2.9 (Figures 7.11(b) and 7.11(c)); the cylinder length l (height) of the cylindrical tank wall must be substituted with l_D which is the length of unstiffened shell between light stiffeners (spacing between stiffeners).[14] When determining the critical circumferential buckling stress $\sigma_{\theta,Rcr}$ in Equation (7.110), boundary conditions BC2-BC2 (Table 7.6) should be used [509]. The influence of axial compressive stresses from end closure pressure and/or from additional external load (roof) should be covered/verified by using Equation (7.89).

7.2.7.6 Buckling of a Light Ring-Stiffened Cylindrical Wall: Ring Local Buckling

The following checks should be performed for light ring stiffeners, in order to verify the distorsional (Figure 7.12(a)) and local (Figure 7.12(b)) buckling mode of the ring itself. Two cases are needed [509]:

1) for ring stiffeners with L, T and H cross-section in Figures 7.17(a), 7.17(b) and 7.17(d), and for external annular plate rings in Figure 7.17(c), their ratios $\frac{b_w}{t_w}$, $\frac{b_f}{t_f}$ and $\frac{b_r}{t_r}$ should comply with the maximum

---

14 This assumption is valid for all the expressions in Section 7.2.9.

width-to-thickness ratios, for compression parts, of Class 3 webs and outstand flanges according to Table 5.2 in UNI ENV 1993-1-1:2005. As an example, considering the external annular plate rings in Figure 7.17(c), should be:

$$\frac{b_r}{t_r} \leq 14\varepsilon = 14\sqrt{\frac{235}{f_{yk}}} \tag{7.56}$$

where f_{yk} is in MPa;

2) for internal annular plate rings (Figure 7.17(c)), it should be proved that:

$$\chi' \geq 1.5\chi_{loc} \qquad \text{but} \qquad \chi' \leq 1 \tag{7.57}$$

where χ_{loc} is the reduction factor for local shell buckling (unstiffened cylindrical wall between ring stiffeners) according to Equations (7.93) and Section 7.2.9. The reduction factor χ' for plate buckling may be evaluated according to Points 4.4/4.5, UNI ENV 1993-1-5:2004 using [275]:

$$\bar{\lambda} = \sqrt{\frac{f_{yk}}{\sigma_{\theta cr}}} \tag{7.58}$$

The critical circumferential plate buckling stress of the internal annular plate, in expression (7.58), may be approximated as:

$$\sigma_{\theta cr} = CE\left(\frac{t_r}{b_r}\right)^2 \tag{7.59}$$

derived by Jumikis and Rotter in [280]. The plate buckling factor[15] C in Equation (7.59) is [601, 605]:

$$C = \frac{\eta_s c_s + \eta_c c_c}{\eta_s + \eta_c} \tag{7.60}$$

where:

$$c_s = 0.385 + 0.452\sqrt{\frac{b_r}{r}} \tag{7.61a}$$

$$c_c = 1.154 + 0.560\frac{b_r}{r} \tag{7.61b}$$

$$\eta_c = 0.43 + \frac{1}{4000}\left(\frac{r}{b_r}\right)^2 \tag{7.61c}$$

$$\eta_s = \left(\frac{t}{t_r}\right)^{2.5} \tag{7.61d}$$

A ring is generally provided at the intersection between two cylinder walls or a wall and a cone of a storage silo or tank to resist a large circumferential compressive force at the intersection. If two different thicknesses are jointing the same ring, Equation (7.61d) becomes:

$$\eta_s = \frac{1}{2}\left[\frac{\sum_{i=1}^{2} t_i^{2.5}}{t_r^{2.5}}\right] \tag{7.62}$$

---

15 Assuming $C = 0.43$ is a conservative approach [509].

Geometric limits to avoid elastic distortional buckling (Figures 7.12(a) and 7.17(b)) of T-section ring-beams have been derived by Teng and Chan in [601] and depicted in Figure 7.19: the linear interpolation rule can be used in the range $0.5 < \frac{t_f}{t_w} < 2$; it should be noted that this range of thickness variation considered is not too restrictive as the annular plate and the stiffener normally have similar thicknesses, anyway if $\frac{t_f}{t_w} > 2$ the curve in Figure 7.19 corresponding to $\frac{t_f}{t_w} = 2$ can be conservatively used.

7.2.7.7 Buckling of a Light Ring-Stiffened Cylindrical Wall: Out-of-Plane Buckling of the Ring

For ring stiffeners with L, T and H cross-section, the out-of-plane torsional buckling of the ring stiffeners (Figure 7.12(c)) should be verified according to:

$$\chi'' \geq 1.5\chi_{\text{loc}} \qquad \text{but} \qquad \chi'' \leq 1 \tag{7.63}$$

where χ_{loc} is the reduction factor for local shell buckling of the unstiffened cylindrical wall between light ring stiffeners according to Equations (7.93) and the suggestion given on Section 7.2.7.5; χ'' is the reduction factor for torsional buckling of compression members derived from Points 6.3.1.2/4, UNI ENV 1993-1-1:2005 using as a member the slenderness parameter:

$$\bar{\lambda} = \sqrt{\frac{N_{\theta\text{rp}}}{N_{\theta\text{cr}}}} \tag{7.64}$$

in which the circumferential force of the ring stiffener in the fully plastic state is $N_{\theta\text{rp}} = A_D f_{yk}$, while the critical torsional buckling force of the ring stiffener is taken from the expression derived by Teng and Rotter in [602]:

$$N_{\theta\text{cr}} = \frac{EJ_{Dz}}{r_0^2}\left[\frac{\left(1 - \frac{b}{r_0}m^2\right)^2 + \frac{GJ_{Dt}m^2}{EJ_{Dz}}\left(1 - \frac{b}{r_0}\right)^2}{\bar{\eta}m^2 - \frac{z_0}{r_0}}\right] \tag{7.65}$$

where coordinate z is measured radially in the plane of the ring section (Figure 7.17) and J_{Dz} is the principal moments of inertia around it; meridional coordinate y is normal to the ring plane and vertically oriented (J_{Dy}); $b = b_r$ for annular plate rings as in Figure 7.17(c) and $b = b_w + \frac{t_f}{2}$ for ring stiffeners with L, T and H cross-section (Figures 7.17(a), 7.17(b), and 7.17(d)); J_{Dt} is the (primary) polar moment of

Figure 7.19 Geometric limits for distortional buckling strength of T-section ringbeams. (Figures 7.12(a) and 7.17(b) for main geometrical properties.) Source: Adapted from [601].

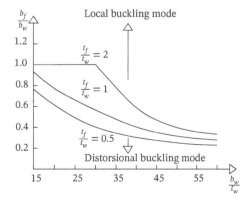

inertia of the cross-section (torsional constant) and GJ_{Dt} is the torsional rigidity; furthermore, the following relations will be used

$$m^2 = \left(\frac{z_0}{r_0 \overline{\eta}} + \frac{r_0}{b} \right) \sqrt{\frac{GJ_{Dt} z_0}{EJ_{Dz} r_0 \overline{\eta}}} \tag{7.66a}$$

$$\overline{\eta} = \frac{J_{Dy} + J_{Dz}}{A_D r_0^2} + \left(\frac{z_0}{r_0} \right)^2 \tag{7.66b}$$

7.2.7.8 Heavy Ring Stiffeners

On principle, the same design checks have to be performed as for the light ring, but modified using the effective cross-sectional properties A_{Re}, J_{Re}, and W_{Re} derived from Equations (7.40)–(7.46) using the following modifications:

- when calculating the effective width values b_{eN} and b_{eM}, using Equations (7.41) and Table 7.7 respectively, the length l_D should be replaced by l_R and the circumferential wave number should be taken as $m = 2$;
- for three (or more) intermediate heavy ring stiffeners, at distance l_{R1}, l_{R2}, ..., the length l_R is the mean value of the lengths of the two (or more) adjacent shell sections $l_R = \frac{l_{R1} + l_{R2}}{2}$;
- for heavy ring stiffeners at the end (base or top) of a ring-stiffened shell (end rings), the length $l_R = \frac{l}{2}$ (Figure 7.16);
- for an end ring, Equation (7.41a) becomes:

$$b_{eN} = b_0 + \frac{1.56 \eta_l \sqrt{rt}}{2} \tag{7.67}$$

and only 50% of the b_{eM} term, derived from Table 7.7, should be introduced into Equations (7.42) and (7.43).

The heavy ring stiffeners are designed using the same design checks performed for the light rings:

- sufficient bending stiffness should be proved by verifying that (limiting stiffness):

$$J_{Re} \geq \frac{p_{Ed} r^3 l_R}{3E} \tag{7.68}$$

which takes up expression (7.37); it is clear that a heavy ring stiffener is more or less designed as a free compression ring with a circumferential buckling wave number $m = 2$ (Figure 7.14(a));
- the flexural global buckling check should be performed as follows (Equation (7.53)):

$$\sigma_{\theta,Rcr} = \left(\frac{N_R}{A_{Re}} \right) + \left(\frac{M_R}{W_{Re}} \right) \leq \frac{f_{yk}}{\gamma_M} \tag{7.69}$$

where:

$$N_R = p_{Ed} r b_{eN} \tag{7.70a}$$

$$M_r = \frac{EJ_{Re}}{r} \left(\frac{\eta_2}{67} \right) \left(\frac{\frac{p_{Ed}}{p_{Ed,cr}}}{1 - \frac{p_{Ed}}{p_{Ed,cr}}} \right) \tag{7.70b}$$

$$p_{Ed,cr} = \frac{3EJ_{Re}}{r^3 l_R} \tag{7.70c}$$

The term η_2 in Equation (7.70b) is equal to 1.5 for cold-formed ring stiffeners and to 1.3 for warm-formed or welded ring stiffeners; expression (7.70c) follows from Equation (7.37) for a free compression ring in its plane;

- the heavy ring local buckling checks should be performed in full analogy to Section 7.2.7.6 for light rings;
- the out-of-plane torsional buckling checks of the heavy ring should be performed in full analogy to Section 7.2.7.7 for light rings.

7.2.8 Stringers

Longitudinal stiffeners, usually called stringers, are most effective, compared to circumferential rings (commonly used to resist circumferential and radial forces), in increasing the axial or bending strength of cylindrical tanks and pipes. Let us consider a circular tank of radius r, height l, and uniform thickness t with n_S the number of equidistant longitudinal stiffeners ("stringers" at Point 1.3.1.8, UNI ENV 1993-1-6:2007) [379], as depicted in Figure 7.20(a) [502]; stringers are placed meridionally and represent the generatrix of the rotational shell. The rules, written by Schmidt and Samuelson in [512], apply:

1) to circular cylindrical shells (Figure 7.20(a)) stiffened by equidistant longitudinal stringers under meridional compressive membrane stress resultants N_x which are constant over the length x and steadily variable (or constant) over the circumference y_m. The membrane force N_x refers to the stress resultant per unit circumferential width on the smeared ("smeared wall", Point A.1, UNI ENV 1993-1-5:2004) wall thickness $t_m = t + \frac{A_S}{b}$ in which $b = \frac{2\pi r}{n_S}$ (Figure 7.20(b)) and A_S is the cross-sectional area of a stiffener.
2) to longitudinally stiffened shells with radially (w) restrained edges providing boundary conditions BC1 or BC2 according to Table 7.6;
3) to longitudinally stiffened shells which comply with the following three geometrical conditions:

$$\frac{A_S}{bt} \leq 2 \tag{7.71a}$$

$$\frac{J_{Sy}}{bt^3} \leq 15 \tag{7.71b}$$

$$\frac{J_{St}}{bt^3} \leq 2.4 \tag{7.71c}$$

where J_{Sy} is the second moment of area of a stiffener about its circumferential axis y (centroid/barycentre) and J_{St} is the uniform torsion constant (primary polar moment of inertia) of a stiffener. For shells that are more heavily stiffened than the restrictions given in Equations (7.71), stringers may be treated conservatively as an assembly of independently buckling columns, according to Point 6.3, UNI ENV 1993-1-1:2005.

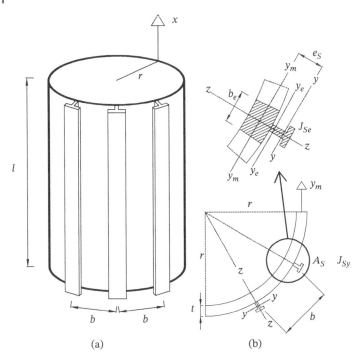

Figure 7.20 Global and local geometry of longitudinally stiffened cylindrical shells using stringers. Source: From [512]/European Convention for Constructional Steelwork.

Possible buckling modes of longitudinally stiffened cylinder shells have been classified by Schmidt and Samuelson in [512] and Singer in [552–554]:

- local buckling or torsional buckling of the stringers as depicted in Figure 7.21(a);
- local buckling of the shell panels between the longitudinal stiffeners (Figure 7.21(b)) that will usually precede global collapse in axially compressed cylindrical shells [551, 556], when the stringers are widely spaced (high value of b [603]);
- global buckling of the stiffened shell (closely spaced stringers usually generates this general instability failure) as depicted in Figure 7.21(c).

7.2.8.1 Local Buckling of the Stringers

The stringers should be verified such that local and torsional buckling of the stiffeners are excluded. For the cross-section shown in Figure 7.22, the following conditions, mainly derived from Point 5.5, UNI ENV 1993-1-1:2005 (section of Class 2), must be fulfilled:

$$\frac{h_S}{s_S} \leq k_1 \sqrt{\frac{E}{f_{yk}}} \tag{7.72}$$

where $k_1 = 0.35$ for a rectangular flange in Figure 7.22(a) and $k_1 = 1.25$ for Figures 7.22(b)–7.22(e);

$$\frac{b_S}{t_S} \leq k_2 \sqrt{\frac{E}{f_{yk}}} \tag{7.73}$$

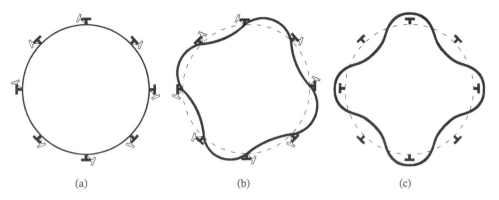

Figure 7.21 Three possible buckling modes: (a) local buckling or torsional buckling of the stringers; (b) local buckling of the shell panels between the longitudinal stiffeners; (c) global buckling of the stiffened shell. Source: Adapted from [551].

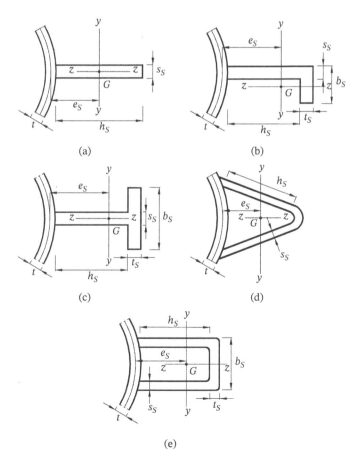

Figure 7.22 Possible geometry of the longitudinal stiffeners in case of (a) flanges, (b) L, (c) T and (d)–(e) hollow sections.

where $k_2 = 0.35$ in Figure 7.22(b), $k_2 = 0.70$ in Figure 7.22(c), and $k_2 = 1.25$ in Figure 7.22(e). According to Point 1004, DIN 18800-3, a limitation on the torsional buckling resistance of an axially compressed stiffener with an open cross-section (Figures 7.22(b)–7.22(c)), is introduced:

$$\frac{J_{Sw}}{J_{St}} \leq 0.19\frac{E}{f_{yk}} \tag{7.74}$$

where J_{Sw} is the polar second moment (warping) of an area of a stiffener about the cylinder middle surface y_m.

7.2.8.2 Local Buckling of the Shell Panels
The buckling strength of the unstiffened cylindrical panel between the longitudinal stringers should be verified according to Koiter's curvature parameter [309]:

$$k_S = \frac{\sqrt[4]{12(1-v^2)}}{2\pi}\frac{b}{\sqrt{rt}} \rightarrow \text{ where } v = 0.3 \rightarrow k_S = 0.289\frac{b}{\sqrt{rt}} \tag{7.75}$$

The influence of k_S, Equation (7.75), on the post-buckling behaviour results in the following three cases:

1) a panel with $k_S \leq 1$ should be designed as an axially compressed plate strip according to Point 4, UNI ENV 1993-1-5:2004 [275];
2) a panel with $k_S \geq 5$ should be designed as an axially compressed unstiffened cylindrical shell according to Point 8, UNI ENV 1993-1-6:2007 and Section 7.2.9;
3) for panels with $1 < k_S < 5$ the design axial (meridional) buckling stress should be determined using the following quadratic interpolation (Equation (7.90a)):

$$\sigma_{x,Rd} = \sigma_{x,Rd}^P\left(1 - \sqrt{\frac{k_S-1}{4}}\right) + \sigma_{x,Rd}^S\sqrt{\frac{k_S-1}{4}} \tag{7.76}$$

where $\sigma_{x,Rd}^P$ is the design buckling stress for a plate strip according to Point 1. (Point 4, UNI ENV 1993-1-5:2004) and $\sigma_{x,Rd}^S$ is the design buckling stress for a full cylindrical shell according to Point 2. (Point 8, UNI ENV 1993-1-6:2007).

7.2.8.3 Global Buckling of the Stiffened Shell
According to the approach proposed at Point 8, UNI ENV 1993-1-6:2007, the global buckling verification should be represented by limiting the design values of membrane stress $N_{x,Ed}$ to the design buckling meridional resistance (Equation (7.90a)):

$$N_{x,Ed} \leq N_{x,Rd} \tag{7.77a}$$

$$N_{x,Rd} = \frac{N_{x,Rk}}{\gamma_M} \tag{7.77b}$$

where $N_{x,Rk}$ is the meridional characteristic buckling stress and $\gamma_M \geq 1.1$ is the partial factors for resistance (Table 2.2, UNI ENV 1993-4-1). Using Equation (7.91a), the characteristic buckling stress resultant is:

$$N_{x,Rk} = \chi_x t_m f_{yk} \tag{7.78}$$

where the global buckling reduction factor χ_x can be derived from Equations (7.93) using the global relative slenderness parameter:

$$\lambda_x = \sqrt{\frac{t_m f_{yk}}{N_{x,Rcr}}} \tag{7.79}$$

The global buckling reduction factor χ_x should be determined, using Equations (7.93), as a function of the nondimensional parameter $\frac{A_S}{bt}$:

- if $\frac{A_S}{bt} \geq 0.2$ for heavily stiffened shells, the elastic imperfection reduction factor $\alpha_x = 0.65$, the plastic range factor $\beta = 0.60$, the interaction exponent of the capacity curve $\eta = 1.0$ and the squash limit relative slenderness $\lambda_{x0} = 0.40$ (Figure 7.24);
- if $\frac{A_S}{bt} \leq 0.06$ for lightly stiffened shells, α_x comes from Equation (7.99), $\beta = 0.60$, $\eta = 1.0$ and $\lambda_{x0} = 0.20$;
- if $0.2 < \frac{A_S}{bt} < 0.06$, for moderately stiffened shells, the buckling parameters $(\alpha_x, \beta, \eta, \lambda_{x0})$ should be linearly interpolated between the two previously defined extremes.

The critical buckling resultant $N_{x,Rcr}$ in Equation (7.79) could be obtained using the following two approaches (the second one more conservative):

1) The first approach is based on the Donnell-Mushtari-Vlasov formulation, extended by Block et al. [47] to include orthotropic cylinder eccentrically stiffened by both stringers and rings [31] in a smeared fashion, and then approximately modified by Miller in [399] to introduce the effective width b_e into the relevant stiffness expressions (Figure 7.20(b)) [512]:

$$N_{x,Rcr} = \min \left[\frac{A_{33} + \frac{A_{12}A_{23}-A_{13}A_{22}}{A_{11}A_{22}-A_{12}^2}A_{13} + \frac{A_{12}A_{13}-A_{11}A_{23}}{A_{11}A_{22}-A_{12}^2}A_{23}}{\left(\frac{n\pi}{l}\right)^2} \right] \tag{7.80}$$

where n denotes the number of longitudinal half waves in the buckle pattern (meridional axis x in Figure 7.20(a)); m the number of circumferential full waves along axis y_m in Figure 7.20(b) (full waves are depicted in Figure 7.14; n and m integers); coefficients A_{ij} can be written as [47]:

$$A_{11} = K_x \left(\frac{n\pi}{l}\right)^2 + K_{x\theta} \left(\frac{m}{r}\right)^2 \tag{7.81a}$$

$$A_{22} = K_\theta \left(\frac{m}{r}\right)^2 + K_{\theta x} \left(\frac{n\pi}{l}\right)^2 \tag{7.81b}$$

$$A_{12} = (K_v + K_{x\theta}) \left(\frac{n\pi}{l}\right) \left(\frac{m}{r}\right) \tag{7.81c}$$

$$A_{13} = \frac{K_v}{r} \left(\frac{n\pi}{l}\right) + K_S \left(\frac{n\pi}{l}\right)^3 \tag{7.81d}$$

$$A_{23} = \frac{K_\theta}{r} \left(\frac{m}{r}\right) \tag{7.81e}$$

$$A_{33} = D_x \left(\frac{n\pi}{l}\right)^4 + D_{x\theta} \left(\frac{n\pi}{l}\right)^2 \left(\frac{m}{r}\right)^2 + D_\theta \left(\frac{m}{r}\right)^4 + \frac{K_\theta}{r^2} \tag{7.81f}$$

The stiffnesses K and D in Equations (7.81) are:

$$K_x = \frac{Et}{1-v^2} \frac{b_e}{b} + \frac{EA_S}{b} \tag{7.82a}$$

$$K_{x\theta} = K_{\theta x} = \frac{Gt}{2}\left(1 + \frac{b_e}{b}\right) \tag{7.82b}$$

$$K_\theta = \frac{Et}{1 - v^2} \tag{7.82c}$$

$$K_v = vK_\theta = v\frac{Et}{1 - v^2} \tag{7.82d}$$

$$K_S = \frac{EA_S e_S}{b} \tag{7.82e}$$

$$D_x = \frac{Et^3}{12(1 - v^2)}\frac{b_e}{b} + \frac{EJ_{Sy}}{b} + \frac{EA_S e_S^2}{b} \tag{7.82f}$$

$$D_{x\theta} = v\frac{Et^3}{6(1 - v^2)} + \frac{Gt^3}{6}\left(1 + \frac{b_e}{b}\right) + \frac{GJ_{St}}{b} \tag{7.82g}$$

$$D_\theta = \frac{Et^3}{12(1 - v^2)} \tag{7.82h}$$

where e_S is the eccentricity of the stiffener axis y from the cylinder middle surface with axis y_m, positive for external stiffener, negative for internal stiffener (Figure 7.20(b)). The effective width b_e [30] in expressions (7.82) may be obtained by using the von Karman-Sechler-Donnell equation [292]:

$$b'_e = \frac{2\pi}{\sqrt{12(1 - v^2)}}t\sqrt{\frac{E}{f_{yk}}} \rightarrow \text{ with } v = 0.3 \rightarrow b'_e = 1.90\, t\sqrt{\frac{E}{f_{yk}}} \le b \tag{7.83}$$

Alternatively and if a more favourable value is sought, b_e should be iteratively determined from the following expression [512]:

$$b''_e = bK_e\sqrt{\frac{Et_m}{N_{x,Rcr}}} \le b \tag{7.84}$$

where K_e is the greater of the two following values:

$$K_e = 0.65\sqrt{\frac{t}{r\sqrt{1 + 0.01\frac{r}{t}}}} \tag{7.85a}$$

$$K_e = 0.275n_S\frac{t}{r} \tag{7.85b}$$

If the iteration procedure ends with a value b''_e that is less than b'_e, the latter value $b'_e = b_e$ from Equation (7.83). The iterative procedure can be summarized as follows:

Step 1 Start with $b''_e = b$.
Step 2 Evaluate the critical values n_{cr} and m_{cr}, which are the integer values n and m able to minimize $N_{x,Rcr}$ in Equation (7.80). If m_{cr} does not meet both of the following conditions:

$$m_{cr} \le \frac{n_S}{3.5} \tag{7.83}$$

$$m_{cr} \ge 4 \tag{7.84}$$

the second approach at Point 2. and Equation (7.87) should be used instead of (7.80). Condition (7.86a) is needed to be sure that the smearing concept, Block et al. [47], can be applied by increasing the number n_S of stringers; in this way b in Figure 7.20 is reduced and the spacing of the discrete stiffeners is small compared to the theoretical buckling wavelength (thanks to this condition, the local buckling of the shell panels between the longitudinal stiffeners in Figure 7.21(b) can be avoided). Condition (7.86b) defines the accuracy in applying the formulation derived from Donnell-Mushtari-Vlasov.

Step 3 Determine a new value for b''_e from expression (7.84) using $N_{x,Rcr}$ as evaluated in Step 2.

Step 4 Steps 2–4 are repeated until convergence (the value of b''_e compared to the one evaluated in the previous iteration).

2) The second approach conservatively evaluates the critical buckling resultant using the approach proposed by Walker and Sridharan [656, 657] as modified by Samuelson et al. [503]:

$$N_{x,Rcr} = \frac{\pi^2 E J_{Se}}{l^2 b} + \Psi t_m \alpha_x \sigma_{xcr} \tag{7.87}$$

where J_{Se} is the second moment of area of an effective stiffener including the effective width b_e of the cylinder wall, in Figure 7.20(b), about its circumferential axis y_e; $\Psi = 1$ for external stiffeners and $\Psi = \frac{1}{1+A_S/(bt)}$ for internal stiffeners; α_x can be derived from Equation (7.99) by substituting $t = t_m$ and $\Delta w_k = \frac{t_m}{Q}\sqrt{\frac{r}{t_m}}$ using $Q = 25$, Class C (normal). The critical meridional buckling stress σ_{xcr} comes from expression (7.98):

$$\sigma_{xcr} = 0.605\frac{Et}{r} \tag{7.88}$$

Equation (7.87) can be subdivided into two contributions: the first one, $\frac{\pi^2 E J_{Se}}{l^2 b}$, is the critical elastic column Euler buckling force of an independently buckling perfect stiffener (hinged at both sides) and the second one, $\Psi t_m \alpha_x \sigma_{xcr}$, is the elastic buckling stress of an imperfect unstiffened cylinder (equal to expression (7.95a)).

7.2.9 Buckling Limit State

According to the explanation given by Professor Schmidt and Professor Rotter in [510], a number of different approaches to verify and design shells against buckling have been proposed in the past, but for the European Standard UNI ENV 1993-1-6, only three approaches are approved for use in the assessment of buckling resistance:

1) Global numerical GMNIA analysis: GMNIA stands for geometrically and materially non-linear analysis with imperfections included. This is a fully numerical non-linear analysis that can be seen as a combination of GMNA (geometrically and materially non-linear analysis) and GNIA (geometrically non-linear analysis with imperfections) analyses.

2) Global numerical MNA/LBA analysis: MNA stands for materially non-linear analysis (elasto-plastic material law is adopted) and LBA for linear elastic bifurcation (eigenvalue) small deflection linear elastic analysis.

3) Buckling stresses: it is based on existing formulas, as given below, to evaluate both the stress state in the shell caused by the loading and the resistance of the structure to those loads.

The limit state of buckling (LS3 at Point 4.1.3, UNI ENV 1993-1-6) should be taken as the condition in which all or part of the tank wall suddenly develops large displacements normal to the shell surface, caused by loss of stability under compressive membrane (meridional σ_x direction in Figure 7.23(a) or circumferential σ_θ in Figure 7.23(b)) or shear membrane stresses ($\tau_{x\theta}$ in Figure 7.23(c)) in the shell wall, leading to the inability to sustain any increase in the stress resultants, possibly causing the total collapse of the structure. The linear elastic membrane stress field (σ_θ, σ_x, $\tau_{x\theta}$) under purely axisymmetric conditions of loading and support, induced by the applied combination of loads, should be taken as the design key[16] values of compressive and shear membrane stresses ($\sigma_{x,Ed}$, $\sigma_{\theta,Ed}$, $\tau_{x\theta,Ed}$) (Point 8.5.1 (1), UNI ENV 1993-1-6). Although buckling is not a purely stress-initiated failure phenomenon, the buckling limit state LS3, should be represented by limiting the design values of membrane stresses (Point 8.5.3 (1), UNI ENV 1993-1-6). The influence of bending effects on the buckling strength may be ignored, provided they arise as a result of meeting boundary compatibility requirements. If more than one of the three buckling-relevant design membrane stress components in Figure 7.23 is present under the actions considered, the following interaction check for the combined membrane stress state should be carried out [493, 502, 669]:

$$\left(\frac{\sigma_{x,Ed}}{\sigma_{x,Rd}}\right)^{k_x} - k_i \left(\frac{\sigma_{x,Ed}}{\sigma_{x,Rd}}\right)\left(\frac{\sigma_{\theta,Ed}}{\sigma_{\theta,Rd}}\right) + \left(\frac{\sigma_{\theta,Ed}}{\sigma_{\theta,Rd}}\right)^{k_\theta} + \left(\frac{\tau_{x\theta,Ed}}{\tau_{x\theta,Rd}}\right)^{k_\tau} \leq 1 \qquad (7.89)$$

where $\sigma_{x,Ed}$, $\sigma_{\theta,Ed}$ and $\tau_{x\theta,Ed}$ are the design values of the acting buckling-relevant meridional, circumferential, shear membrane stress, respectively. The design buckling resistance should be obtained from:

$$\sigma_{x,Rd} = \frac{\sigma_{x,Rk}}{\gamma_M} \qquad (7.90a)$$

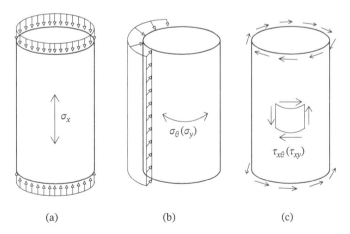

(a) (b) (c)

Figure 7.23 Cylindrical stresses and action loads for the design against buckling: (a) axial compression, (b) circumferential (hoop) compression, and (c) shear.

---

16 The key values of membrane stresses should be taken as the maximum value of each stress at the axial coordinate in the structure.

$$\sigma_{\theta,Rd} = \frac{\sigma_{\theta,Rk}}{\gamma_M} \tag{7.90b}$$

$$\tau_{x\theta,Rd} = \frac{\tau_{x\theta,Rk}}{\gamma_M} \tag{7.90c}$$

where the partial factor for resistance to buckling $\gamma_M \geq 1.1$ should be taken from Table 2.2, UNI ENV 1993-4-2 for tanks, unless the National Annex defines a different value; $\sigma_{x,Rk}$, $\sigma_{\theta,Rk}$ and $\tau_{x\theta,Rk}$ are the meridional, hoop, and shear characteristic buckling stresses, respectively. The three characteristic buckling stresses account for plasticity effects and geometric imperfections [139]:

$$\sigma_{x,Rk} = \chi_x f_{yk} \tag{7.91a}$$

$$\sigma_{\theta,Rk} = \chi_\theta f_{yk} \tag{7.91b}$$

$$\tau_{x\theta,Rk} = \chi_\tau \frac{f_{yk}}{\sqrt{3}} \qquad \text{Von Mises} \tag{7.91c}$$

Furthermore, the buckling interaction exponent parameters, k_x, k_θ, k_τ and k_i in Equation (7.89), may be estimated as a function of the buckling reduction factors χ_x, χ_θ and χ_τ according to Annex D in Point D.1.6 (1), UNI ENV 1993-1-6 [493]:

$$k_x = 1.25 + 0.75\chi_x \tag{7.92a}$$

$$k_\theta = 1.25 + 0.75\chi_\theta \tag{7.92b}$$

$$k_\tau = 1.75 + 0.25\chi_\tau \tag{7.92c}$$

$$k_i = \left(\chi_x \cdot \chi_\theta \right)^2 \tag{7.92d}$$

The buckling reduction factor χ_x, χ_θ and χ_τ (Point 8.5.2, UNI ENV 1993-1-6), that just for a simplification in the notation are represented as $\chi_{(x,\theta,\tau)}$, should be determined as a function of the relative slenderness of the shell λ_x, λ_θ and λ_τ ($\lambda_{(x,\theta,\tau)}$), and define the form of the capacity curve[17] in Figure 7.24 [487, 489]:

$$\chi_{(x,\theta,\tau)} = 1 \qquad \text{if} \quad \lambda_{(x,\theta,\tau)} \leq \lambda_{(x0,\theta0,\tau0)} \tag{7.93a}$$

$$\chi_{(x,\theta,\tau)} = 1 - \beta \left(\frac{\lambda_{(x,\theta,\tau)} - \lambda_{(x0,\theta0,\tau0)}}{\lambda_{(xp,\theta p,\tau p)} - \lambda_{(x0,\theta0,\tau0)}} \right)^\eta \qquad \text{if} \quad \lambda_{(x0,\theta0,\tau0)} \leq \lambda_{(x,\theta,\tau)} \leq \lambda_{(xp,\theta p,\tau p)} \tag{7.93b}$$

$$\chi_{(x,\theta,\tau)} = \frac{\alpha_{(x,\theta,\tau)}}{\lambda^2_{(x,\theta,\tau)}} \qquad \text{if} \quad \lambda_{(x,\theta,\tau)} \geq \lambda_{(xp,\theta p,\tau p)} \tag{7.93c}$$

where the value of the plastic limit relative slenderness $\lambda_{(xp,\theta p,\tau p)}$, below which the effects of material yield significantly affect the buckling process (Figure 7.24), should be determined from:

$$\lambda_{(xp,\theta p,\tau p)} = \sqrt{\frac{\alpha_{(x,\theta,\tau)}}{1 - \beta}} \tag{7.94}$$

---

17 The form of the capacity curve is thus independently defined for every geometry and load case, by the choice of values for the parameters $\alpha_{(x,\theta,\tau)}$, β, η and $\lambda_{(x0,\theta0,\tau0)}$.

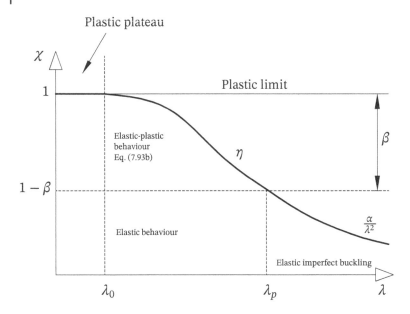

Figure 7.24 Capacity curve of a tank shell as a function of the following factors: $\alpha_{(x,\theta,\tau)}$, $\lambda_{(x0,\theta0,\tau0)}$, β and η; plastic plateau, elastic-plastic interaction, and elastic imperfect buckling. Source: Adapted from [487].

Equation (7.93c) describes the elastic buckling stress accounting for geometric imperfection, as depicted in Figure 7.24. In this case, where the behaviour is entirely elastic, the characteristic buckling stresses (7.91), may alternatively be determined directly from (Point 8.5.2 (4), Note 2, UNI ENV 1993-1-6):

$$\sigma_{x,Rk} = \alpha_x \sigma_{x,Rcr} \tag{7.95a}$$

$$\sigma_{\theta,Rk} = \alpha_\theta \sigma_{\theta,Rcr} \tag{7.95b}$$

$$\tau_{x\theta,Rk} = \alpha_\tau \tau_{x\theta,Rcr} \tag{7.95c}$$

where $\alpha_{(x,\theta,\tau)}$ are the elastic imperfection reduction factors.

The relative shell slenderness parameters λ_x, λ_θ and λ_τ, corresponding to the three stress components[18] $\sigma_{x,Rcr}$, $\sigma_{\theta,Rcr}$ and $\tau_{x\theta,Rcr}$, are:

$$\lambda_x = \sqrt{\frac{f_{yk}}{\sigma_{x,Rcr}}} \tag{7.96a}$$

$$\lambda_\theta = \sqrt{\frac{f_{yk}}{\sigma_{\theta,Rcr}}} \tag{7.96b}$$

$$\lambda_\tau = \sqrt{\frac{f_{yk}}{\sqrt{3}\tau_{x\theta,Rcr}}} \tag{7.96c}$$

---

18 Each of these stress components represents the smallest bifurcation limit load determined assuming the idealized conditions of elastic isotropic material behaviour, perfect geometry, perfect load application, and support and absence of residual stresses (LBA).

The elastic critical buckling stresses $\sigma_{x,Rcr}$, $\sigma_{\theta,Rcr}$ and $\tau_{x\theta,Rcr}$, the elastic imperfection reduction factor $\alpha_{(x,\theta,\tau)}$, the plastic range factor β, the interaction exponent η of the capacity curve and the squash limit relative slenderness $\lambda_{(x0,\theta0,\tau0)}$, are given in UNI ENV 1993-1-6, corresponding to three cases [491]: meridional axial compression buckling, circumferential hoop compression buckling, and shear buckling as a function of the dimensionless length parameter ω:

$$\omega = \frac{l}{r}\sqrt{\frac{r}{t}} = \frac{l}{\sqrt{rt}} \tag{7.97}$$

where l is the cylinder length between defined boundaries, r the radius of cylinder middle surface and t the thickness of the shell.

According to Points D.1.2.2, D.1.3.2 and D.1.4.2, UNI ENV 1993-1-6:2007, cylinders need not be checked against buckling when:

1) $\frac{r}{t} \leq 0.03\frac{E}{f_{yk}}$ (plastic plateau Figure 7.24) in the case of meridional buckling;
2) $\frac{r}{t} \leq 0.21\sqrt{\frac{E}{f_{yk}}}$ in the case of circumferential buckling;
3) $\frac{r}{t} \leq 0.16\left(\frac{E}{f_{yk}}\right)^{0.67}$ in the case of shear buckling.

In addition to the meridional (elastic and elastic-plastic), circumferential and shear buckling, two more verifications should be added [69, 70, 125, 193, 558]:

1) buckling check of cylinder wall under wind pressure [276, 321, 344] (Point D.1.3.2 (4)-(5), UNI ENV 1993-1-6:2007) taking into account, as suggested by Greiner et al. in [194, 197, 198], that the effects vary strongly with the aspect ratio of the cylinder, hence a distinction between stocky, intermediate and slender shells should be made:
   - $C_\theta \frac{r}{l}\sqrt{\frac{r}{t}} = \frac{C_\theta}{\omega}\frac{r}{t} \geq 15$ for stocky cylindrical shells [197, 474, 620, 621];
   - $C_\theta \frac{r}{l}\sqrt{\frac{r}{t}} = \frac{C_\theta}{\omega}\frac{r}{t} < 15$ and $\frac{l}{r} \leq 20$ for intermediate cylindrical shells [196];
   - $50 \leq \frac{r}{t} \leq 400$ and $\frac{l}{r} \geq 20$ for slender cylindrical shells [513–516].

   The factor C_θ accounts for the effect of the boundaries and varies between 1.0 and 1.5 (as in Equation (7.110));
2) buckling check of cylinder wall (silos containing granular material) [51, 473] with an aspect ratio of $2 < \frac{l}{(2r)} < 6$ subjected to eccentric discharge [480, 483, 484, 495, 604]. Four different pressure patterns and their effects from the Australian code (AS 3774-1996), the German code (DIN 1055-6), the ISO code (ISO 11697) and Part 4 of Eurocode 1 (UNI ENV 1991-4:2006), are compared by Song and Teng in [559].

7.2.9.1 Meridional Buckling
The critical meridional buckling stress is derived from the classical Donnell-Mushtari-Vlasov formulation (without the coefficient C_x) [5, 61, 129, 131, 169, 318, 386, 486, 557]:

$$\sigma_{x,Rcr} = 0.605EC_x\frac{t}{r} \tag{7.98}$$

The term $0.605\frac{t}{r}$, originally obtained independently by Timoshenko [609], Lorenz [345, 346], and South-well [560], is widely referred to as the classical non-dimensional elastic critical stress [54, 64, 148, 294, 357, 493, 612, 686]. The factor C_x covers the special features of short and long cylinders as a function of the dimensionless length parameter ω:

- a long cylinder with $\omega > 0.5\frac{r}{t}$ globally buckles as a compressed Euler column [32]:

$$C_x = \max\left\{1 + \frac{0.2}{C_{xb}}\left[1 - 2\omega\frac{t}{r}\right]; 0.60\right\}$$

where C_{xb} is a parameter depending on the boundary conditions at both ends of the tanks applied to the normal (radial) displacement w, meridional displacement u and meridional rotation β_ϕ:

- $C_{xb} = 6$ if both boundary conditions BC1 (Table 7.6) are applied to both ends;
- $C_{xb} = 3$ if BC1 (Table 7.6) is applied to one end and BC2 (Table 7.6) to the other end;
- $C_{xb} = 1$ if both boundary conditions BC2 (Table 7.6) are applied to both ends;

- for short cylinders with $\omega \leq 1.7$ the critical meridional buckling stress approaches the plate buckling stress of an indefinitely wide plate strip:

$$C_x = 1.36 - \frac{1.83}{\omega} + \frac{2.07}{\omega^2}$$

- for a medium-length cylinder with $1.7 \leq \omega \leq 0.5\frac{r}{t}$, the factor $C_x = 1.0$ because Equation (7.98) results in the classical linear Donnell-Mushtari-Vlasov shell buckling expression applied to medium-length cylinders under uniform axial compression $0.605E\frac{t}{r}$.

The meridional elastic imperfection factor α_x is [487]:

$$\alpha_x = \frac{0.62}{1 + 1.91\left(\dfrac{\Delta w_k}{t}\right)^{1.44}} \tag{7.99}$$

where the characteristic imperfection amplitude $\Delta w_k = \frac{t}{Q}\sqrt{\frac{r}{t}}$ is a function of the non-dimensional fabrication quality parameter Q depending on the specified fabrication tolerance quality: Class A (excellent) $Q = 40$, Class B (high) $Q = 25$, Class C (normal) $Q = 16$ (Appendix D.1.2.2, UNI ENV 1993-1-6:2007). In order to derive the stability reduction factors χ, the elastic-plastic stability interaction is defined in terms of three parameters, as previously described: the plastic range factor $\beta = 0.60$, the interaction exponent $\eta = 1.0$ and the meridional squash limit slenderness $\lambda_{x0} = 0.20$.

The term α_x is denoted as the "unpressurized" elastic imperfection factor; the presence in the tank of an internal pressure modifies the effect of the geometrical imperfections.[19] Hence, α_x (Equation (7.99)), used in the evaluation of the plastic limit relative slenderness (Equation (7.94)), should be replaced by the "pressurized" elastic imperfection factor α_{xp}, that corresponds to the smaller of the following two values:

$$\alpha_{xp} = \min\left(\alpha_{xpe}; \alpha_{xpp}\right) \tag{7.100}$$

where α_{xpe} is a factor covering pressure-induced elastic stabilization; this means that internal pressure causes increased strength (elastic stabilization or elastic strength gain). Factor α_{xpp} is a factor covering

---

19 This question has been deeply investigated and explored with reference to Figure 5.14 and Equations (5.61) mainly using as a reference the research done by Scharf [506] and also by Rammerstorfer and Fischer in [166, 469, 470, 507].

pressure-induced plastic destabilization, this means that internal pressure causes decreasing strength (plastic destabilization or plastic strength loss). It is clear from Figure 7.25 that the effect of strength gains under elastic buckling conditions and the strength losses due to plastic buckling are almost completely independent for a thin cylinder. The smaller of the two values α_{xpe} and α_{xpp} controls the design, "diamond buckling" in Figure 7.26 [294, 325, 502, 612] or "elephant foot buckling" in Figure 7.27 [139, 377, 488] respectively.

The factor α_{xpe} should be obtained from:

$$\alpha_{xpe} = \alpha_x + (1 - \alpha_x)\left[\frac{\bar{p}_s}{\bar{p}_s + \frac{0.3}{\sqrt{\alpha_x}}}\right] \tag{7.101}$$

where:

$$\bar{p}_s = \frac{p_s}{\sigma_{x,Rcr}}\frac{r}{t} \tag{7.102}$$

and $\sigma_{x,Rcr}$ comes from Equation (7.98); α_x is derived from expression (7.99) and p_s is the smallest[20] design value of local internal pressure at the location of the point being assessed, guaranteed to coexist with the meridional compression; Point 5.3.2.2 (1), UNI ENV 1993-4-1:2007, Note 3, clarifies that where the design stress resultants are being evaluated to verify adequate resistance to the buckling limit state under stored loads, in general, the stored material properties should be chosen to maximize the axial compression and the condition of discharge with patch loads in EN 1991-4 should be chosen. However, where the internal pressure is beneficial in increasing the buckling resistance (diamond buckling), only the filling pressures (for a consistent set of material properties) should be adopted in conjunction with the discharge axial forces, since the beneficial pressures (Figure 7.25) may fall to the filling values locally even though the axial compression derives from the discharge condition. In this context, Schmidt and Rotter in [511] at Point 10.2.5.2 suggest applying the partial factor $\gamma_F = 1.5$ for the axial compression, but the partial

Figure 7.25 Membrane stress state inside the von Mises envelope when the failures mode occurs: influence of circumferential tension σ_θ from internal pressure on the meridional buckling stress σ_x [482, 488, 493].

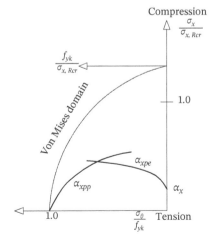

---

20 The smallest value is a request that comes from the evidence given in Figure 7.25 where high pressure induces elastic stabilization.

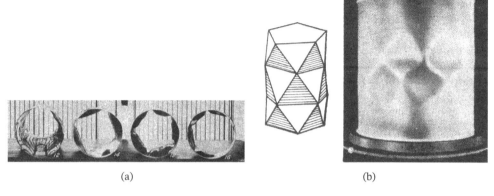

Figure 7.26 Examples of "diamond buckling". Source: (a) From [325]/Courtesy of the University of Illinois at Urbana-Champaign Archives; (b) From [185]/Courtesy of Professor Gioncu.

Figure 7.27 Examples of "elephant foot buckling": (a) Alaska earthquake (27 March 1964, Magnitude 9.2); (b) San Fernando earthquake, California (9 February 1971, Magnitudo 6.6) (from Karl V. Steinbrugge Collection, University of California, Berkeley).

factor $\gamma_F = 1$ to the internal pressure when calculating α_{xpe}. The factor α_{xpe} should not be applied to long cylinders where $\omega > 0.5\frac{r}{t}$, while it can be applied to medium-length cylinder $1.7 \leq \omega \leq 0.5\frac{r}{t}$ and short cylinder $\omega \leq 1.7$ and $C_x = 1$.

The factor α_{xpp} suggested by Rotter in [481] is:

$$\alpha_{xpp} = \left[1 - \left(\frac{\bar{p}_g}{\lambda_x^2}\right)^2\right]\left[1 - \frac{1}{1.12 + s^{1.5}}\right]\left[\frac{s^2 + 1.21\lambda_x^2}{s(s+1)}\right] \tag{7.103}$$

where:

$$\bar{p}_g = \frac{p_g}{\sigma_{x,Rcr}}\frac{r}{t} \tag{7.104}$$

and p_g is the largest design value of local internal pressure at the location of the point being assessed, and possibly coexistent with the meridional compression (as depicted in Figure 7.25 [486, 488] where internal pressure causes decreasing strength, the largest internal pressure that can possibly coexist with the meridional compression must be introduced); λ_x comes from Equation (7.96a) and $s = r/(400 \cdot t)$. Particularly, Point 5.3.2.2 (1), UNI ENV 1993-4-1:2007, Note 2, suggests that the stored solid properties should be chosen to maximize the internal pressure and the condition of discharge using patch loads in EN 1991-4.

Recommendations "Seismic Design of Storage Tanks, New Zealand National Society for Earthquake Engineering" (Edition December 1986 or revised version November 2009), in order to verify elastic buckling in membrane compression, instead of expression (7.101), suggests at Point 5.5.2, Edition 2009 (or Point 4.5.2, Edition 1986) the following relation:

$$\frac{\sigma_m}{\sigma_{c1}} \leq 0.19 + 0.81 \frac{\sigma_p}{\sigma_{c1}} \tag{7.105}$$

where the unknown variable σ_m is the vertical membrane compression stress for tank with internal pressure; $\sigma_{c1} = 0.6E\frac{t}{r}$ is the classical nondimensional elastic Timoshenko-Lorenz-Southwell critical stress [345, 346, 560, 609] (already described in Equation (7.98)); the remaining terms are:

$$\sigma_p = \sigma_{c1}\sqrt{\left[1 - \left(1 - \frac{\bar{p}}{5}\right)^2\left[1 - \left(\frac{\sigma_0}{\sigma_{c1}}\right)^2\right]\right]} \leq \sigma_{c1} \tag{7.106a}$$

$$\bar{p} = \frac{p}{\sigma_{c1}}\frac{r}{t} \leq 5 \tag{7.106b}$$

where p has the meaning of p_s in Equation (7.102). The coefficient σ_0 in relation (7.106a) is:

$$\sigma_0 = \begin{cases} f_y\left(1 - \frac{\lambda^2}{4}\right) & \lambda^2 = \frac{f_y}{\bar{\sigma}\cdot\sigma_{c1}} \leq 2 \\ \bar{\sigma}\sigma_{c1} & \lambda^2 = \frac{f_y}{\bar{\sigma}\cdot\sigma_{c1}} \geq 2 \end{cases} \tag{7.107}$$

where:

$$\bar{\sigma} = 1 - \Psi\frac{\Delta w_k}{t}\left[\sqrt{\left(1 + \frac{2}{\Psi\frac{\Delta w_k}{t}}\right)} - 1\right] \tag{7.108}$$

in which $\Psi = 1.24$, f_y is the nominal value of yield strength (Table 3.1, Point 3, UNI ENV 1993-1-1:2005) and $\frac{\Delta w_k}{t}$ is the ratio of maximum imperfection amplitude to wall thickness; $\Delta w_k = \frac{t}{Q}\sqrt{\frac{r}{t}}$ is a function of the non-dimensional fabrication quality parameter Q depending on the specified fabrication tolerance quality: Class A (excellent) $Q = 40$, Class B (high) $Q = 25$, Class C (normal) $Q = 16$ (Appendix D.1.2.2, UNI ENV 1993-1-6:2007).

Recommendations "Seismic Design of Storage Tanks, New Zealand National Society for Earthquake Engineering" (Edition December 1986 or revised version November 2009), in order to verify elastic-plastic buckling, instead of expression (7.103), suggests at Point 5.5.2, Edition 2009 (or 4.5.2 Edition 1986) the following relation:

$$\sigma_m \le \sigma_{c1} \left[1 - \left(\frac{pr}{tf_y} \right)^2 \right] \left[1 - \frac{1}{1.12 + \bar{r}^{1.5}} \right] \left[\frac{\bar{r} + \frac{f_y}{250}}{\bar{r} + 1} \right] \tag{7.109}$$

where $\bar{r} = \frac{1}{400} \frac{r}{t}$ and p has the meaning of p_g in Equation (7.104).

7.2.9.2 Circumferential Buckling

The critical circumferential buckling stress $\sigma_{\theta,Rcr}$ depends strongly from the shell segment length, and three cases can be obtained:

- for long (slender) tanks which are defined by $\frac{\omega}{C_\theta} > 1.63 \frac{r}{t}$, the critical circumferential buckling stress should be obtained from:

$$\sigma_{\theta,Rcr} = E \left(\frac{t}{r} \right)^2 \left[0.275 + 2.03 \left(\frac{C_\theta}{\omega} \frac{r}{t} \right)^4 \right] \tag{7.110}$$

where C_θ covers the influences of the boundary conditions at both ends of the tanks applied to the normal (radial) displacement w, meridional displacement u and meridional rotation β_ϕ:

- $C_\theta = 1.5$ if both boundary conditions BC1 (Table 7.6) are applied to both ends;
- $C_\theta = 1.25$ if BC1 (Table 7.6) is applied to one end and BC2 (Table 7.6) to the other end;
- $C_\theta = 1$ if both boundary conditions BC2 (Table 7.6) are applied to both ends;
- $C_\theta = 0.6$ if BC1 (Table 7.6) is applied to one end and BC3 (Table 7.6) to the other end;
- $C_\theta = 0$ if BC2 (Table 7.6) is applied to one end and BC3 (Table 7.6) to the other end or if both boundary conditions BC3 (Table 7.6) are applied to both ends.

Equation (7.110) when $\omega \to \infty$ becomes equal to $\sigma_{\theta,Rcr} = 0.275 E \left(\frac{t}{r} \right)^2$ which corresponds to expression (7.37) that is the critical pressure of an isolated ring under uniform external pressure with $(1 - v^2)$ and $v = 0.3$ [246];

- for short (squat) tanks which are defined by $\frac{\omega}{C_\theta} < 20$ the critical circumferential buckling stress has been originally derived from Batdorf in [29] using Donnell's equations [129]:

$$\sigma_{\theta,Rcr} = 0.92 E \frac{C_{\theta s}}{\omega} \frac{t}{r} \tag{7.111}$$

where $C_{\theta s}$ covers the influences of the boundary conditions at both ends of the tanks applied to the normal (radial) displacement w, meridional displacement u and meridional rotation β_ϕ:

- $C_{\theta s} = 1.5 + 10/\omega^2 - 5/\omega^3$ if both boundary conditions BC1 (Table 7.6) are applied to both ends;
- $C_{\theta s} = 1.25 + 8/\omega^2 - 4/\omega^3$ if BC1 (Table 7.6) is applied to one end and BC2 (Table 7.6) to the other end;
- $C_{\theta s} = 1.0 + 3/\omega^{1.35}$ if both boundary conditions BC2 (Table 7.6) are applied to both ends;
- $C_{\theta s} = 0.6 + 1.0/\omega^2 - 0.3/\omega^3$ if BC1 (Table 7.6) is applied to one end and BC3 (Table 7.6) to the other end;

- for medium-length tanks which are defined by $20 \le \frac{\omega}{C_\theta} \le 1.63 \frac{r}{t}$, the critical circumferential buckling stress should be obtained from:

$$\sigma_{\theta,Rcr} = 0.92 E \frac{C_\theta}{\omega} \frac{t}{r} \tag{7.112}$$

The circumferential elastic imperfection factor α_θ should be taken as a function of the fabrication tolerance quality class: Class A (excellent) $\alpha_\theta = 0.75$, Class B (high) $\alpha_\theta = 0.65$, Class C (normal) $\alpha_\theta = 0.50$. The plastic range factor, the interaction exponent, and the circumferential squash limit slenderness should be respectively taken as $\beta = 0.60$, $\eta = 1.0$, and $\lambda_{\theta 0} = 0.40$.

7.2.9.3 Shear Buckling

The critical shear buckling stress was originally derived from the linear Donnell-Mushtari-Vlasov shell buckling expression [18, 493] and can be obtained from:

$$\tau_{x\theta, Rcr} = 0.75 EC_\tau \frac{t}{r} \sqrt{\frac{1}{\omega}} \qquad (7.113)$$

where the factor C_τ covers different slendernesses of the tank through the parameter ω:

- for long cylinders which are defined by $\omega > 8.7\frac{r}{t}$, the factor $C_\tau = \frac{1}{3}\sqrt{\omega\frac{t}{r}}$;
- for short cylinders which are defined by $\omega < 10$, the factor $C_\tau = \sqrt{1 + \frac{42}{\omega^3}}$;
- for medium-length cylinders which are defined by $10 \leq \omega \leq 8.7\frac{r}{t}$ the factor $C_\tau = 1.0$.

The shear elastic imperfection factor α_τ should be taken as a function of the fabrication tolerance quality class: Class A (excellent) $\alpha_\tau = 0.75$, Class B (high) $\alpha_\tau = 0.65$, Class C (normal) $\alpha_\tau = 0.50$. The plastic range factor, the interaction exponent and the shear squash limit slenderness should be respectively taken as $\beta = 0.60$, $\eta = 1.0$, and $\lambda_{\tau 0} = 0.40$.

7.3 Requirements for Concrete Tanks

The design of reinforced concrete tanks requires attention not only to strength ultimate requirements, but also to crack control and durability [237]. Comment C3.1 to Point 3.1, NZS 3106 (2009) clearly states that serviceability rather than ultimate limit state considerations will commonly govern the design of liquid concrete retaining structures. ACI 350-06 "Code Requirements for Environmental Engineering Concrete Structures" extends the basic ACI 318-19 with additional serviceability requirements for crack width control due to flexure, shrinkage of concrete, and temperature movement. The commentary of ACI 350-06 makes it clear that ACI 318-19 requirements alone will not produce watertight structures. ACI 350-06 serviceability requirements include [237]:

- reduced working load stresses, and requirements for size and spacing of reinforcement;
- increased minimum reinforcement for temperature and shrinkage movement, which is dependent on the grade of reinforcing steel and the length between shrinkage dissipating joints;
- waterstop requirements at all joints;
- concrete mix design requirements;
- increased cover requirements compared to ACI 318-19.

In the ultimate and serviceability limit design of the concrete components, the designer is in many ways less constrained compared to the steel designer of metallic tanks, if only for the possibility of using the rules for ordinary reinforced concrete frame structures. Anyway, in the following, some particular rules of the commonly used design approaches for concrete tanks are provided [471]. Furthermore, design and detailing in reinforced concrete may be in accordance with the following recommendations used throughout this section:

- UNI ENV 1991-4:2006 "Actions on structures - Part 4: Silos and tanks" provides general principles and actions for the structural design of silos and tanks, used for the storage of particulate solids and for the storage of liquids. This code should be used in conjunction with EN 1990 and EN 1991.
- UNI EN 1992-3:2006 "Design of concrete structures - Part 3: Liquid retaining and containment structures" covers additional rules to those in UNI EN 1992-1-1:2004 for the design of structures constructed from plain or lightly reinforced concrete, reinforced concrete or prestressed concrete for the containment of liquids or granular solids. This part gives principles and application rules for the design of those elements of structure which directly support the stored liquids or materials (i.e. the directly loaded walls of tanks, reservoirs or silos, plates, junctions).
- ACI (American Concrete Institute), Concrete Shell Structures Practice and Commentary, (ACI 334.1R-92) together with ACI (American Concrete Institute), Building Code Requirements for Structural Concrete (ACI 318-19) and Commentary (ACI 318R-19) and the Uniform Building Code, Vol.2, Structural Engineering Design Provisions. They all provide information on the design, analysis, and construction of concrete thin shells (such as domes (surfaces of revolution) and cylindrical shells) and folded plates (prismatic, nonprismatic, or faceted).
- ACI (American Concrete Institute), Testing reinforced concrete structures for water-tightness, (ACI 350.1R, Title N. 90-S35) and Control of Cracking in Concrete Structures, (ACI 224R-01) provide guidance on water-tightness criteria and water-tightness testing of cast-in-place reinforced concrete structures such as containments, reservoirs, tanks and basins.
- IS (Indian Standard):3370 year 1965/1967 (updated version 2004), "Code of practice for concrete structures for storage of liquids": Part I is about general requirements on materials, control of cracking for water-tightness and junctions; Part II gives design rules for plates and walls (design tables are in Part IV); Part III is on prestressed concrete structures.
- "Circular concrete tanks without prestressing", Portland Cement Association (1993) and "Design of liquid-containing concrete structures for earthquake forces" Portland Cement Association (2002) both are guides for the design and detailing of concrete liquid containing structures for earthquake (and static) forces according to the model building codes. They both cover rectangular and circular tanks with non-flexible and flexible wall-to-base slab connections.
- BS (British Standard) 8007:1987: "Code of practice for design of concrete structures for retaining aqueous liquids" provides recommendations for the design and construction of normal reinforced and prestressed concrete vessels used for the containment of aqueous liquids.

7.3.1 Serviceability Limit State

The serviceability limit state, which includes leakage, durability, and deflection, invariably governs the concrete tank design.

7.3.1.1 Leakage

Starting from leakage as a first field of investigation, Point 7.3.1, UNI EN 1992-3:2006 suggests classifying liquid retaining structures in relation to the degree of protection against leakage required. Table 7.8 gives

Table 7.8 Fluid tightness classification for concrete retaining and containment structures.

Tightness Class	Leakage requirements
0	Some degree of leakage acceptable, or leakage of liquids irrelevant
1	Leakage to be limited to a small amount; some surface staining or damp patches acceptable
2	Leakage to be minimal; appearance not to be impaired by staining
3	No leakage permitted

the classification, noticing that all concrete will permit the passage of small quantities of liquids and gasses by diffusion. Depending on the classification of the element of the tank from Table 7.8, appropriate limits to cracking should be selected, paying due regard to the required function of the structure and the stored material.[21] Here are the requirements from Point 7.3.1(111), UNI EN 1992-3:2006:

Tightness Class 0 the provisions at Point 7.3.1, UNI EN 1992-1-1:2004 may be adopted.

Tightness Class 1 any cracks which can be expected to pass through the full thickness of the section should be limited to w_{k1}. The recommended w_{k1} values for structures retaining water are defined as a function of the ratio of the hydrostatic pressure, h_D (in metres of water pressure units $m_{H2O} = 10000 \, N/m^2$) to the wall thickness of the containing structure, h. For $h_D/h \leq 5$, $w_{k1} = 0.2$ mm, while for $h_D/h \geq 35$, $w_{k1} = 0.05$ mm. For intermediate values, a linear interpolation between 0.2 and 0.05 may be used (similar values at Point 5.1.1, NZS 3106 - 2009). The provisions at Point 7.3.1, UNI EN 1992-1-1:2004 apply where the full thickness of the section is not fully cracked and where the conditions at Points 7.3.1(112/113), UNI EN 1992-3:2006, described just below, are fulfilled.

Tightness Class 2 cracks which may be expected to pass through the full thickness of the section should generally be avoided unless appropriate measures (e.g. liners, water stoppers or water bars) have been incorporated.

Tightness Class 3 generally, special measures (e.g. liners or prestress) will be required to ensure watertightness.

According to Point 7.3.1(112) in UNI EN 1992-3:2006, in order to provide adequate performance[22] for structures of Tightness Class 2 and 3, the design value of the depth of the compression zone should be at least x_{min} (the lesser of 50 mm and $0.2h$ where h is the element thickness) calculated for the quasi-permanent combination of actions. Where a section is subjected to alternate actions, cracks should be considered to pass through the full thickness of the section unless it can be shown that some part of the section thickness will always remain in compression. This thickness of concrete in compression should normally be at least x_{min} under all appropriate combinations of actions. The action effects may be calculated on the assumption of linear elastic material behaviour. The resulting stresses in a section

---

21 Silos holding dry materials may generally be designed as Tightness Class 0, however, it may be appropriate for Tightness Class 1, 2 or 3 to be used where the stored material is particularly sensitive to moisture.

22 Performance here means that cracks do not pass through the full width of a section of the tank (Table 7.8).

should be calculated assuming that concrete in tension is ignored. If the provisions for Tightness Class 1 in Table 7.8 are met, then cracks through which water flows may be expected to heal in members which are not subjected to significant changes of loading or temperature during service. In the absence of more reliable information, healing may be assumed where the expected range of strain at a section under service conditions is less than $150 \cdot 10^{-6}$ (Point 7.3.1(113), UNI EN 1992-3:2006).

Where the minimum reinforcement given at Point 7.3.2, UNI EN 1992-1-1:2004 is provided in the tank's section, Figures 7.28(a) and 7.28(b) give values of maximum bar diameters and bar spacings, respectively, for various design crack widths w_{k1} for sections totally in tension. The maximum bar diameter given by Figure 7.28(a) should be modified using the following expression:

$$\phi_s = \phi_s^\star \left(\frac{f_{ct,eff}}{2.9} \right) \frac{h}{10(h-d)} \tag{7.114}$$

where ϕ_s is the modified maximum bar diameter; $\phi_s^\star$ is the maximum bar diameter obtained from Figure 7.28(a); h is the overall thickness of the member and d is the depth to the centroid of the outer layer of reinforcement from the opposite face of the concrete; $f_{ct,eff}$ [MPa] is the mean value of the tensile

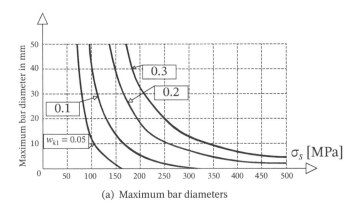

(a) Maximum bar diameters

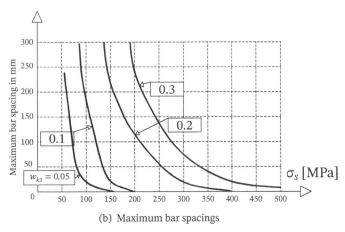

(b) Maximum bar spacings

Figure 7.28 Maximum bar diameters and spacings for crack control in members subjected to axial tension.

strength of the concrete effective at the time when the cracks may first be expected to occur ($f_{\text{ct,eff}} = f_{\text{ctm}}$ if cracking is expected earlier than 28 days where f_{ctm} is the mean value of the tensile strength of the concrete). For cracking caused dominantly by restraint, the bar sizes given in Figure 7.28(a) should not be exceeded where the steel stress is the value obtained immediately after cracking (which is called σ_s and it is the absolute value of the maximum stress permitted in the reinforcement immediately after formation of the crack; it may be taken as the yield strength of the reinforcement f_{yk}). For cracks caused dominantly by loading, either the maximum bar sizes from Figure 7.28(a) or the maximum bar spacings from Figure 7.28(b) may be complied with. The steel stress should be calculated on the basis of a cracked section under the relevant combination of actions. For intermediate values of design crack width, values may be interpolated. Where it is desirable to minimize the formation of cracks due to restrained imposed deformations resulting from temperature change or shrinkage, this may be achieved for Tightness Class 1 tank structures by ensuring that the resulting tensile stresses do not exceed the available tensile strength $f_{ctk,0.05}$ of the concrete, and, for Tightness Class 2 or 3 tank structures, where a liner is not used, by ensuring that the whole section remains in compression. These conditions may be achieved by:

1) limiting the temperature rise due to hydration of the cement;
2) removing or reducing restraints;
3) reducing the shrinkage of the concrete;
4) using concrete with a low coefficient of thermal expansion;
5) using concrete with a high tensile strain capacity (only for Tightness Class 1);
6) application of prestressing.

Point 24.3, ACI 318R-19 suggests that an improvement in crack control is obtained where the reinforcement is well distributed over the zone of maximum concrete tension. Several bars at moderate spacing are much more effective in controlling cracking than one or two larger bars of equivalent area. The maximum spacing s of bonded reinforcement closest to the tension face should be lesser of (Appendix C, ACI 318-19):

$$380 \left(\frac{280}{f_s} \right) - 2.5c_c \quad \text{and} \quad 300 \left(\frac{280}{f_s} \right) \tag{7.115}$$

where c_c is the least distance from the surface of deformed reinforcement to the tension face and f_s in MPa is the tensile stress in reinforcement at service loads, excluding pre-stressing reinforcement. This provision for spacing is intended to limit surface cracks to a width that is generally acceptable in practice but may vary widely in a given structure.

Under the seismic action relevant to the damage limitation state, crack widths should be verified against the limit values specified in Points 4 and 7, UNI EN 1992-1-1:2004, taking into account the appropriate environmental exposure class and the sensitivity of the steel to corrosion. In the case of lined concrete tanks, transient concrete crack widths should not exceed a value that might induce local deformation in the liner exceeding 50% of its ultimate uniform elongation (Point 4.5.1.2.1 (2), EC8-4).

7.3.1.2 Durability

A durable containment structure should meet the requirements of serviceability, strength, and stability throughout its design working life, without significant loss of utility or excessive unforeseen maintenance. In order to achieve the required design working life of the structure, adequate measures should be taken to protect each structural element against the relevant environmental actions. Abrasion of the inner face of the walls of a tank or silo may cause contamination of the stored material or lead to significant loss of cover. Three mechanisms of abrasion may occur:

1) mechanical attack due to the filling and discharging process;
2) physical attack due to erosion and corrosion with changing temperature and moisture conditions;
3) chemical attack due to reaction between the concrete and the stored material.

Appropriate measures should be taken to ensure that the elements subject to abrasion will remain serviceable for the design working life. Corrosion protection of steel reinforcement depends mainly on density, quality, and thickness of concrete cover. The cover density and quality are achieved by controlling the maximum water/cement ratio and minimum cement content and may be related to a minimum strength class of concrete. The minimum concrete cover,[23] depending on the exposure classes related to environmental conditions in accordance with EN 206-1 (2006), may be derived from Point 4.4.1.2, UNI EN 1992-1-1:2004. Exposure conditions are chemical and physical conditions to which the shell structure is exposed in addition to the mechanical actions and are classified according to the following list (Point 4.1, UNI EN 1992-1-1:2004):

1) Class designation X0 (no risk of corrosion or attack) for concrete inside buildings with very low air humidity.
2) Class designation XC1 to XC4 (corrosion induced by carbonation) for concrete inside buildings with low, medium, and high air humidity or for concrete surfaces subject to long-term water contact or for concrete permanently submerged in water.
3) Class designation XD1 to XD3 (corrosion induced by chlorides) for concrete surfaces exposed to airborne chlorides, swimming pools or bridges; concrete components exposed to industrial waters containing chlorides or car park slabs.
4) Class designation XS1 to XS3 (corrosion induced by chlorides from sea water) for (marine) structures near to or on the coast.
5) Class designation XF1 to XF4 (freeze/thaw attack) for horizontal/vertical concrete surfaces exposed to rain and freezing, roads and bridge decks (exposed to de-icing agents).
6) Class designation XA1 to XA3 (chemical attack) in case of natural soils and ground water.

A quite similar classification can be also found at Point 19.3.1, ACI 318-19 where four exposure categories that affect the requirements for concrete to ensure adequate durability are defined:

1) Exposure Category F applies to concrete exposed to moisture and cycles of freezing and thawing, with or without de-icing chemicals.
2) Exposure Category S applies to concrete in contact with soil or water containing deleterious amounts of water-soluble sulfate ions.
3) Exposure Category W applies to concrete in contact with water.
4) Exposure Category C applies to non-pre-stressed and pre-stressed concrete exposed to conditions that require additional protection against corrosion of reinforcement.

---

23 According to Point 4.4.1.1, UNI EN 1992-1-1:2004, the concrete cover is the distance between the surface of the reinforcement closest to the nearest concrete surface (including links and stirrups and surface reinforcement where relevant) and the nearest concrete surface.

Furthermore, as suggested at Point C7, NZS 3106 (2009), the air space below roof soffits is often at 100% humidity and the slab/roof is exposed internally to severe corrosive conditions. Thus the roof slab should be designed with the same crack control philosophy as for the rest of the structure/wall.

7.3.1.3 Deformability

The deflection/deformation of a slab, shell or wall should not be such that it adversely affects its proper functioning or appearance. Appropriate limiting values of deflection taking into account the nature of the structure, of the finishes, partitions, glazing, cladding, services or fixings and the function of the structure should be established. In some industrial applications, it may be required to ensure the proper functioning of machinery or apparatus supported by the structure, or to avoid ponding on flat roofs (Point 7.4.1, UNI EN 1992-1-1:2004). Hence, according to Point R6.6.3.2.2, ACI 318-19, analyses of deflections, time-dependent deflections, vibrations, and building periods are needed at various service (unfactored) load levels to determine the performance of the structure in service. Some values can be derived from Point 7.4.1, UNI EN 1992-1-1:2004 where it is highlighted that the appearance and general utility of the structure could be impaired when the calculated slab/shell or cantilever subjected to quasi-permanent loads exceeds span/250. Furthermore, deflections that could damage adjacent parts of the system/structure (or non-structural elements) should be limited to span/500 that is normally an appropriate limit for quasi-permanent loads. Similar values can be also found at Point 24.2, ACI 318-19 where elastic deflections, calculated considering the effects of cracking and reinforcement on member stiffness, vary in a range between span/180 and span/480 depending on the type of member (general/flat roofs or floors), boundary condition (supporting or not and eventually attached to non-structural elements) and type of deflection (immediate or time-dependent). Point 7.3.1.1, ACI 318-19 suggests that for solid non-pre-stressed slabs not supporting or attached to partitions or other construction likely to be damaged by large deflections, overall slab thickness h should not be less than the limits in Table 7.9, unless the calculated deflection limits (between span/180 and span/480) are satisfied.

7.3.2 Ultimate Limit State

The ultimate limit state (i.e. the moment resistance) of reinforced or prestressed concrete cross-sections can be evaluated following these assumptions taken from Point 6.1, UNI EN 1992-1-1:2004:

- plane sections remain plane;
- the strain in bonded reinforcement or bonded prestressing tendons, whether in tension or in compression, is the same as that in the surrounding concrete;

Table 7.9 Minimum thickness of solid non-pre-stressed one-way slabs, according to Point 7.3.1.1, ACI 318-19.

Boundary conditions	Minimum thickness h
Simply supported	span/20
One end continuous	span/24
Both ends continuous	span/28
Cantilever	span/10

- the tensile strength of the concrete is ignored;
- the stresses in the concrete in compression are derived from the design stress/strain relationship given at Point 3.1.7, UNI EN 1992-1-1:2004:

$$\sigma_c = f_{cd} \left[1 - \left(1 - \frac{\varepsilon_c}{\varepsilon_{c2}} \right)^n \right] \quad \text{for} \quad 0 \leq \varepsilon_c \leq \varepsilon_{c2} \tag{7.116a}$$

$$\sigma_c = f_{cd} \quad \text{for} \quad \varepsilon_{c2} \leq \varepsilon_c \leq \varepsilon_{cu2} \tag{7.116b}$$

where ε_{c2} is the strain reaching the maximum strength, ε_{cu2} is the ultimate strain, f_{cd} is the design value of concrete compressive strength; all the strength and deformation characteristics for concrete (including the exponent n) can be derived in Point 3.1, UNI EN 1992-1-1:2004;
- the stresses in the reinforcing or prestressing steel are derived from the design curves given in Points 3.2 and 3.3, UNI EN 1992-1-1:2004;
- the initial strain in pre-stressing tendons is taken into account when assessing the stresses in the tendons.

At the ultimate limit state the maximum action effects (membrane forces and bending moments, circumferential or meridional, and membrane shear) induced in the seismic (and static) design situation should be less or equal to the resistance of the shell evaluated as in the persistent or transient design situations (Point 3.5.2.2, EC8-4). This includes all types of failure modes (i.e. bending with axial force, shear for in-plane or radial shear, punching, torsion) and calculation of resistances and verifications that must be carried out in accordance with UNI EN 1992-1-1:2004 and UNI EN 1992-3:2006.

7.4 Detailing and Particular Rules

7.4.1 Walls

Typically in the design of reinforced concrete wall members (cylindrical or rectangular) the tensile strength of concrete is ignored and any significant cracking in a tank containing liquid is unacceptable [128]. For this reason, it must be assured that the stress in the concrete, from maximum ring tension T_{max} or maximum moment M_{max}, is kept at a minimum to prevent excessive cracking. Two procedures can be suggested to estimate the minimum value of the wall thickness t_{min}:

- in the case of a circular cylindrical reinforced concrete tank of thickness t_{min} the expression proposed by Domel and Gogate in [128] can be used:

$$t_{min} = 0.0003 \cdot T_{max} \tag{7.117}$$

where the thickness t_{min} is in centimetres and T_{max} in kN;
- in the case of a rectangular reinforced concrete tank of thickness t_{min} (for a single vertical wall) the expression proposed by Calvi and Nascimbene in [66] can be adopted:

$$d = \sqrt{\frac{M_{max}}{0.8 \, \xi \, b(1 - 0.4\xi)f_{cd}}} \tag{7.118a}$$

$$t_{min} = d + c \tag{7.118n}$$

where d is the depth to the centroid of the outer layer of reinforcement from the opposite face of the concrete and c the concrete cover; f_{cd} is the design value of the concrete compressive strength; b is the unit length of a single wall strip; the normalized neutral axis depth $\xi = \frac{x}{d}$ and x the neutral axis depth.

This section refers mainly to reinforced concrete walls in cylindrical or rectangular tanks with a length to thickness ratio of 4 or more (as indicated at Point 9.6.1, UNI EN 1992-1-1:2004) and in which the reinforcement is taken into account in the strength analysis. The amount and proper detailing of reinforcement may be derived from a more refined strut-and-tie model (described in detail in Point 6.5, UNI EN 1992-1-1:2004). For walls subjected predominantly to out-of-plane bending, the rules for slabs apply as clearly stated in Section 7.4.2.

Both horizontal and vertical shear reinforcement are required for all walls (Point R11.6.1, ACI 318-19). The distributed reinforcement is identified as being oriented parallel to either the longitudinal or transverse axis of the wall. Regarding **vertical reinforcement**, suggestions are given in Point 9.6.2, UNI EN 1992-1-1:2004.

- the area of the vertical reinforcement should lie between $A_{s,vmin} = 0.002A_c$ and $A_{s,vmax} = 0.04A_c$ where A_c is the cross-sectional area of concrete;
- where the minimum area of reinforcement $A_{s,vmin}$ controls the design, half of this area should be located on each face of the wall;
- the distance between two adjacent vertical bars should not exceed 3 times the wall thickness or 400 mm, whichever is the lesser (in agreement with Point 11.7.2.1, ACI 318-19);
- in the case of cylindrical tanks for flexural reinforcement, the minimum tension reinforcement ratio in potential plastic hinge zones of walls and floor slabs should be $\rho_{min} = \frac{\sqrt{f'_c}}{4f_y}$, where f'_c is the characteristic compressive cylinder strength of concrete at 28 days and f_y the yield strength of the reinforcement (Point 6.4.2, NZS 3106 (2009)).

Regarding **horizontal reinforcement**, suggestions are given in Point 9.6.3:

- horizontal reinforcement running parallel to the faces of the wall (and to the free edges) should be provided at each surface and it should not be less than $A_{s,hmin}$ which ever is the greater between 25% of the vertical reinforcement or $0.001A_c$;
- the spacing between two adjacent horizontal bars should not be greater than 400 mm;
- in the case of cylindrical tanks, the minimum hoop reinforcement ratio at any wall section should be $\rho_{min} = 0.05\frac{f'_c}{f_y}$ where f'_c is the characteristic compressive cylinder strength of concrete at 28 days and f_y the yield strength of the reinforcement (Point 6.4.1, NZS 3106 (2009)). Furthermore, Point C6.4.1, NZS 3106 (2009) suggests that sufficient reinforcement is required providing a ductile hoop response by ensuring that the cracked strength is greater than the uncracked strength, so that reinforcement yielding will spread around the tank perimeter rather than be concentrated at a few crack locations.

Regarding **transverse reinforcement**, suggestions are given at Point 9.6.4:

- in any part of a wall where the total area of the vertical reinforcement in the two faces exceeds $0.02A_c$, transverse reinforcement in the form of links (or stirrups) should be provided in accordance with the requirements for classical columns (Point 9.5.3, UNI EN 1992-1-1:2004);
- the diameter of the transverse reinforcement (links, stirrups, loops, or helical spiral reinforcement) should not be less than 6 mm or one quarter of the maximum diameter of the longitudinal bars,

whichever is the greater. The diameter of the wires of welded mesh fabric for transverse reinforcement should not be less than 5 mm; the transverse reinforcement (links, stirrups, loops, or helical spiral reinforcement) should be anchored adequately.

- the spacing of the transverse reinforcement along the column should not exceed the least of the following three distances: 20 times the minimum diameter of the longitudinal bars, the lesser dimension of the wall section and 400 mm;
- the maximum spacing just required above should be reduced by a factor 0.6: (a) in sections within a distance equal to the larger dimension of the wall cross-section above or below a beam or slab and need not be taken greater than 4 times the thickness of the wall; (b) near lapped joints, if the maximum diameter of the longitudinal bars is greater than 14 mm. A minimum of 3 bars evenly placed in the lap length is required.
- where the direction of the longitudinal bars changes (i.e. at changes in wall size), the spacing of transverse reinforcement should be calculated, taking account of the lateral forces involved. These effects may be ignored if the change of direction is less than or equal to 1 in 12.
- every longitudinal bar or bundle of bars placed in a corner should be held by transverse reinforcement. No bar within a compression zone should be further than 150 mm from a restrained bar.
- where the main reinforcement is placed nearest to the wall faces, transverse reinforcement should also be provided in the form of links with at least of 4 per m² of wall area.

At the ultimate limit state, according to Point 11.4.2.1, ACI 318-19, walls should be designed for the maximum factored moment that can accompany the factored axial force for each applicable load combination and for the maximum in-plane and out-of-plane factored shear. In addition to the minimum reinforcement required (or the reinforcement evaluated), at least two 16 mm (diameter) bars in walls having two layers of reinforcement in both directions and one 16 mm (diameter) bar in walls having a single layer of reinforcement in both directions should be provided around window, door, and similarly sized openings. Such bars should be anchored to develop the yield strength for the reinforcement in tension at the corners of the openings (Point 11.7.5.1, ACI 318-19).

7.4.2 Slabs

A slab is a member for which the minimum panel dimension is not less than 5 times the overall slab thickness as suggested by Point 5.3.1 (4), UNI EN 1992-1-1:2004 (ACI 302.1R-04 and ACI 302.1R-15). In the scientific literature [17, 155, 185, 231, 335, 687, 688], the shell or slab thickness h is assumed to be small with respect to the main other size L (or the curvature radii); thus a slab/shell can be classified as thin when the ratio $\frac{h}{L} < \frac{1}{20}$. A slab subjected to dominantly uniformly distributed loads may be considered to be one-way spanning if either (Point 5.3.1, UNI EN 1992-1-1:2004):

- it possesses two free (unsupported) and sensibly parallel edges;
- it is the central part of a sensibly rectangular slab supported on four edges with a ratio of the longer to shorter span greater than 2.

For non-pre-stressed cast-in-situ slabs without interior beams spanning between supports on all sides, having a maximum ratio of long-to-short span of 2, overall slab thickness h should not be less than the following limits, where L_{max} is the clear span in the long direction, measured face-to-face of supports and f_{yk} the yield strength for non-prestressed reinforcement (Point 8.3.1.1, ACI 318-19 without drop panels):

- $\frac{L_{max}}{33}$ when $f_{yk} = 280$ MPa in exterior slabs without edge beams; $\frac{L_{max}}{36}$ in exterior slabs with beams between columns along exterior edges; $\frac{L_{max}}{36}$ in interior slabs;
- $\frac{L_{max}}{30}$ when $f_{yk} = 420$ MPa in exterior slabs without edge beams; $\frac{L_{max}}{33}$ in exterior slabs with beams between columns along exterior edges; $\frac{L_{max}}{33}$ in interior slabs (Figure 7.29)[24];
- $\frac{L_{max}}{27}$ when $f_{yk} = 550$ MPa in exterior slabs without edge beams; $\frac{L_{max}}{30}$ in exterior slabs with beams between columns along exterior edges; $\frac{L_{max}}{30}$ in interior slabs.

Furthermore, Point 9.2, UNI EN 1992-1-1:2004 gives detailed design rules for one-way and two-way solid slabs for which $b, l_{eff} \geq 5h$ where h is the thickness, b is the length of the clear span in the short direction and l_{eff} the effective span:

$$l_{eff} = l_n + a_1 + a_2 \tag{7.119}$$

where l_n is the clear distance between the faces of the supports; values for a_1 and a_2, at each end of the span, may be determined from the appropriate a_i values in Figure 7.30 where t is the width of the supporting element. Some suggestions, derived from Point 9.3.1, UNI EN 1992-1-1:2004, regarding flexural and shear reinforcement are given in the following.

7.4.2.1 Flexural Reinforcement of Slabs

For the minimum and the maximum steel percentages in the main direction of the slab, the following rules apply:

- the area of longitudinal tension reinforcement should not be taken as less than $A_{s,min} = 0.26 \frac{f_{ctm}}{f_{yk}} b_t d$ but not less than $0.0013\, b_t d$ where f_{ctm} is the mean value of axial tensile strength of concrete, d is the depth

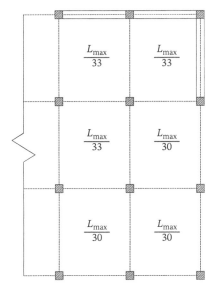

Figure 7.29 Minimum thickness in slabs (without drop panels on top of the columns or column capitals), according to Point 8.3.1.1, ACI 318-19 using $f_{yk} = 420$ MPa and L_{max} as the clear span in the long direction.

---

24 Notes on ACI 318-08 Building Code Requirements for Structural Concrete with Design Applications, Point 9.5.3.

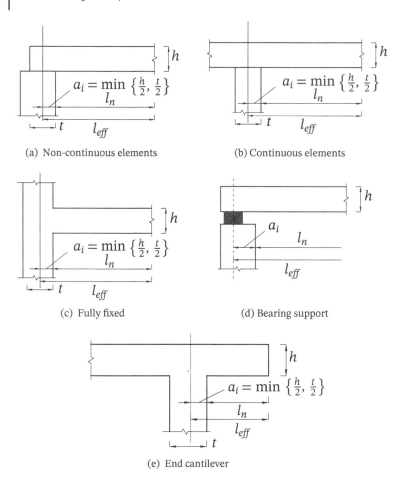

$$a_i = \min \left\{ \tfrac{h}{2}, \tfrac{t}{2} \right\}$$

(a) Non-continuous elements

$$a_i = \min \left\{ \tfrac{h}{2}, \tfrac{t}{2} \right\}$$

(b) Continuous elements

$$a_i = \min \left\{ \tfrac{h}{2}, \tfrac{t}{2} \right\}$$

(c) Fully fixed

(d) Bearing support

$$a_i = \min \left\{ \tfrac{h}{2}, \tfrac{t}{2} \right\}$$

(e) End cantilever

Figure 7.30 Effective span length l_{eff} in the case of different support conditions.

to the centroid of the outer layer of reinforcement from the opposite face of the concrete and b_t denotes the mean width of the tension zone;

- the cross-sectional area of tension or compression reinforcement should not exceed $A_{s,max} = 0.04A_c$ outside lap locations where A_c is the cross-sectional area of concrete.

Secondary transverse reinforcement of not less than 20% of the principal reinforcement should be provided in one-way slabs. In areas near supports, transverse reinforcement to principal top bars is not necessary where there is no transverse bending moment. Furthermore, the spacing of bars should not exceed s_{max}: the recommended value from Point 9.3.1.1 (3), UNI EN 1992-1-1:2004 for the principal reinforcement is $3h \leq 400$ mm, for the secondary reinforcement $3.5h \leq 450$ mm. In areas with concentrated loads or areas of maximum moment, the above provisions become respectively, for the principal reinforcement $2h \leq 250$ mm and for the secondary reinforcement $3h \leq 400$ mm.

Some suggestions are herein introduced related to curtailment of longitudinal tension reinforcement (Point 9.2.1.3 (1)–(3), UNI EN 1992-1-1:2004):

- sufficient reinforcement should be provided in all sections to resist the envelope of the acting tensile force, including the effect of inclined cracks in webs and flanges; for members with shear reinforcement, the additional tensile force ΔF_{td} in the longitudinal bars should be calculated according to the following expression [622]:

$$\Delta F_{td} = 0.5 V_{Ed}(\cot \theta - \cot \alpha) \tag{7.120}$$

where V_{Ed} is the design value of the applied shear force; θ is the angle between the concrete compression strut and the slab axis perpendicular to the shear force; α is the angle between shear reinforcement and the slab axis perpendicular to the shear force; $\frac{M_{Ed}}{0.9d} + \Delta F_{td}$ should be taken to be not greater than $\frac{M_{Ed,max}}{0.9d}$ where $M_{Ed,max}$ is the maximum moment along the slab. For members without shear reinforcement ΔF_{td} may be estimated by shifting the moment curve a distance $a_l = d$. This "shift rule" may also be used as an alternative for members with shear reinforcement, where $a_l = \frac{0.9d}{2}(\cot \theta - \cot \alpha)$.

- the resistance of bars within their anchorage lengths may be taken into account, assuming a linear variation of force, but as a conservative simplification this contribution may be ignored.

Some suggestions are herein introduced related to anchorage of bottom reinforcement at end supports (Point 9.2.1.4 (1)–(3), UNI EN 1992-1-1:2004):

- the area of bottom reinforcement provided at supports with little or no end fixity assumed in design, should be at least $\frac{1}{4}$ of the area of steel provided in the span; in a one-way slab, at simple supports, at least $\frac{1}{3}$ of the maximum positive moment reinforcement should extend along the slab bottom into the support, except for pre-cast slabs where such reinforcement should extend at least to the centre of the bearing length (Point 7.7.3.8.1, ACI 318-19); always in a one-way slab, at other supports, at least $\frac{1}{4}$ of the maximum positive moment reinforcement should extend along the slab bottom into the support at least 15 cm (Point 7.7.3.8.2, ACI 318-19);
- the tensile force to be anchored may be determined according to Equation (7.120) (members with shear reinforcement) including the contribution of the axial force if any, or according to the following shift rule:

$$F_{Ed} = |V_{Ed}|\frac{a_l}{z} + N_{Ed} \tag{7.121}$$

where N_{Ed} is the axial force, to be added to or subtracted from the tensile force; $z = 0.9d$ and $a_l = \frac{z}{2}(\cot \theta - \cot \alpha)$;

- the anchorage length is l_{bd} (Figures 7.31 and 7.34(a)) and is measured from the line of contact between the slab and the support; it must be determined according to the following (Point 8.4.4 (1), UNI EN 1992-1-1:2004):

$$l_{bd} = \alpha_1 \alpha_2 \alpha_3 \alpha_4 \alpha_5 l_{b,rqd} \geq l_{b,min} \tag{7.122}$$

where α_1, α_2, α_3, α_4 and α_5 are coefficients given in Table 7.10: α_1 is for the effect of the form of the bars assuming adequate cover (Figure 7.34); α_2 is for the effect of concrete minimum cover (Figure 7.35); α_3 is for the effect of confinement by transverse reinforcement (not welded); α_4 is for the influence of one or more welded transverse bars (with diameters $\phi_t > 0.6\phi$ placed not less than 5ϕ inside length

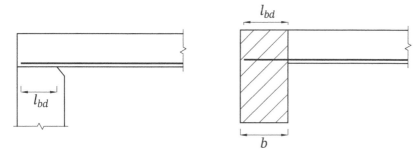

(a) Direct support where slab is supported by wall or column

(b) Indirect support where slab is intersecting another supporting beam or slab

Figure 7.31 Anchorage of bottom reinforcement at end supports used in the evaluation of the anchorage length l_{bd} in expression (7.122).

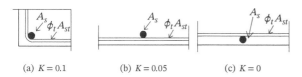

(a) $K = 0.1$ (b) $K = 0.05$ (c) $K = 0$

Figure 7.32 K values for slabs (and beams) according to expressions given in Table 7.10.

$l_{b,rqd}$ [418]) along the design anchorage length l_{bd}; α_5 is for the effect of the pressure transverse to the plane of splitting along the design anchorage length l_{bd}. The product $(\alpha_2\alpha_3\alpha_5) \geq 0.7$ and $l_{b,rqd}$ is the basic required anchorage length to anchor the force $A_s\sigma_{sd}$ in a straight bar assuming constant bond stress equal to f_{bd} (Point 8.4.3 (2), UNI EN 1992-1-1:2004):

$$l_{b,rqd} = \frac{\phi}{4}\frac{\sigma_{sd}}{f_{bd}} \tag{7.123}$$

where σ_{sd} is the design stress of the bar at the position from where the anchorage length is measured; the design value of the ultimate bond stress for ribbed bars f_{bd} may be taken as $f_{bd} = 2.25\,\eta_1\eta_2 f_{ctd}$ (Point 8.4.2 (2)) where $f_{ctd} = f_{ctk,0.05}/\gamma_c$ (5% fractile and γ_c is the partial factors for concrete for ultimate limit states); η_1 is a coefficient related to the quality of the bond condition and the position of the bar during concreting (Figure 7.33): hence $\eta_1 = 1.0$ when good bond conditions are obtained and $\eta_1 = 0.7$ for all other cases and for bars in structural elements built with slip-forms, unless it can be shown that good bond conditions exist; η_2 is related to the bar diameter, $\eta_2 = 1.0$ if $\phi \leq 32$ mm and $\eta_2 = (132 - \phi)/100$ if $\phi > 32$ mm [418]. Furthermore, in Equation (7.122), $l_{b,min}$ is the minimum anchorage length and it should be greater than max $\{0.3l_{b,rqd};\ 10\phi;\ 100\text{mm}\}$ for anchorages in tension and $l_{b,min} > $ max $\{0.6l_{b,rqd};\ 10\phi;\ 100\text{mm}\}$ for anchorages in compression. As a simplified alternative to expression (7.122) the tension anchorage of certain shapes shown in Figure 7.34 may be provided as an equivalent anchorage length $l_{b,eq}$ (Point 8.4.4 (2)). This equivalent length may be taken as: $\alpha_1 l_{b,rqd}$ for shapes shown in Figures 7.34(b)–7.34(d) (Table 7.10 is needed to obtain values for α_1); $\alpha_4 l_{b,rqd}$ for shapes shown in Figure 7.34(e) (Table 7.10 is needed to obtain values for α_4); the length $l_{b,rqd}$ should be evaluated using Equation (7.123).

Table 7.10 Values for coefficients $\alpha_1, \alpha_2, \alpha_3, \alpha_4$ and α_5 used in the evaluation of the design anchorage length l_{bd} in Equation (7.122).

		Reinforcement bar	
	Type of anchorage	In tension	In compression
Shape of bars	Straight	$\alpha_1 = 1.0$	$\alpha_1 = 1.0$
	Other than straight (Figures 7.34(b), 7.34(c) and 7.34(d))	$\alpha_1 = 0.7$ if $c_d > 3\phi$ otherwise $\alpha_1 = 1.0$ (Figure 7.35 for c_d)	$\alpha_1 = 1.0$
Concrete cover	Straight	$\alpha_2 = 1 - 0.15(c_d - \phi)/\phi$ $\geq 0.7 \leftarrow c_d = 3\phi$ $\leq 1.0 \leftarrow c_d = \phi$	$\alpha_2 = 1.0$
	Other than straight (Figures 7.34(b), 7.34(c) and 7.34(d))	$\alpha_2 = 1 - 0.15(c_d - 3\phi)/\phi$ ≥ 0.7 ≤ 1.0	$\alpha_2 = 1.0$
Confinement by transverse reinforcement not welded	All types	$\alpha_3 = 1 - K\lambda$ ≥ 0.7 ≤ 1.0	$\alpha_3 = 1.0$
Confinement by welded transverse reinforcement	All types; position and size in Figure 7.34(e)	$\alpha_4 = 0.7$	$\alpha_4 = 0.7$
Confinement by transverse pressure	All types	$\alpha_5 = 1 - 0.04p$ ≥ 0.7 ≤ 1.0	-

where:

$\lambda = \left(\sum A_{st} - \sum A_{st,\min} \right) / A_s$

$\sum A_{st} = nA_{st}$ where n is the number of transverse reinforcements along

the design anchorage length l_{bd} and A_{st} is the cross-sectional area of a single bar;

$\sum A_{st,\min}$ is the cross-sectional area of the minimum transverse reinforcement

$= 25\% A_s$ for beam elements and 0 for slabs;

A_s area of a single anchored bar with maximum bar diameter;

K values shown in Figure 7.32;

p is the transverse pressure [MPa] at ultimate limit state along l_{bd}.

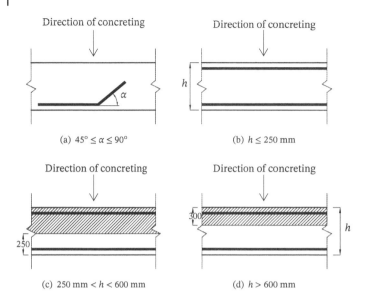

Direction of concreting

Direction of concreting

h

(a) $45° \leq \alpha \leq 90°$

(b) $h \leq 250$ mm

Direction of concreting

Direction of concreting

250

300

h

(c) 250 mm $< h <$ 600 mm

(d) $h >$ 600 mm

Figure 7.33 Bond conditions: (a) and (b) good bond conditions for all bars; (c) and (d) bars placed in the hatched zone: poor bond conditions, while bars placed in the unhatched zone: good bond conditions.

Some suggestions are herein introduced related to anchorage of bottom reinforcement at intermediate supports (Point 9.2.1.5 (1)–(2), UNI EN 1992-1-1:2004):

- the area of reinforcement given in Point 9.2.1.4 (1) (and related to the anchorage of the bottom reinforcement at end supports) applies;
- the anchorage length should not be less than 10 ϕ (for straight bars) or not less than the diameter of the mandrel d_m (for hooks and bends with bar diameters at least equal to 16 mm) or twice the diameter of the mandrel (in other cases) (Figure 7.36(a)).

Some suggestions are herein introduced related to reinforcement in slabs near supports, corners, and free edges (Point 9.3.1.2/3/4, UNI EN 1992-1-1:2004 and ACI 318-19):

- in simply supported slabs, half the calculated span reinforcement should continue up to the support and be anchored therein in accordance with Section 7.4.2.1;
- where partial fixity occurs along an edge of a slab, but is not taken into account in the analysis, the top reinforcement should be capable of resisting at least 25% of the maximum moment in the adjacent span. This reinforcement should extend at least 0.2 times the length of the adjacent span, measured from the face of the support (Figure 7.36(b)); furthermore, it should be continuous across internal supports and anchored at end supports. At an end support the moment to be resisted may be reduced to 15% of the maximum moment in the adjacent span.
- unrestrained corners of slabs tend to lift when loaded; if this lifting tendency is restrained by edge walls or beams, bending moments result in the slab. If the detailing arrangements at a support/corner are such that lifting of the slab at the corner is restrained, suitable reinforcement should be provided. Point 8.7.3.1.2, ACI 318-19 suggests that a corner reinforcement (top and bottom) have to be provided for a

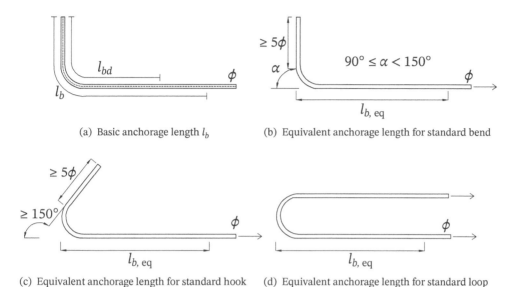

(a) Basic anchorage length l_b

(b) Equivalent anchorage length for standard bend

(c) Equivalent anchorage length for standard hook

(d) Equivalent anchorage length for standard loop

(e) Equivalent anchorage length for welded transverse bar

Figure 7.34 Anchorage in case of curved bar.

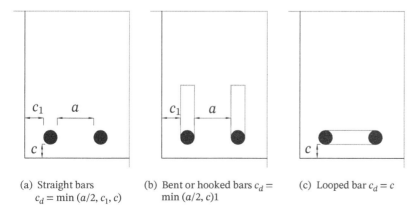

(a) Straight bars
$c_d = \min(a/2, c_1, c)$

(b) Bent or hooked bars $c_d = \min(a/2, c)1$

(c) Looped bar $c_d = c$

Figure 7.35 Minimum value of the cover c_d for slabs (and beams).

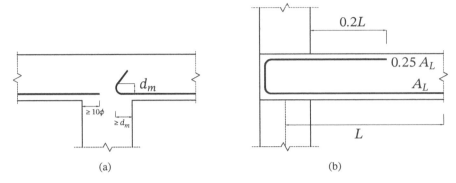

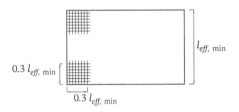

Figure 7.36 (a) Anchorage at intermediate supports (d_m is the diameter of the mandrel); (b) reinforcement in slabs' near supports.

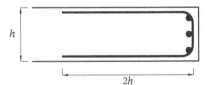

Figure 7.37 Slab corner reinforcement, according to DIN 1045-1.

Figure 7.38 Edge reinforcement for a slab, according to Point 9.3.1.4, UNI EN 1992-1-1:2004.

distance in each direction from the corner equal to one-fifth the longer span ($0.2\,l_{max}$ where l_{max} is the longer span of the slab); Point 13.3.2 (6), DIN 1045-1 suggests using a corner reinforcement (top and bottom) not less than $0.3\,l_{eff,min}$ (Figure 7.37); [335];
- along a free (unsupported) edge, a slab should normally contain longitudinal and transverse reinforcement, generally arranged as shown in Figure 7.38; furthermore, the normal reinforcement provided for a slab may act as an edge reinforcement.

7.4.2.2 Shear Reinforcement of Slabs
Some suggestions are herein introduced related to shear reinforcement in slabs (Points 9.2.2 and 9.3.2, UNI EN 1992-1-1:2004 and Point 8.7.6, ACI 318-19) [418]:

- a slab in which shear reinforcement is provided should have a depth of at least 200 mm (Point 9.3.2, UNI EN 1992-1-1:2004) or greater (250 mm) according to Point R8.7.6, ACI 318-19;
- the shear reinforcement should form an angle between 45° and 90° to the longitudinal axis of the structural element; it is essential that shear reinforcement engage longitudinal reinforcement at both the top and bottom of the slab, as shown for typical details in Figure 7.39;

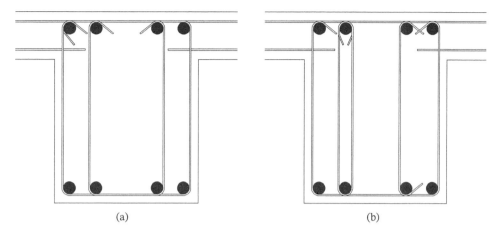

Figure 7.39 Single or multiple-leg stirrup-type (closed bar) slab shear reinforcement from Point R8.7.6, ACI 318-19 and Point 9.2.2, UNI EN 1992-1-1:2004.

- the shear reinforcement may consist of a combination of links enclosing the longitudinal tension reinforcement and the compression zone if is possible (usually in a vertical tank's walls); bent-up bars or cages, ladders which are cast in without enclosing the longitudinal reinforcement but are properly anchored in the compression and tension zones; at least 50% of the necessary shear reinforcement should be in the form of links;
- the ratio of shear reinforcement is given by:

$$\rho_w = \frac{A_{sw}}{sb_w \sin \alpha} \tag{7.124}$$

where: ρ_w is the shear reinforcement ratio which should not be less than $\rho_{w,\min} = \frac{0.08\sqrt{f_{ck}}}{f_{yk}}$; A_{sw} is the area of shear reinforcement within length s; s is the spacing of the shear reinforcement measured along the longitudinal axis of the member; b_w is the breadth of the web of the member; α is the angle between the shear reinforcement and the longitudinal axis;
- the maximum longitudinal spacing between shear assemblies should not exceed $s_{l,\max} = 0.75d(1 + \cot \alpha)$ (Point 8.7.6.3, ACI 318-19 suggests $0.5d$); the maximum longitudinal spacing of bent-up bars should not exceed $s_{b,\max} = 0.6d(1 + \cot \alpha)$ (d is the effective depth of the cross-section);
- the transverse spacing of the legs in a series of shear links should not exceed $s_{t,\max} = 0.75d \leq 600$ mm;
- improved performance, compared with a leg of a stirrup having bends at the ends, can be obtained using a stud head that exhibits smaller slip and, thus, results in smaller shear crack widths. The improved performance results in increased limits for shear strength and spacing between peripheral lines of headed shear stud reinforcement.

7.4.3 Joints

In the design practice for reinforced concrete corner connection joints, it is desirable to avoid local failure in the corner connection itself [494]. Knee joints subjected to negative moment (moment tends to close the corner; it is also called closing bending moment) or positive moment (moment tends to open the corner; it is also called opening bending moment) need reasonable reinforcing details giving an acceptable structural performance. The importance of the detailing on the behaviour of corners in reinforced concrete containment structures has been often pointed out in many research studies [62, 63, 178, 308, 335, 426, 584]. To ensure ductility, the design of a joint ought to be such that the failure of the adjoining members precedes failure of the joint. Some suggestions are given for corners subjected to closing moment as well as opening corners, in the following. Anyway, here some basic guiding principles are introduced:

- determine location and direction of all internal forces and the corresponding load path that satisfies equilibrium;
- use adequately anchored reinforcement wherever a tensile force is required for equilibrium and never rely on the concrete's ability to carry tension;
- use only ductile reinforcement when the reinforcement is required for strength;
- include adequate quantities of reinforcement for crack control and waterproofing;
- ensure steel details are practical and that steel can be fixed and concrete can be satisfactorily placed and compacted around complex details with adequate cover;
- ensure details are economical.

Where a tank's walls are connected monolithically at a corner (wall to wall or wall to roof or wall to base plate) and are subjected to moments and shears which tend to open or close the corner, a strut and tie system as covered in Point 5.6.4, UNI EN 1992-1-1:2004 (or Chapter 23 in ACI 318-19) is an appropriate design approach. Strut-and-tie models consist of struts representing compressive stress fields, then ties representing the reinforcement, and of the connecting nodes. The forces in the elements of a strut-and-tie model should be determined by maintaining the equilibrium with the applied loads in the ultimate limit state.

7.4.3.1 Connections Subjected to Negative Moment

The most likely cause of premature failure [475] is due to bearing under the bend of the tension steel passing around the outside of the corner. Providing that the radius of this bend is gradual (a diameter of the mandrel greater than 15 times the diameter of the bar and less than 1.6 times the wall thickness [335]) and that sufficient anchorage is provided for the lapping bars, no problems should arise. Some examples of reinforcement arrangements in closing bending moment connection nodes subject to negative moment M^- are shown in Figure 7.40 [62, 308, 334, 335, 494]. In addition to the primary tie reinforcement, confinement or skin reinforcement may be needed to control cracking in the connection region, especially for connection of large members.

7.4.3.2 Connections Subjected to Positive Moment

Usually negative corner moments cause less detailing problems compared to positive corner moments M^+. Figure 7.41 shows the stress distribution along the diagonal in a corner subjected to positive moment. The bending stress, σ_x, exhibits a peak tension at the inside of the corner which explains why corner cracks occur for quite small loads [32, 426]:

$$\sigma_x = \frac{M}{A\left(r_0 - r_n\right)} \frac{y}{r_n - y} \tag{7.125}$$

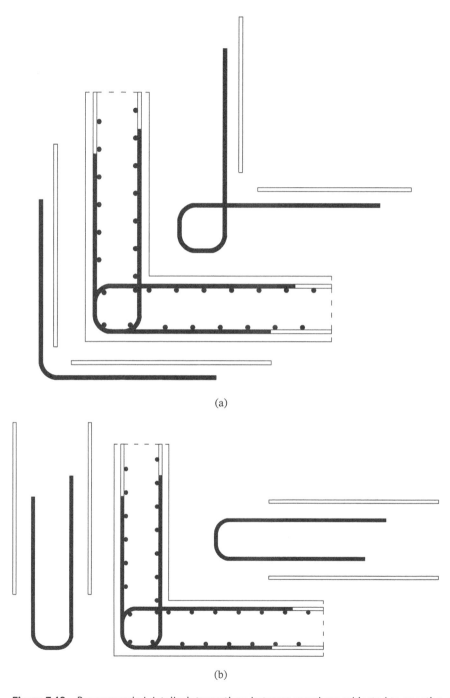

(a)

(b)

Figure 7.40 Recommended details: intersections between members subjected to negative moment.

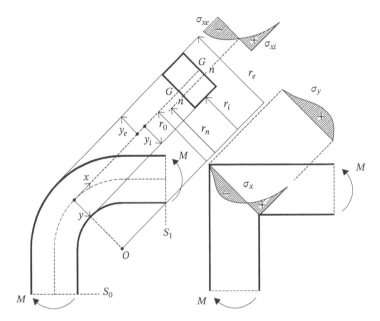

Figure 7.41 Tensile stress distribution across joint sections with a positive moment applied.

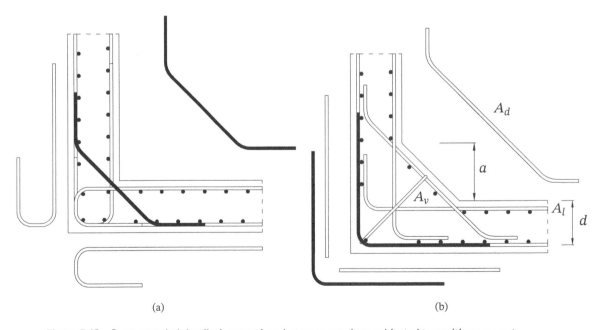

(a)   (b)

Figure 7.42 Recommended details: intersections between members subjected to positive moment.

Stresses corresponding to the internal and external fibres of the section have the following expression:

$$\sigma_{xi} = \frac{My_i}{A\left(r_0 - r_n\right)r_i} \tag{7.126a}$$

$$\sigma_{xe} = \frac{My_e}{A\left(r_0 - r_n\right)r_e} \tag{7.126b}$$

The tensile stresses σ_y cause a diagonal crack across the corner which results in sudden failure unless reinforcement is provided. Both tensile stresses may be considered parabolically distributed perpendicular to the joint diagonal. Providing that the radius of this bend is gradual, a diameter of the mandrel greater than 10 times the diameter of the bar is suggested [335]. Reinforcement should be provided as a loop in the corner region or as two overlapping U bars in combination with inclined links $A_v = \sqrt{2}A_l$ [475, 476] (Figure 7.42). For large opening moments a diagonal bar, as shown in Figure 7.42(a) (area A_d and a diameter equal to 2/3 of the diameter of the longitudinal bar A_l [335]) and links to prevent splitting should be considered. As an alternative, the relation proposed by Reynolds and Steedman in [475] can be used:

$$A_d = \frac{\sqrt{2}\left(M^+ - 0.6 \cdot 0.9\, d\, f_{yd}A_l\right)}{0.9(a + d)f_{yd}} \tag{7.127}$$

Special attention must be paid to bending and fixing the diagonal links, which must be designed to resist all the force in the main tension bars. Care must also be taken to provide adequate cover to these main bars where they pass round the inner corner [475].

Appendix A

Dimensionless Design Charts

A.1 Introduction

This Appendix presents dimensionless design charts (Figures A.1–A.6) for the calculation of hoop force, N_θ^{max}, vertical bending moment, M_y^{max}, for convective, impulsive (with rigid wall container assumption), and hydrostatic (or vertical), pressure distributions acting on a cylindrical storage tank with a vertical axis (charts are based on Points C4.3.1 and Appendix A in NZSEE-09). The following charts are related to tank radius R, height H, and wall thickness s for a range of $\frac{H}{R} = 0.2, 0.5, 1.0, 2.0, 5.0$ and $\frac{R}{s} = 50, 100, 200, 500,$ and 1000. Values of the dimensionless hoop force $N_\theta^\star$ and bending moment $M_y^\star$ are given along the height of the tank's wall $\zeta = \frac{z}{H}$.

Seismic Design and Analysis of Tanks, First Edition. Gian Michele Calvi and Roberto Nascimbene.
© 2023 John Wiley & Sons, Inc. Published 2023 by John Wiley & Sons, Inc.

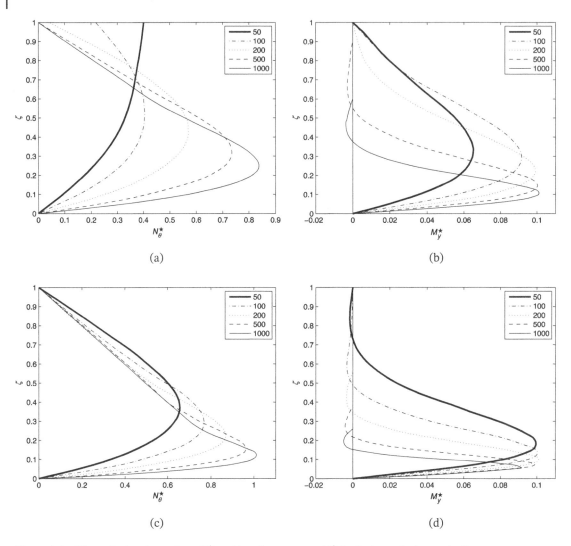

Figure A.1 Dimensionless hoop force $N_\theta^\star$ and bending moment $M_y^\star$ for hydrostatic (or vertical) pressure distributions: (a)–(b) $\frac{H}{R} = 0.2$ and (c)–(d) $\frac{H}{R} = 0.5$.

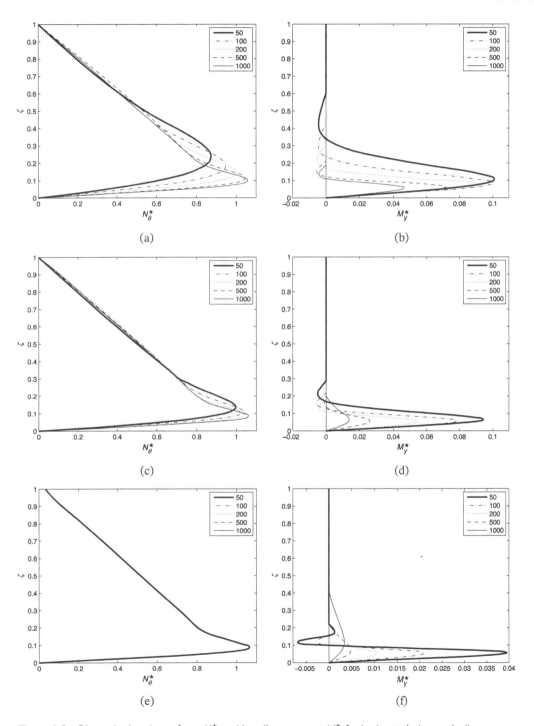

Figure A.2 Dimensionless hoop force $N_\theta^\star$ and bending moment $M_y^\star$ for hydrostatic (or vertical) pressure distributions: (a)–(b) $\frac{H}{R} = 1.0$, (c)–(d) $\frac{H}{R} = 2.0$, and (e)–(f) $\frac{H}{R} = 5.0$.

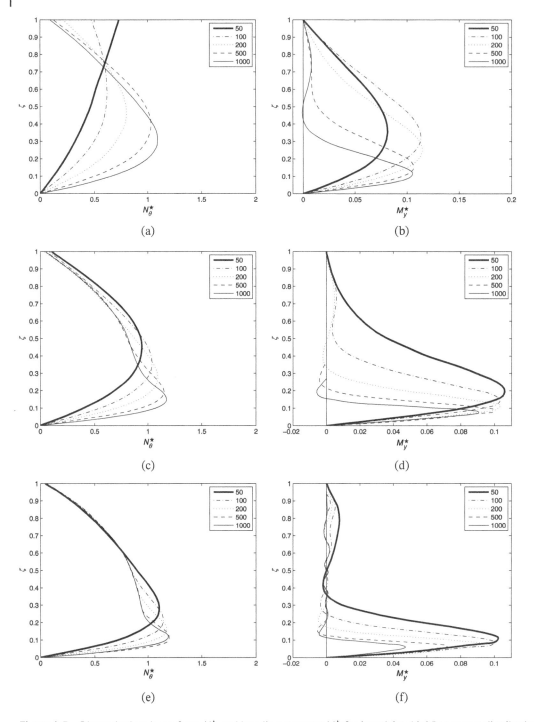

Figure A.3 Dimensionless hoop force $N_\theta^\star$ and bending moment $M_y^\star$ for impulsive (rigid) pressure distributions: (a)–(b) $\frac{H}{R} = 0.2$, (c)–(d) $\frac{H}{R} = 0.5$ e, (e)–(f) $\frac{H}{R} = 1.0$.

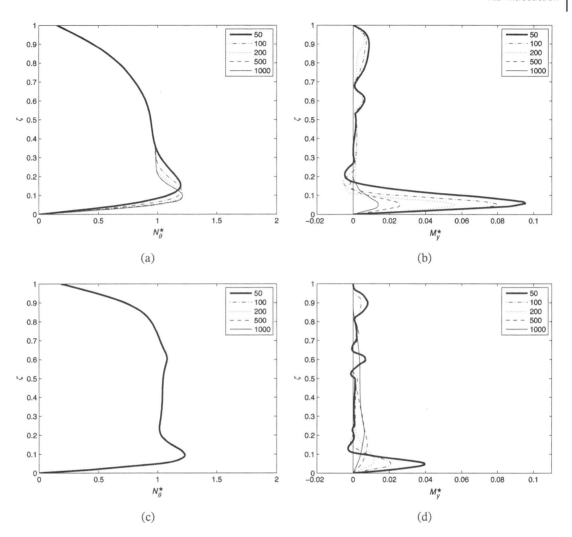

Figure A.4 Dimensionless hoop force $N_\theta^\star$ and bending moment $M_y^\star$ for impulsive (rigid) pressure distributions: (a)–(b) $\frac{H}{R} = 2.0$ and (c)–(d) $\frac{H}{R} = 5.0$.

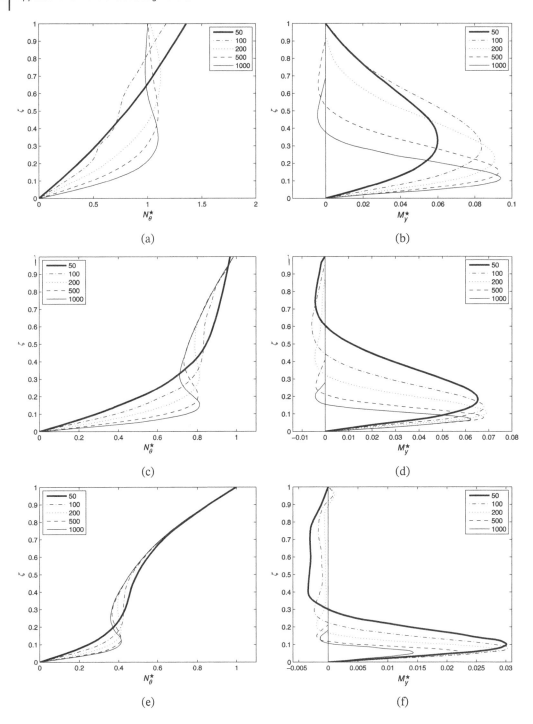

Figure A.5 Dimensionless hoop force $N_\theta^\star$ and bending moment $M_y^\star$ for convective (1st-mode) pressure distributions: (a)–(b) $\frac{H}{R} = 0.2$, (c)–(d) $\frac{H}{R} = 0.5$ e, (e)–(f) $\frac{H}{R} = 1.0$.

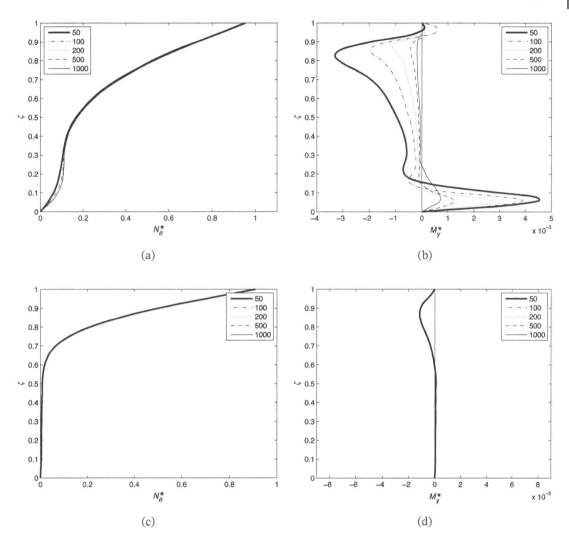

Figure A.6 Dimensionless hoop force $N_\theta^\star$ and bending moment $M_y^\star$ for convective (1st−mode) pressure distributions: (a)−(b) $\frac{H}{R} = 2.0$ and (c)−(d) $\frac{H}{R} = 5.0$.

Appendix B

Codes, Manuals, Recommendations, Guidelines, Reports

B.1 Introduction

This Appendix lists various international codes, standards, manuals, guidelines, and recommendations, used in this book as references for analysis and design.

ACI (American Concrete Institute), Prediction of Creep, Shrinkage, and Temperature Effects in Concrete Structures, (ACI 209R-92, Reapproved 2008).

ACI (American Concrete Institute), Control of Cracking in Concrete Structures, (ACI 224R-01).

ACI (American Concrete Institute), Guide for Concrete Floor and Slab Construction, (ACI 302.1R-04 and ACI 302.1R-15).

ACI (American Concrete Institute), Building Code Requirements for Structural Concrete (ACI 318-19) and Commentary (ACI 318R-19).

ACI (American Concrete Institute), Concrete Shell Structures Practice and Commentary, (ACI 334.1R-92).

ACI (American Concrete Institute), Design and Construction of Circular Prestressed Concrete Structures, (ACI 344, Title N. 67-40).

ACI (American Concrete Institute), Specification for Cast-in-Place Concrete Pipe, (ACI 346-09).

ACI (American Concrete Institute), Testing Reinforced Concrete Structures for Watertightness, (ACI 350.1R, Title N. 90-S35).

ACI (American Concrete Institute), Seismic Design of Liquid-Containing Concrete Structures and Commentary, (ACI 350.3-06).

ACI (American Concrete Institute), Guide for the Analysis, Design and Construction of Concrete-Pedestal Water Towers, (ACI 371R-98).

ACI (American Concrete Institute), Guide for the Analysis, Design and Construction of Elevated Concrete and Composite Steel-Concrete Water Storage Tanks, (ACI 371R-08).

ACI (American Concrete Institute), Design and Construction of Circular Wire- and Strand-Wrapped Prestressed Concrete Structures, (ACI 372R-13).

ACI (American Concrete Institute), Design and Construction of Circular Prestressed Concrete Structures with Circumferential Tendons, (ACI 373R-97).

Seismic Design and Analysis of Tanks, First Edition. Gian Michele Calvi and Roberto Nascimbene.
© 2023 John Wiley & Sons, Inc. Published 2023 by John Wiley & Sons, Inc.

AIJ (Architectural Institute of Japan), Design Recommendation for Storage Tanks and their Supports with Emphasis on Seismic Design, 2010 Edition.

ALA (American Lifelines Alliance), Guidelines for the Design of Buried Steel Pipe, FEMA-ASCE, 2001 (with addenda through February 2005).

ALA (American Lifelines Alliance), Seismic Design and Retrofit of Piping Systems, FEMA-ASCE, 2002.

ALA (American Lifelines Alliance), Guidelines for Implementing Performance Assessments of Water Systems, Volume I-II (Commentary), FEMA-National Institute of Building Sciences, 2005.

ALA (American Lifelines Alliance), Seismic Guidelines for Water Pipelines, FEMA-National Institute of Building Sciences, 2005.

ANSI/API Specification 5L, Specification for Line Pipe, (ISO 3183:2007, Petroleum and Natural Gas Industries - Steel Pipe for Pipeline Transportation Systems), 2007 (Errata, January 2009 and Addendum, February 2009).

ANSI/ASCE 15-93, "Standard practice for direct design of buried precast concrete pipe using standard installations (SIDD)", approved 18 December 1993.

API 12B (2009) Specification for Bolted Tanks for Storage of Production Liquids. American Petroleum Institute Standards, Washington, DC.

API 620 (2012) Design and Construction of Large, Welded, Low-Pressure Storage Tanks. American Petroleum Institute Standards, Washington, DC.

API 650 (2012) Welded Tanks for Oil Storage. American Petroleum Institute Standards, Washington, DC.

API 1105 Bulletin on Construction Practices for Oil and Products Pipe Lines. American Petroleum Institute Standards, Washington, DC.

API 1117 (2008) Recommended Practice for Movement in In-service Pipelines, (Errata, December 2008 and Addendum, August 2009).

AS 3774-1996 "Loads on bulk solids containers".

ASCE (American Society of Civil Engineers), Buried Flexible Steel Pipe. Design and Structural Analysis, ASCE Manuals and Reports of Engineering Practice, n. 119 of 2009.

ASCE (American Society of Civil Engineers), Design of Cylindrical Concrete Shell Roofs, ASCE Manuals of Engineering Practice, n. 31 of 1952.

ASCE (American Society of Civil Engineers), Guidelines for the Seismic Design of Oil and Gas Pipeline Systems, Committee on Gas and Liquid Fuel Lifelines, 1984.

ASCE (American Society of Civil Engineers), Minimum Design Loads for Buildings and Other Structures (ASCE 7-2010).

ASCE (American Society of Civil Engineers), Seismic Design Criteria for Structures, Systems, and Components in Nuclear Facilities, ASCE/SEI 43-05.

ASME B31.1-2007 (American Society of Mechanical Engineers), Power Piping, ASME Code for Pressure Piping.

ASME BPVC-2017 (American Society of Mechanical Engineers), Boiler and Pressure Vessel Code, BPVC17.

ASTM A416/A416M-2006 Standard Specification for Steel Strand, Uncoated Seven-Wire for Prestressed Concrete.

ATC-6 Seismic Design Guidelines for Highway Bridges, Applied Technology Council, California (Published 1981, 210 pages).

ATC-32 Improved Seismic Design Criteria for California Bridges: Provisional Recommendations, Applied Technology Council, California (Published 1996, 215 pages).

ATC-40 Seismic Evaluation and Retrofit of Concrete Buildings, Applied Technology Council, California (Published, 1996, 612 pages)

AWWA D100-11 Welded Carbon Steel Tanks for Water Storage, American Water Works Association, Colorado.

AWWA D102-06 Coating Steel Water-Storage Tanks, American Water Works Association, Colorado.

AWWA D103-09 Factory-Coated Bolted Carbon Steel Tanks for Water Storage, American Water Works Association, Colorado.

AWWA D110-04 Wire and Strand-Wound, Circular, Prestressed Concrete Water Tanks, American Water Works Association, Colorado.

AWWA D115-06 Tendon-Prestressed Concrete Water Tanks, American Water Works Association, Colorado.

AWWA M42 Steel Water-Storage Tanks, Manual of Water Supply Practice, American Water Works Association, Colorado.

AWWA M45 Fiberglass Pipe Design, Manual of Water Supply Practice, American Water Works Association, Colorado.

BS 5500:1997 Specification for Unfired Fusion Welded Pressure Vessels.

BS 5896:1980 Specification for High Tensile Steel Wire and Strand for the Prestressing of Concrete.

BS 8007:1987 Code of Practice for Design of Concrete Structures for Retaining Aqueous Liquids.

BS 8010:2004 Code of Practice for Pipelines. Steel Pipelines on Land.

BS 8110-1:1997 Structural Use of Concrete. Part 1: Code of Practice for Design and Construction.

BS 8110-2:1985 Structural Use of Concrete. Part 2: Code of Practice for Special Circumstances.

BS 8110-3:1985 Structural Use of Concrete. Part 3: Design Charts for Singly Reinforced Beams, Doubly Reinforced Beams and Rectangular Columns.

CEB-FIP Model Code 1990, Design Code.

CEB Bulletin N. 208 "Fire Design of Concrete Structures in Accordance with CEB-FIP Model Code 90".

CEN (1996) European Prestandard ENV 1998: Eurocode 8 - Design Provisions for Earthquake Resistance of Structures. Comité Européen de Normalisation, Brussels.

Circ. LL. PP. 5 Maggio 1966, n. 2136, Istruzioni sull'impiego delle tubazioni in acciaio saldate nelle costruzioni degli acquedotti.

Circ. LL. PP. 20 Marzo 1986, n. 27291, Istruzioni relative alle *Norme tecniche per le tubazioni.*

Circ. LL. PP. 4 Luglio 1996, Istruzioni per l'applicazione delle *Norme tecniche relative ai criteri generali per la verifica di sicurezza delle costruzioni e dei carichi e sovraccarichi.*

Circ. LL. PP. 10 Aprile 1997, Istruzioni per l'applicazione delle *Norme tecniche per le costruzioni in zone sismiche.*

Concrete Pipe Design Manual. Revised to Include Standard Installation (2000), American Concrete Pipe Association (www.concrete-pipe.org).

COVENIN 3623:2000, *Diseño sismorresistente de tanques metálicos*, Norma Venezolana.

CNR 10012/85 "Istruzioni per la valutazione delle azioni sulle costruzioni".

DIN 1045-1 (2001), *Concrete, reinforced and prestressed concrete structures - Part.1 Design and construction.*

DIN 1055-6 (1987-05), *Design loads for buildings: loads in silo bins.*

DIN 18800-3 (2008) *Steel structures. Part 3: stability-buckling of plates.*

DIN 18800-4 (2008) *Steel structure. Part 4: stability-analysis of safety against buckling of shells.*

D. Min. LL. PP. 12 Dicembre 1985, *Norme tecniche per le tubazioni.*

D. Min. LL. PP. 3 Dicembre 1987, *Norme tecniche per la progettazione, esecuzione e collaudo delle costruzioni prefabbricate.*

D. Min. LL. PP. 9 Gennaio 1996, *Norme tecniche per il calcolo, l'esecuzione ed il collaudo delle strutture in cemento armato, normale e precompresso e per le strutture metalliche.*

D. Min. LL. PP. 16 Gennaio 1996, *Norme tecniche relative ai criteri generali per la verifica di sicurezza delle costruzioni e dei carichi e sovraccarichi.*

D. Min. LL. PP. 16 Gennaio 1996, *Norme tecniche per le costruzioni in zone sismiche.*

DOE-STD-1020-2012, "Natural Phenomena Hazards Analysis and Design Criteria for DOE Facilities", U.S. Department of Energy, Washington, D.C. 20585.

EPRI NP-6041-SL, "A Methodology for Assessment of Nuclear Power Plant Seismic Margin (Revision 1)", Research Project 2722-23, Electric Power Research Institute, August 1991.

European Technical Approval No ETA-06/0006. VSL Post-Tensioning System.

FEMA 233 (Federal Emergency Management Agency), Earthquake Resistant Construction of Gas and Liquid Fuel Pipeline Systems Serving, or Regulated by, the Federal Government, Earthquake Hazard Reduction Series 67, NISTIR 4795, 1992.

FEMA 273 (Federal Emergency Management Agency), NEHRP (National Earthquake Hazards Reduction Program) Guidelines for the Seismic Rehabilitation of Buildings, 1997.

FEMA 274 (Federal Emergency Management Agency), NEHRP (National Earthquake Hazards Reduction Program) Commentary on the Guidelines for the Seismic Rehabilitation of Buildings, 1997.

FEMA 450 (Federal Emergency Management Agency), NEHRP (National Earthquake Hazards Reduction Program) Recommended Provisions for Seismic Regulations for New Buildings and Other Structures. Part 1: Provisions and Part 2: Commentary, 2003.

fib Bulletin 12, (2001), *Punching of structural concrete slabs*, 314 pages.

fib Bulletin 13, (2001), *Nuclear containments*, 121 pages.

fib Bulletin 31, (2005), *Post-tensioning in buildings*, 109 pages.

FIP/3/2, Gennaio 1978, *Recommendations for the design of prestressed concrete tanks*, 44 pages.

FIP/3/3, Marzo 1978, *The design and construction of prestressed concrete reactor vessels*, 79 pages.

FM Approvals Class Number 4020, "Steel Tanks for Fire Protection", Maj 2011.

Guidelines for the Seismic Design of Oil and Gas Pipeline Systems, ASCE Technical Council on Lifeline Earthquake Engineering, 1987.

HSE (Health and Safety Executive), Fluid Structure Interaction Effects on and Dynamic Response of Pressure Vessels and Tanks Subjected to Dynamic Loading, prepared by The Steel Construction Institute, Research Report RR527, 2007.

IASS Working Group N. 5, "Recommendations for Reinforced Concrete Shells and Folded Plates", International Association for Shell and Spatial Structures, Madrid, Spain, 1979, 66 pagine.

IAEA-TECDOC-1347, Consideration of External Events in the Design of Nuclear Facilities Other than Nuclear Power Plants, with Emphasis on Earthquakes, International Atomic Energy Agency, March 2003.

IBC (International Building Code) 2012 Edition. International Code Council (ICC).

IITK-GSDMA (2007) "Guidelines for Seismic Design of Liquid Storage Tanks, Previsions with Commentary and Explanatory Examples" issued by the NICEE in India, National Information Centre of Earthquake Engineering, Indian Institute of Technology, Kanpur.

IITK-GSDMA (2007) "Guidelines for Seismic Design of Buried Pipelines, Previsions with Commentary and Explanatory Examples" issued by the NICEE in India, National Information Centre of Earthquake Engineering, Indian Institute of Technology, Kanpur.

International Building Code, IBC 2009.

IS:456 2000, "Plain and reinforced concrete. Code of Practice".

IS:1893 (1984) "Criteria for earthquake resistant design of structures" (Fourth Revision).

IS:3370 (Part I) 1965, revised in 2004, "Code of practice for concrete structures for storage of liquids: general requirements".

IS:3370 (Part II) 1965, revised in 2004, "Code of practice for concrete structures for storage of liquids: reinforced concrete structures".

IS:3370 (Part III) 1967, revised in 2004, "Code of practice for concrete structures for storage of liquids: prestressed concrete structures".

IS:3370 (Part IV) 1967, revised in 2004, "Code of practice for concrete structures for storage of liquids: design tables".

ISO 11697:1995 *Bases for design of structures - loads due to bulk materials.*

ISO 13623:2009 *Petroleum and natural gas industries - Pipeline transportation systems.*

ISO 13847:2000 *Petroleum and natural gas industries - Pipeline transportation systems - Welding of pipelines.*

JSCE (Japan Society of Civil Engineers), Earthquake Resistant Design Codes in Japan, 2000.

NCh2369 2003 Edition, "Diseño sismico de estructuras e instalaciones (industriales)" (Seismic design for industrial structures and facilities), Chile.

Norme Tecniche per le Costruzioni, D. Min. 14 gennaio 2008, pubblicato sulla Gazzetta Ufficiale n. 29 del 4 febbraio 2008.

Notes on ACI 318-08 Building Code Requirements for Structural Concrete with Design Applications; Ed. M.E. Kamara, L.C. Novak and B.G. Rabbat, PCA.

Nuclear Reactors and Earthquake, United States Atomic Energy Commission, TID-7024 (22nd Edition, August 1963). Prepared by Lockhead Aircraft Corporation and Holmes & Narver Inc. for the Division of Reactor Development (TID-7024).

NZS 1170.5 2004, "Structural design actions, Part 5: Earthquake actions".

NZS 3101 2006, "Concrete Structures Standard. Part 1: The design of concrete structures".

NZS 3101 2006, "Concrete Structures Standard. Part 2: Commentary on the design of concrete structures".

NZS 3106 (2009), "Design of concrete structures for the storage of liquids" (Superseding NZS 3106:1986).

NZS 4203 1992, "Code of practice for general structural design and design loading for buildings".

O.P.C.M. n. 3274 20/03/2003 "Norme tecniche per il progetto, la valutazione e l'adeguamento, sismico degli edifici" come modificato dall'O.P.C.M. 3431 del 3 Maggio 2005.

O.P.C.M. n. 3274 20/03/2003, Allegato 4, "Norme tecniche per il progetto sismico di opere di fondazione e di sostegno dei terreni".

O.P.C.M. n. 3431 3/05/2005, "Ulteriori modifiche ed integrazioni all'O.P.C.M. n. 3274".

Phase I - Report on Folded Plate Construction, Proceedings ASCE V. 89, ST6, Dec. 1963, pp. 365–406.

Portland Cement Association (1993), "Circular concrete tanks without prestressing".

Portland Cement Association (1998), "Rectangular concrete tanks", Javeed A. Munshi, 5th Edition.

Portland Cement Association (2002), "Design of liquid-containing concrete structures for earthquake forces", Javeed A. Munshi.

prEN 10138:2006 *Prestressing Steels-Part 1: General requirements & Part 3: Strand.*

PRCI (Pipeline Research Council International), Guidelines for the Seismic Design and Assessment of Natural Gas and Liquid Hydrocarbon Pipelines, (L51927, PR-268-9823), 2004.

Prestressed Concrete Cylinder Pipe Guide Specifications, Ameron International Water Trasmission Group (www.ameronpipe.com).

Seismic Design of Storage Tanks (December 1986 or November 2009 Edition), Recommendations of a Study Group of the New Zealand National Society for Earthquake Engineering, New Zealand National Society for Earthquake Engineering (M.J.N. Priestley, Editor and Chairman).

UNI 7616:1976 *Raccordi di polietilene ad alta densità per condotte di fluidi in pressione. Metodi di prova.*

UNI 9032:2008 *Tubazioni di resine termoindurenti rinforzate con fibre di vetro (PRFV) con o senza cariche - Linee guida per la definizione dei requisiti per l'impiego.*

UNI 9099:1989 *Tubi di acciaio impiegati per tubazioni interrate o sommerse. Rivestimento esterno di polietilene applicato per estrusione.*

UNI EN 545:2007 *Tubi, raccordi e accessori di ghisa sferoidale e loro assemblaggi per condotte d'acqua - Requisiti e metodi di prova.*

UNI EN 969:2009 *Tubi, raccordi e accessori di ghisa sferoidale e loro assemblaggi per condotte di gas - Prescrizioni e metodi di prova.*

UNI EN 1452-2:2001 *Sistemi di tubazioni di materia plastica per adduzione - Policloruro di vinile non plastificato (PVC-U) - Tubi.*

UNI EN 1473:2007 *Installazioni ed equipaggiamenti per il gas naturale liquefatto (GNL) - Progettazione delle installazioni di terra.*

UNI EN 1594:2009 *Trasporto e distribuzione di gas - Condotte per pressione massima di esercizio maggiore di 16 bar - Requisiti funzionali.*

UNI EN 1770:2000 *Prodotti e sistemi per la protezione e riparazione delle strutture di calcestruzzo. Metodi di prova. Determinazione del coefficiente di dilatazione termica.*

UNI EN 14015:2006 *Specification for the design and manifacture of site built, vertical, cylindrical, flat–bottomed, above ground, steel tanks for the storage of liquids at ambient temperature and above.*

UNI EN 1990:2002 *Eurocodice - Criteri generali di progettazione strutturale.*

UNI EN 1991-1-1:2002 *Actions on structures - Part 1-1: General actions - Densities, self-weight, imposed loads for buildings.*

UNI EN 1991-1-5:2004 *Actions on structures - Part 1-5: General actions - Thermal actions.*

UNI EN 1991-1-7:2006 *Parte 1-7: Azioni in generale - Azioni eccezionali.*

UNI ENV 1991-4:2006 *Azioni sulle strutture. Parte 4: Azioni su silos e serbatoi.*

UNI EN 1992-1-1:2004 *Design of concrete structures - Part 1-1: General rules and rules for buildings.*

UNI EN 1992-1-2:2005 *Progettazione delle strutture di calcestruzzo - Parte 1-2: Regole generali - Progettazione strutturale contro l'incendio.*

UNI EN 1992-2:2006 *Design of Concrete Structures - Part 2: Concrete Bridges - Design and Detailing Rules.*

UNI EN 1992-3:2006 *Design of Concrete Structures. Part 3: Liquid Retaining and Containment Structures.*

UNI ENV 1993-1-1:2005 *Progettazione delle strutture di acciaio. Parte 1-1: Regole generali e regole per gli edifici.*

UNI EN 1993-1-3:2007 *Progettazione delle strutture di acciaio. Parte 1-3: Regole generali e regole supplementari per l'impiego dei profilati e delle lamiere sottili piegati a freddo.*

UNI ENV 1993-1-5:2004 *Progettazione delle strutture di acciaio. Parte 1-5: Elementi strutturali a lastra.*

UNI ENV 1993-1-6:2007 *Design of Steel Structures. Part 1-6: Strength and Stability of Shell Structures.*

UNI ENV 1993-4-1:2007 *Progettazione delle strutture di acciaio. Parte 4-1: Silos.*

UNI ENV 1993-4-2:2007 *Progettazione delle strutture di acciaio. Parte 4-2: Serbatoi.*

UNI ENV 1993-4-3:2007 *Progettazione delle strutture di acciaio. Parte 4-3: Condotte.*

UNI EN 1997-1:2005 *Geotechnical Design - Part 1: General Rules.*

UNI EN 1998-1:2004 *Design of Structures for Earthquake Resistance - Part 1: General Rules, Seismic Actions and Rules for Building.*

EN 1998-2:2005 *Design of Structures for Earthquake Resistance - Part 2: Bridges.*

UNI EN 1998-4:2006 *Progettazione delle strutture per la resistenza sismica - Parte 4: Silos, serbatoi e condotte.*

UNI EN 1998-5:2020 *Design of Structures for Earthquake Resistance Part 5: Geotechnical Aspects, Foundations, Retaining and Underground Structures.*

UNI EN 10025-1:2005 *Prodotti laminati a caldo di acciaio non legati per impieghi strutturali. Condizioni tecniche generali di fornitura.*

UNI EN 10208:2009 *Tubi di acciaio per condotte di fluidi combustibili - Condizioni tecniche di fornitura.*

UNI EN 10224:2006 *Tubi e raccordi di acciaio non legato per il convogliamento di acqua e di altri liquidi acquosi - Condizioni tecniche di fornitura.*

UNI EN 12007:2004 *Trasporto e distribuzione di gas - Condotte con pressione massima di esercizio non maggiore di 16 bar - Raccomandazioni funzionali generali.*

UNI EN 12201-1:2004 *Sistemi di tubazioni di materia plastica per la distribuzione dell'acqua - Polietilene (PE) - Generalità.*

UNI EN 12201-2:2004 *Sistemi di tubazioni di materia plastica per la distribuzione dell'acqua - Polietilene (PE) - Tubi.*

UNI EN 12452:2007 *Serbatoi fissi cilindrici di acciaio saldato, per gas di petrolio liquefatto (GPL), prodotti in serie, di capacità geometrica fino a 13 m^3 per installazione fuori terra - Progettazione e fabbricazione.*

UNI EN 12493:2004 *Serbatoi di acciaio saldato per gas di petrolio liquefatto (GPL) - Cisterne per trasporto su strada - Progettazione e costruzione.*

UNI EN 12732:2005 *Trasporto e distribuzione di gas - Saldatura delle tubazioni di acciaio - Requisiti funzionali.*

UNI EN 14075:2007 *Serbatoi fissi cilindrici di acciaio saldato, per gas di petrolio liquefatto (GPL), prodotti in serie, di capacità geometrica fino a 13 m^3 per installazione interrata - Progettazione e fabbricazione.*

UNI EN 14620:2006 *Progettazione e fabbricazione di serbatoi di acciaio verticali, cilindrici, a fondo piatto, costruiti in sito, per lo stoccaggio di gas liquefatti refrigerati operanti a temperature tra 0 °C e -165 °C. Parte 1-5.*

UNI EN 206-1:2006 *Concrete - Part 1: Specification, Performance, Production and Conformity.*

Uniform Building Code, Vol. 2, Structural Engineering Design Provisions, 1997 Edition, International Conference of Building Officials, Whittier, CA, 1997, 492 pagine.

UNI ISO 5256:1987 *Tubi ed accessori di acciaio impiegati per tubazioni interrate o immerse. Rivestimento esterno ed interno a base di bitume o di catrame.*

References

1 Abalı, E. and Uçkan, E. (2010). Parametric analysis of liquid storage tanks base isolated by curved surface sliding bearings, *Soil Dynamics and Earthquake Engineering*, **30**, 21–31.

2 Abramson, H.N. (1966). The dynamic behavior of liquids in moving containers, Technical Report NASA SP-106, Southwest Research Institute, San Antonio, TX.

3 Abramson, H.N. (1969). Slosh suppression, Report NASA SP-8031, Southwest Research Institute, San Antonio, TX.

4 Abramson, H.N. and Garza, L.R. (1964). Some measurements of the effects of ring baffles in cylindrical tanks, *Journal of Spacecraft and Rockets*, **1**(5), 560–562.

5 Adams, N.J.I. (1991). Seismic design rules for flat bottom cylindrical liquid storage tanks, *International Journal of Pressure Vessels and Piping*, **49**(1), 61–95.

6 Ahari, M.N., Eshghi, S., and Ashtiany, M.G. (2009). The tapered beam model for bottom plate uplift analysis of unanchored cylindrical steel storage tanks, *Engineering Structures*, **31**(3), 623–632.

7 Akiyama, N. and Yamaguchi, H. (1988). Experimental study on lift-off behavior of flexible cylindrical tank, in: *Proceedings of the 9th World Conference on Earthquake Engineering*, Kyoto, Tokyo, Paper Number 10-4-9, Vol. VI, 655–660.

8 Alampalli, S. and Elgamel, A.W. (1990). Dynamic response of retaining walls including supported soil backfill: A computational model, Earthquake Engineering Research Institute, in: *Proceedings of the 4th U.S. National Conference on Earthquake Engineering*, **3**, 623–632.

9 Alembagheri, M. and Estekanchi, H.E. (2011). Seismic assessment of unanchored steel storage tanks by endurance time method, *Earthquake Engineering and Engineering Vibration*, **10**(4), 591–604.

10 Almroth, B.O. and Holmes, A.M.C. (1972). Buckling of shells with cutouts, experiment and analysis, *International Journal of Solids Structures*, **8**, 1057–1071.

11 Aly, A.M., Nguyen, M.T., and Lee, S-W. (2015). Numerical analysis of liquid sloshing using the incompressible smoothed particle hydrodynamics method, *Advances in Mechanical Engineering*, **7**(2), Article ID 765741.

12 Anumod, A.S., Harinarayanan, S., and Usha, S. (2014). Finite element analysis of steel storage tank under seismic load, *International Journal of Engineering Research and Applications (IJERA)*. Special Issue, January 24 and 25, "Trends and Recent Advances in Civil Engineering (TRACE)", 47–54.

13 Armenio, V. and La Rocca, M. (1996). On the analysis of sloshing of water in rectangular containers: Numerical study and experimental validation, *Ocean Engineering*, **23**(8), 705–739.

Seismic Design and Analysis of Tanks, First Edition. Gian Michele Calvi and Roberto Nascimbene.
© 2023 John Wiley & Sons, Inc. Published 2023 by John Wiley & Sons, Inc.

14 Aron, H. (1874). Das Gleichgewicht und die Bewegung einer unendlich dünnen, beliebig gekrümmten elastischen Schale, *Journal für die Reine und Angewandte Mathematik*, **78**, 136–174.

15 Arze, E. (1969). Seismic failure and repair of an elevated water tank, in: *Proceedings of the 4th World Conference on Earthquake Engineering*, Vol. III, pp. B6–57 to B6–69.

16 ASCE (2004). *Primer on Seismic Isolation*, sponsored by Task Committee on Seismic Isolation, Eds. Andrew W. Taylor and Takeru Igusa. ASCE.

17 Ashwell, D.G. (1976). *Finite elements for Thin Shells and Curved Members*, New York: John Wiley & Sons Ltd.

18 Athiannan, K. and Palaninathan, R. (2004). Buckling of cylindrical shells under transverse shear, *Thin-Walled Structures*, **42**(9), 1307–1328.

19 Auli, W., Fischer, F.D., and Rammerstorfer, F.G. (1985). Uplifting of earthquake-loaded liquid-filled tanks, in: *Proceedings Pressure Vessels and Piping (PVP) Conference*, 98–7, 71–85.

20 AWWA (American Water Works Association) (2010). *Steel Water Storage Tanks. Design, Construction, Maintenance, and Repair*, Technical Editor Stephen W. Meier, New York: McGraw Hill.

21 Azabi, T.M. (2014). Behaviour of reinforced concrete conical tanks under hydrostatic loading, MSc. dissertation, School of Graduate and Postdoctoral Studies, Western University, London, ON, Canada.

22 Azzuni, E. and Guzey, S. (2015). Comparison of the shell design methods for cylindrical liquid storage tanks, *Engineering Structures*, **101**, 621–630.

23 Balendra, T., Ang, K.K., Paramasivam, P., and Lee, S.L. (1982). Seismic design of flexible cylindrical liquid storage tanks, *Earthquake Engineering and Structural Dynamics*, **10**(3), 477–496.

24 Balendra, T. and Nash, W.A. (1976). Earthquake analysis of a cylindrical liquid storage tank with a dome by finite element method, A computer program distributed by NISEE/Computer Applications, Department of Civil Engineering, University of Massachusetts, Amherst.

25 Bandyopadhyay, K., Cornell, A., Costantino, C., Kennedy, R., Miller, C., and Veletsos, A. (1995). *Seismic Design and Evaluation Guidelines for the Department of Energy High-Level Waste Storage Tanks and Appurtenances*, BNL 52361 (Rev. 10/95) UC-406 and UC-510, Engineering Research and Application Division Department of Advanced Technology, Brookhaven National Laboratory, Associated University, Inc., Upton, New York.

26 Bandyopadhyay, K.K. (1991). Overview of seismic panel activities, BNL 46540 (DE91 018726). in: *Proceedings of the 3rd DOE Natural Phenomena Hazards Mitigation Conference*, 423–429.

27 Barton, D.C. and Parker, J.V. (1987). Finite element analysis of the seismic response of anchored and unanchored liquid storage tanks, *Earthquake Engineering and Structural Dynamics*, **15**(3), 299–322.

28 Basu, B. and Gupta, V.K. (1996). A note on damage-based inelastic spectra, *Earthquake Engineering and Structural Dynamics*, **25**(5), 421–433.

29 Batdorf, S.B. (1947). A simplified method of elastic-stability analysis for thin cylindrical shells, NACA Report No. 874.

30 Bažant, Z.P. and Cedolin, L. (1991). *Stability of Structures: Elastic, Inelastic, Fracture and Damage Theories*, Oxford: Oxford University Press.

31 Becker, H. (1958). General instability of stiffened cylinders, National Advisory Committee for Aeronautics, NACA Technical Note No. 4237.

32 Belluzzi, O. (1941, 1943, 1961, 1955). *Scienza delle costruzioni*, Vols I, II, III, IV, Bologna: Zanichelli (in Italian).

33 Benson, D.J. (1989). An efficient, accurate, simple ALE method for nonlinear finite element programs, *Computer Methods in Applied Mechanics and Engineering*, **72**(3), 305–350.

34 Berahman, F. and Behnamfar, F. (2007). Seismic fragility curves for un-anchored on-grade steel storage tanks: Bayesian approach, *Journal of Earthquake Engineering*, **11**(2), 166–192.

35 Berahman, F. and Behnamfar, F. (2009). Probabilistic seismic demand model and fragility estimates for critical failure modes of un-anchored steel storage tanks in petroleum complexes, *Probabilistic Engineering Mechanics*, **24**(4), 527–536.

36 Bielak, J. (1975). Dynamic behaviour of structures with embedded foundations, *Earthquake Engineering and Structural Dynamics*, **3**(3), 259–274.

37 Bielak, J. (1976). Modal analysis for building-soil interaction, *ASCE Journal of the Engineering Mechanics Division*, **102**(5), 771–786.

38 Billimoria, H.D. and Hagstrom, J. (1978). Stiffness coefficients and allowable loads for nozzles in flat-bottom storage tanks, *American Society of Mechanical Engineers (ASME) Journal of Pressure Vessel Technology*, **100**(4), 389–399.

39 Billington, D.P. (1965). *Thin Shell Concrete Structures*, 1st Edition, New York: McGraw Hill.

40 Billington, D.P. (1982). *Thin Shell Concrete Structures*, 2nd Edition, New York: McGraw Hill.

41 Biswal, K.C., Bhattacharyya, S.K., and Sinha, P.K. (2003a). Dynamic characteristics of liquid filled rectangular tank with baffles, *Journal of the Institution of Engineers (India)*, **84**, 145–148.

42 Biswal, K.C., Bhattacharyya, S.K., and Sinha, P.K. (2003b). Free-vibration analysis of liquid-filled tank with baffles, *Journal of Sound and Vibration*, **259**(1), 177–192.

43 Biswal, K.C., Bhattacharyya, S.K., and Sinha, P.K. (2004). Dynamic response analysis of a liquid-filled cylindrical tank with annular baffle, *Journal of Sound and Vibration*, **274**(1), 13–37.

44 Biswal, K.C. and Nayak, S.K. (2013). Slosh dynamics of liquid filled baffled tank under seismic excitations, in: *New Developments in Structural Engineering and Construction*, Eds. S. Yazdani and A. Singh, ISEC-7, Honolulu, June 18–23, 2013.

45 Blackler, M.J. (1986). Stability of silos and tanks under internal and external pressure, PhD dissertation, School of Civil and Mining Engineering, The University of Sydney.

46 Bleich, H.H. (1956). Longitudinal forced vibrations of cylindrical fuel tanks, *Journal of Jet Propulsion*, **26**(2), 109–111.

47 Block, D.L., Card, M.F., and Mikulas, M.M. (1965). *Buckling of Eccentrically Stiffened Orthotropic Cylinders*, NASA TN D-2960.

48 Bo, L. and Jia-xiang, T. (1994). Vibration studies of base-isolated liquid storage tanks, *Computers & Structures*, **52**(5), 1051–1059.

49 Bodner, S.R. (1957). General instability of a ring-stiffened, circular cylindrical shell under hydrostatic pressure, *Journal of Applied Mechanics*, **24**, 269–277.

50 Bozorgmehrnia, S., Ranjbar, M.M., and Madandoust, R. (2013). Seismic behavior assessment of concrete elevated water tanks, *Journal of Rehabilitation in Civil Engineering*, **1–2**, 69–79.

51 Brown, C.J. and Nielsen, J. (1998). *Silos. Fundamentals of Theory, Behaviour and Design*, Boca Raton, Florida: CRC Press.

52 Brunesi, E. (2014). Influence of structural openings on the buckling strength of cylindrical steel tanks, PhD thesis, IUSS Pavia.

53 Brunesi, E., Nascimbene, R., Pagani, M., and Beilic, D. (2014). Seismic performance of storage steel tanks during the May 2012 Emilia, Italy, earthquakes, *Journal of Performance of Constructed Facilities*, DOI: 10.1061/(ASCE)CF.1943-5509.0000628.

54 Brush, D.O. and Almroth, B.O. (1975). *Buckling of Bars, Plates and Shells*, New York: McGraw Hill.

55 Buchanan, H. (1965). Optimization of slosh baffle geometry, in: *NASA TM X-533389 Aero-Astrodynamics Research Review No 3*, Chapter VII Structural Dynamics, 95–100.

56 Buchanan, H. and Lott, L. (1966). Effect of Reynolds number on slosh damping by flat ring baffles, NASA TM X-53559.

57 Burris, L., Steunenberg, R.K. and Miller, W.E. (1986). The application of electrorefining for recovery and purification of fuel discharge from the integral fast reactor, paper presented at CONF-861146-14 (DE87 004746), Annual AIChE Meeting, Miami, Florida.

58 Bushnell, D. (1980). Computerized analysis of shells - governing equation, Technical Report AFWAL-81-3048, Lockheed Applied Mechanics Laboratory, Palo Alto Research Laboratory, Palo Alto, California.

59 Bushnell, D. (1984). Computerized analysis of shells - governing equation, *Computers & Structures*, **18**(3), 471–536.

60 Bustamante, J.I. and Flores, A. (1966). Water pressure on dams subjected to earthquakes, *ASCE Journal of the Engineering Mechanics Division*, **92**(5), 115–127.

61 Cai, M., Holst, J.M.F.G., and Rotter, J.M. (2002). Buckling strength of thin cylindrical shells under localized axial compression, paper presented at 15th ASCE Engineering Mechanics Conference (EM 2002). Columbia University, New York.

62 Calavera, J.R. (1993). *Manual de detalles constructivos en obras de hormigón armado*, Intemac Ed.

63 Calavera, J.R., González, E.V., Fernández, J.G., and Valenciano, F.C. (1999). *Manual de ferralla*, Intemac Ed.

64 Calladine, C.R. (1983). *Theory of Shell Structures*, Cambridge: Cambridge University Press.

65 Calugaro, V. and Mahin, S.A. (2009). Experimental and analytical studies of fixed-base and seismically isolated liquid storage tanks, paper presented at 3rd International Conference on Advanced in Experimental Structural Engineering, San Francisco.

66 Calvi, G.M. and Nascimbene, R. (2011). *Progettare i gusci*, Pavia: IUSS Press (in Italian).

67 Cambra, F.J. (1982). Earthquake response considerations of broad liquid storage tanks, Research Report California Institute of Technology, Pasadena, California, EERL 82–25.

68 Cano, J.H., Giuliani, H., Zaragoza, A., and Sisterna, C. (1996). Theoretical experimental study of cylindrical steel tanks, paper presented at the 11th World Conference on Earthquake Engineering, Acapulco, Mexico, Paper Number 1672.

69 Carson, J.W. (2000). Silo failures: case histories and lessons learned, in: *Proceedings of 3rd Israel Conference for Conveying and Handling of Particulate Solids*, Vol. 1, 4.1–4.11.

70 Carson, J.W. and Holmes, T. (2003). Silo failures: why do they happen?, *Task Quarterly*, **7**(4), 499–512.

71 Case, K.M. and Parkinson, W.C. (1957). Damping of surface waves in an incompressible liquid, *Journal of Fluid Mechanics*, **2**(2), 172–184.

72 Cestelli-Guidi, C. (1980). *Geotecnica e tecnica delle fondazioni*, Vol. II, 6th Edition, Milan: Editore Ulrico Hoepli (in Italian).

73 Chalhoub, M.S. and Kelly, J.M. (1988). Theoretical and experimental studies of cylindrical water tanks in base isolated structures, Report, Earthquake Engineering Research Center, College of Engineering, University of California, Berkeley, California, EERC 88-07.

74 Chalhoub, M.S. and Kelly, J.M. (1990). Shake table test of cylindrical water tanks in base-isolated structures, *Journal of Engineering Mechanics*, **116**(7), 1451–1472.

75 Chandrasekaran, A.R. and Krishna, J. (1954). Water towers in seismic zones, in: *Proceedings of the 3rd World Conference on Earthquake Engineering*, Vol. IV, 161–171.

76 Chen, B-F. (2005). Viscous fluid in tank under coupled surge, heave, and pitch motions, *Journal of Waterway, Port, Coastal, and Ocean Engineering*, **131**(5), 239–256.

77 Chen, B-F. and Nokes, R. (2005). Time-independent finite difference analysis of fully non-linear and viscous fluid sloshing in a rectangular tank, *Journal of Computational Physics*, **209**, 47–81.

78 Chen, G. (1984). Why the "Elephant's Foot" phenomenon of liquid storage tank happened, in: *Proceedings of the 8th World Conference on Earthquake Engineering*, Vol. VII, 445–452.

79 Chen, J.Z. and Kianoush, M.R. (2004). Response of concrete liquid containing structures in different seismic zones, paper presented at the 13th World Conference on Earthquake Engineering, Vancouver, Canada, Paper Number 1441.

80 Chen, J.Z. and Kianoush, M.R. (2005). Seismic response of concrete rectangular tanks for liquid containing structures, *Canadian Journal of Civil Engineering*, **32**, 739–752.

81 Chikazawa, Y., Koshizuka, S., and Oka, Y. (1999). Numerical analysis of sloshing with large deformation of elastic walls and free surfaces using MPS method, *Nihon Kikai Gakkai Ronbunshu, B Hen/Transactions of the Japan Society of Mechanical Engineers*, Part B, **65**(637), 2954–2960.

82 Cho, E.K. and Chang, S.P. (1993). Seismic response of cylindrical liquid storage tanks, paper presented at the 12th International Conference on Structural Mechanics in Reactors Technology, SMiRT-12, Ed. K. Kussmaul, Stuttgart, Germany, Paper SD103/3, 219–224.

83 Cho, K.H., Kim, M.K., Lim, Y.M., and Cho, S.Y. (2004). Seismic response of base-isolated liquid storage tanks considering fluid-structure-soil interaction in time domain, *Soil Dynamics and Earthquake Engineering*, **24**(11), 839–852.

84 Cho, J.R. and Lee, H.W. (2004). Numerical study on liquid sloshing in baffled tank by nonlinear finite element method, *Computer Methods in Applied Mechanics and Engineering*, **193**(23–26), 2581–2598.

85 Cho, J.R. and Lee, S.Y. (2003). Dynamic analysis of baffled fuel-storage tanks using the ALE finite element method, *International Journal for Numerical Methods in Fluids*, **41**(2), 185–208.

86 Chopra, A.K. (1967a). Reservoir-dam interaction during earthquakes, *Bulletin of the Seismological Society of America*, **57**(4), 675–687.

87 Chopra, A.K. (1967b). Hydrodynamic pressures on dams during earthquakes, *ASCE Journal of the Engineering Mechanics Division*, **93**(EM 6), 205–223.

88 Chopra, A.K. (1995). *Dynamics of Structures: Theory and Applications to Earthquake Engineering*, Englewood Cliffs, NJ: Prentice Hall.

89 Chopra, A.K. and Goel, R.K. (1999). Capacity-demand-diagram methods based on inelastic design spectrum, *Earthquake Spectra*, **15**(4), 637–656.

90 Choudhury, D., Subba Rao, K.S., and Ghosh, S. (2002), Passive earth pressure distribution under seismic condition, paper presented at the 15th Engineering Mechanics Conference of ASCE (EM2002). Columbia University, New York.

91 Choun, Y-S. and Yun, C-B. (1996). Sloshing characteristics in rectangular tanks with a submerged block, *Computers & Structures*, **61**(3), 401–413.

92 Choun, Y-S. and Yun, C-B. (1999). Sloshing analysis of rectangular tanks with a submerged structure by using small-amplitude water wave theory, *Earthquake Engineering and Structural Dynamics*, **28**(7), 763–783.

93 Christovasilis, I.P. and Whittaker, A.S. (2008). Seismic analysis of conventional and isolated LNG tanks using mechanical analogs, *Earthquake Spectra*, **24**(3), 599–616.

94 Cifuentes, A.O. (1984). System identification of hysteretic structures, PhD thesis, Technical Report California Institute of Technology, Pasadena, California, EERL 84–04.

95 Clough, D.P. (1977). Experimental evaluation of seismic design methods for broad cylindrical tanks, Research Report, California Institute of Technology, Pasadena, California, EERL 77–10.

96 Clough, D.P. and Clough, R.W. (1978). Earthquake simulator studies of cylindrical tanks, *Nuclear Engineering and Design*, **46**(2), 367–380.

97 Clough, R.W. (1979). Experimental seismic study of cylindrical tanks, *ASCE Journal of the Structural Division*, **105**(ST12), 2565–2590.

98 Clough, R.W. and Clough, D.P. (1977). Seismic response of flexible cylindrical tanks, paper presented at the 4th International Conference on Structural Mechanics in Reactor Technology, SMiRT-4, San Francisco, Paper K5/1.

99 Clough, R.W. and Niwa, A. (1979). Static tilt tests of a tall cylindrical liquid storage tank, Research Report, California Institute of Technology, Pasadena, California, EERL 79–06.

100 Clough, R.W. and Penzien, J. (1994). *Dynamics of Structures*, 2nd Edition, New York: McGraw Hill.

101 Cole, H.A. Jr. (1966a). On a fundamental damping law for fuel sloshing, Report, NASA TN D-3240.

102 Cole, H.A. Jr. (1966b). Baffle thickness effects in fuel sloshing experiments, Report, NASA TN D-3716.

103 Cole, H.A. Jr. and Gambucci, B.J. (1961). Measured two-dimensional damping effectiveness of fuel-sloshing baffles applied to ring baffles in cylindrical tanks, Report, NASA TN D-694.

104 Collatz, L. (1960). *The Numerical Treatment of Differential Equations*, Berlin: Springer-Verlag.

105 Collins, I.F. (1973). A note on the interpretation of Coulomb's analysis of the thrust on a rough retaining wall in terms of the limit theorems of plasticity theory, *Géotechnique*, **23**(3), 442–447.

106 Comincioli, V. (2005). *Analisi numerica; metodi, modelli, applicazioni*, APOGEO.

107 Compagnoni, M.E, Curadelli, O., and Ambrosini, D. (2018). Experimental study on the seismic response of liquid storage tanks with sliding concave bearings, *Journal of Loss Prevention in the Process Industries*, **55**, 1–9.

108 Constantinou, M.C. (1998). Application of seismic isolation systems in Greece, paper presented at '98 Structural Engineers World Congress, Paper T175-3.

109 Conte, S.D. and de Boor, C. (1980). *Elementary Numerical Analysis. An Algorithmic Approach*, 3rd Edition, New York: McGraw Hill International Editions.

110 Crandall, S.H. (1956). *Engineering Analysis. A Survey of Numerical Procedures*, New York: McGraw Hill.

111 Crouse, C.B. (2000). Energy dissipation in soil–structure interaction, paper presented at the 12th World Conference on Earthquake Engineering, New Zealand, Paper Number 0366.

112 Dar, A. and Baughman, P.D. (2011). Seismic margin assessment of unanchored flat-bottom liquid-storage tank on flexible foundation, in: *Proceedings of the 21st International Conference on Structural Mechanics in Reactor Technology*, SMiRT-21, Eds B.K. Dutta, P.V. Durgaprasad, and R.K. Singh, New Delhi, India, Division-VII: Paper ID# 578.

113 Daysal, H. and Nash, W.A. (1984). Soil structure interaction effects on the seismic behavior of cylindrical liquid storage tanks, in: *Proceedings of the 8th World Conference on Earthquake Engineering*, Vol. V, 223–229.

114 Deylami, A. and Sarrafzadeh, M.R. (1996). Seismic analysis of cylindrical liquid storage tank, paper presented at the 11th World Conference on Earthquake Engineering, Acapulco, Mexico, Paper Number 522.

115 Di Carluccio, A., Fabbrocino, G., and Manfredi, G. (2009). Risposta sismica di serbatoi atmosferici per olio combustibile, paper presented at XIII Convegno ANIDIS, Bologna (in Italian).

116 Di Carluccio, A., Fabbrocino, G., Salzano, E., and Manfredi, G. (2008). Analysis of pressurized horizontal vessels under seismic excitation, paper presented at 14th World Conference on Earthquake Engineering, Beijing, China.

117 Diaconu, D., Manolovici, M., Iticovici, M., Marinescu, S., Cârlan, S., and Soroceanu, I. (1977). Seismic response of elevated water-towers, with tronconic tanks and central tube, taking into account the water swinging effect, in: *Proceedings of the 6th World Conference on Earthquake Engineering*, Topic 9, Vol. III, 2816–2822.

118 DiGrado, B.D. and Thorp, G.A. (2004). *The Aboveground Steel Storage Tank Handbook*, Hoboken, NJ: John Wiley & Sons.

119 Dodge, F.T. (1966). Analytical representation of lateral sloshing by equivalent mechanical models, in: *The Dynamic Behaviour of Liquids in Moving Containers*, Ed. H.N., Abramson, NASA SP-106, 198–223.

120 Dodge, F.T. (1971). Engineering study of flexible baffles for slosh suppression, Report, NASA CR-1880.

121 Dodge, F.T. (2000). The new "dynamic behaviour of liquids in moving containers", San Antonio, TX: Southwest Research Institute.

122 Dodge, F.T. and Kana, D.D. (1966a). Moment of inertia and damping of liquids in baffled cylindrical tanks, *Journal of Spacecraft and Rockets*, **3**(1), 153–155.

123 Dodge, F.T. and Kana, D.D. (1966b). Moment of inertia and damping of liquids in baffled cylindrical tanks, Report, NASA CR-383.

124 Doğangün, A., Durmus, A., and Ayvaz, Y. (1997). Earthquake analysis of flexible rectangular tanks using the Lagrangian fluid finite element, *European Journal of Mechanics – A/Solids*, **16**(1), 165–182.

125 Dogangun, A., Karaca, Z., Durmus, A., and Sezen, H. (2009). Cause of damage and failures in silo structures, *Journal of Performance of Constructed Facilities*, **23**(2), 65–71.

126 Doğangün, A. and Livaoğlu, R. (2004). Hydrodynamic pressures acting on the walls of rectangular fluid containers, *Structural Engineering and Mechanics*, **17**(2), 203–214.

127 Doğangün, A. and Livaoğlu, R. (2008). A comparative study of the seismic analysis of rectangular tanks according to different codes, paper presented at the 14th World Conference on Earthquake Engineering, Beijing, China.

128 Domel, A.W. and Gogate, A.B. (1993). *Circular Concrete Tanks Without Prestressing*, Washington, DC: Portland Cement Association.

129 Donnell, L.H. (1933). Stability of thin-walled tubes under torsion, NACA Report No. 479.

130 Donnell, L.H. (1934). A new theory for the buckling of thin cylinders under axial compression and bending, *Transactions of the ASME Series E*, **56**, 795–806.

131 Donnell, L.H. (1976). *Beams, Plates and Shells*, New York: McGraw Hill.

132 Dotoli, R., Lisi, D., Bardaro, D., Perillo, M., and Tomasi, M. (2007). Sloshing response of a LNG storage tank subjected to seismic loading, paper presented at 6th European LS-DYNA Users Conference, Gothenburg, Sweden.

133 Dutta, S.C., Jain, S.K., and Murty, C.V.R. (2000a). Assessing the seismic torsional vulnerability of elevated tanks with RC frame-type staging, *Soil Dynamics and Earthquake Engineering*, **19**, 183–197.

134 Dutta, S.C., Jain, S.K., and Murty, C.V.R. (2000b). Alternate tank staging configurations with reduced torsional vulnerability, *Soil Dynamics and Earthquake Engineering*, **19**, 199–215.

135 Dutta, S.C., Jain, S.K., and Murty, C.V.R. (2001). Inelastic seismic torsional behaviour of elevated tanks, *Journal of Sound and Vibration*, **242**(1), 151–167.

136 Dutta, S.C., Murty, C.V.R., and Jain, S.K. (1996). Torsional failure of elevated water tanks: the problem and some solutions, paper presented at the 11th World Conference on Earthquake Engineering, Acapulco, Mexico, Paper Number 287.

137 Ebeling, R.M. and Morrison, E.E. (1993). The seismic design of waterfront retaining structures, Technical Report ITL-92-11, NCEL TR-939, Port Hueneme, California: Office of Navy Technology and Department of the Army, Naval Civil Engineering Laboratory.

138 Ebrahimian, M., Noorian, M.A,. and Haddadpour, H. (2014). Equivalent mechanical model of liquid sloshing in multi-baffled containers, *Engineering Analysis with Boundary Elements*, **47**, 82–95.

139 Edlund, B.L.O. (2007). Buckling of metallic shells: buckling and postbuckling behaviour of isotropic shells, especially cylinders, *Structural Control and Health Monitoring*, **14**(4), 693–713.

140 Edwards, N.W. (1969). A procedure for dynamic analysis of thin-walled cylindrical liquid storage tanks subjected to lateral ground motions, PhD thesis, University of Michigan.

141 Eidinger, J.M. (2012). Performance of water systems during the Maule M_w 8.8 earthquake of 27 February 2010, *Earthquake Spectra*, **28**(S1), S605–S620.

142 El-Bkaily, M. and Peek, R. (1998). Plastic buckling of unanchored roofed tanks under dynamic loads, *Journal of Engineering Mechanics*, **124**(6), 648–657.

143 El Damatty, A.A., El-Attar, M., and Korol, R.M. (1999). Simple design procedure for liquid-filled steel conical tanks, *ASCE Journal of Structural Engineering*, **125**(8), 879–890.

144 El Damatty, A.A., Korol, R.M., and Mirza, F.A. (1997a). Stability of imperfect steel conical tanks under hydrostatic loading, *ASCE Journal of Structural Engineering*, **123**(6), 703–712.

145 El Damatty, A.A., Korol, R.M., and Mirza, F.A. (1997b). Stability of elevated liquid-filled conical tanks under seismic loading. Part I - Theory, *Earthquake Engineering and Structural Dynamics*, **26**(12), 1191–1208.

146 El Damatty, A.A., Korol, R.M., and Mirza, F.A. (1997c). Stability of elevated liquid-filled conical tanks under seismic loading. Part II - Applications, *Earthquake Engineering and Structural Dynamics*, **26**(12), 1209–1229.

147 El Damatty, A.A. and Sweedan, A.M.I. (2006). Equivalent mechanical analog for dynamic analysis of pure conical tanks, *Thin-Walled Structures*, **44**(4), 429–440.

148 Elishakoff, I. (2000). Elastic stability: from Euler to Koiter there was none like Koiter, *Meccanica*, **35**, 375–380.

149 El-Zeiny, A.A-W. (1995). Nonlinear time-dependent seismic response of unanchored liquid storage tanks, PhD dissertation, University of California, Irvine.

150 Epstein, H.I. (1976). Seismic design of liquid-storage tanks, *ASCE Journal of the Structural Division*, **102**(ST9), 1659–1673.

151 Estekanchi, H.E., Vafai, A., and Sadeghazar, M. (2004). Endurance time method for seismic analysis and design of structures, *Scientia Iranica*, **11**(4), 361–370.

152 Fajfar, P. (2000). A nonlinear analysis method for performance-based seismic design, *Earthquake Spectra*, **16**(3), 573–592.

153 Faltinsen, O.M., Rognebakke, O.F., Lukovsky, I.A., and Timokha, A.N. (2000). Multidimensional modal analysis of nonlinear sloshing in a rectangular tank with finite water depth, *Journal of Fluid Mechanics*, **407**, 201–234.

154 Faltinsen, O.M. and Timokha, A.N. (2001). An adaptive multimodal approach to nonlinear sloshing in a rectangular tank, *Journal of Fluid Mechanics*, **432**, 167–200.

155 Favre, R., Jaccoud, J.-P., Koprna, M., and Radojicic, A. (1990). *Progettare in calcestruzzo armato - Piastre, muri, pilastri e fondazioni*, Milan: Hoepli (in Italian).

156 Fenves, G. and Chopra, A.K. (1985). Simplified earthquake analysis of concrete gravity dams: separate hydrodynamic and foundation interaction effects, *Journal of Engineering Mechanics*, **111**(6), 715–735.

157 Fiore, A., Rago, C., Vanzi, I., Greco, R., and Brisighella, B. (2018). Seismic behavior of a low-rise horizontal cylindrical tank, *International Journal of Advanced Structural Engineering*, **10**, 143–152.

158 Fischer, D. (1979). Dynamic fluid effects in liquid-filled flexible cylindrical tanks, *Earthquake Engineering and Structural Dynamics*, **7**(6), 587–601.

159 Fischer, D. (1981). Ein Vorschlag zur erdbebensicheren Bemessung von flüssigkeitsgefüllten zylindrischen Tankbauwerken, *Der Stahlbau*, **1**, 13–20.

160 Fischer, D.F. (1979). Explicit evaluation of the apparent fluid mass at the vibration of fluid filled cylindrical tanks, in: *Transactions of the 5th International Conference on Structural Mechanics in Reactor Technology*, Berlin, Germany, Vol. K(b). Paper K12/8.

161 Fischer, D.F. and Rammerstorfer, F.G. (1982). The stability of liquid-filled cylindrical shells under dynamic loading, in: *Buckling of Shells: Proceedings of a State-of-the-Art Colloquium, Universität Stuttgart, Germany, May 6-7*, Ed. E. Ramm, Berlin: Springer Verlag, 569–597.

162 Fischer, D.F. and Rammerstorfer, F.G. (1983). Local instabilities of liquid filled cylindrical shells under earthquake excitation, in: *Proceedings of the 7th International Conference on Structural Mechanics in Reactor Technology, SMiRT-7*, Paper K4/8, 303–312.

163 Fischer, D.F. and Rammerstorfer, F.G. (1984). Stability of liquid storage tanks under earthquake excitation, in: *Proceedings of the 8th World Conference on Earthquake Engineering*, Vol. V, 215–222.

164 Fischer, D.F. and Rammerstorfer, F.G. (1999). A refined analysis of sloshing effects in seismically excited tanks, *International Journal of Pressure Vessels and Piping*, **76**(10), 693–709.

165 Fischer, D.F., Rammerstorfer, F.G., and Auli, W. (1985). Strength and stability of uplifting earthquake - loaded liquid-filled tanks, in: *Proceedings of the 8th International Conference on Structural Mechanics in Reactor Technology, SMiRT-8*, Paper BK 2/1^, 475–480.

166 Fischer, D.F., Rammerstorfer, F.G,. and Scharf, K. (1991). Earthquake-resistant design of anchored and unanchored liquid storage tanks under three-dimensional earthquake excitation, *Structural Dynamics Recent Advances*, Ed. G.I. Schüeller, Berlin: Springer-Verlag, 317–371.

167 Fischer, D.F. and Seeber, R. (1988). Dynamic response of vertically excited liquid storage tanks considering liquid-soil interaction, *Earthquake Engineering and Structural Dynamics*, **16**(3), 329–342.

168 Fischer, D.F., Seeber, R., and Rammerstorfer, F.G. (1987). Refined algorithms for the analysis of earthquake loaded tanks, in *Structures and Stochastics Methods*, Ed. A.S. Cakmak, Oxford: Computational Mechanics Publications, Vol. 45, 59–72.

169 Flügge, W. (1960). *Stresses in Shells*, Berlin: Springer-Verlag.

170 García, S.M. (1969). Earthquake response analysis and aseismic design of cylindrical tanks, in: *Proceedings of the 4th World Conference on Earthquake Engineering*, Vol. II, B4–169 to B4–182.

171 Garza, L.R. (1966). A comparison of theoretical and experimental pressures and forces acting on a ring baffle under sloshing conditions, Report, NASA CR-385.

172 Garza, L.R. and Abramson, H.N. (1963). Measurements of liquid damping provided by ring baffles in cylindrical tanks, Report, NASA-CR-52070, Document ID. 19630013631, Contract Number NAS8-1555.

173 Gates, N.C. (1977). The earthquake response of deteriorating systems, PhD thesis, Technical Report, California Institute of Technology, EERL 77–03.

174 Gavrilyuk, I., Hermann, M., Lukovsky, I., Solodun, O., and Timokha, A. (2008). Natural sloshing frequencies in rigid truncated conical tanks, *Engineering Computations*, **25**(6), 518–540.

175 Gavrilyuk, I.P., Lukovsky, I.A., and Timokha, A.N. (2005). Linear and nonlinear sloshing in a circular conical tank, *Fluid Dynamics Research*, **37**(6), 399–429.

176 Gavrilyuk, I.P., Lukovsky, I.A., Trotsenko, Y., and Timokha, A.N. (2006). Sloshing in a vertical circular cylindrical tank with an annular baffle. Part 1, Linear fundamental solutions, *Journal of Engineering Mathematics*, **54**(1), 71–88.

177 Gavrilyuk, I.P., Lukovsky, I.A., Trotsenko, Y., and Timokha, A.N. (2006). Sloshing in a vertical circular cylindrical tank with an annular baffle. Part 2. Nonlinear resonant waves, *Journal of Engineering Mathematics*, **57**(1), 57–78.

178 Gaylord, E.H. Jr., Gaylord, C.N., and Stallmeyer, J.E. (1997). *Structural Engineering Handbook*, 4th Edition. New York: McGraw Hill.

179 Gazetas, G. (1983). Analysis of machine foundation vibrations: state of the art, *Soil Dynamics and Earthquake Engineering*, **2**(1), 2–42.

180 Gazetas, G. and Tassoulas, J.L. (1987). Horizontal stiffness of arbitrarily shaped embedded foundations, *Journal of Geotechnical Engineering*, **113**(5), 440–457.

181 Gedikli, A. and Ergüven, M.E. (1999). Seismic analysis of a liquid storage tank with a baffle, *Journal of Sound and Vibration*, **223**(1), 141–155.

182 Gedikli, A. and Ergüven, M.E. (2003). Evaluation of sloshing problem by the variational boundary element method, *Engineering Analysis with Boundary Elements*, **27**(9), 935–943.

183 Ghaemmaghami, A.R. and Kianoush, M.R. (2010). Effect of wall flexibility on dynamic response of concrete rectangular liquid storage tanks under horizontal and vertical ground motions, *ASCE Journal of Structural Engineering*, **136**(4), 441–451.

184 Ghali, A. (1979). *Circular Storage Tanks and Silos*, London: Spon Press.

185 Gioncu, V. (1979). *Thin Reinforced Concrete Shells*, Hoboken, NJ: John Wiley & Sons.

186 Gnitko, V., Naumemko, Y., and Strelnikova, E. (2017). Low frequency sloshing analysis of cylindrical containers with flat and conical baffles, *International Journal of Applied Mechanics and Engineering*, **22**(4), 867–881.

187 González, E., Almazán, J., Beltrán, J., Herrera, R,. and Sandoval, V. (2013). Performance of stainless steel winery tanks during the 02/27/2010 Maule earthquake, *Engineering Structures*, **56**, 1402–1418.

188 Gould, P.L. (1998). *Analysis of Shells and Plates*, Englewood Cliffs, NJ: Prentice Hall.

189 Graham, E.W. and Rodriguez, A.M. (1952). The characteristics of fuel motion which affect airplane dynamics, *Journal of Applied Mechanics*, **19**(3), 381–388.

190 Green, R.A., Olgun, C.G., and Cameron, W.I. (2008). Response and modeling of cantilever retaining walls subjected to seismic motions, *Computer-Aided Civil and Infrastructure Engineering*, **23**, 309–322.

191 Gregoriou, V.P., Tsinopoulos, S.V., and Karabalis, D.L. (2005). Seismic analysis of base isolated liquefied natural gas tanks, in: *Proceedings of the 5th GRACM International Congress on Computational Mechanics*, 305–312.

192 Greiner, R. (1987). Ring-stiffened cylindrical shells under external pressure: an extended proposal for design recommendations, in: *Proceedings of International Colloquium on Stability of Plate and Shell Structures*, 457–466.

193 Greiner, R. (1995). A concept for the classification of steel containments due to safety considerations, in: *Proceedings of the IABSE Colloquium*, Ed. B. Simpson, 65–75.

194 Greiner, R. (1998). Cylindrical shells: wind loading, in: *Silos: Fundamentals of Theory, Behaviour and Design*, Eds C.J. Brown and J. Nielsen, London: Spon Press, 378–399.

195 Greiner, R. (2004). Cylindrical shells under uniform external pressure, in: *Buckling of Thin Metal Shells*, Eds. J.G Teng and J.M. Rotter, London: Spon Press, 154–174.

196 Greiner, R. and Derler, P. (1995). Effect of imperfections on wind-loaded cylindrical shells, *Thin-Walled Structures*, **23**(1–4), 271–281.

197 Greiner, R. and Guggenberger, W. (2004). Tall cylindrical shells under wind pressure, in: *Buckling of Thin Metal Shells*, Eds. J.G Teng and J.M. Rotter, London: Spon Press, 198–206.

198 Greiner, R., Guggenberger, W., and Schneider, W. (2008). Cylindrical shells under wind loading, in: *Buckling of Steel Shells - European Design Recommendations*, 5th Edition, ECCS Technical Committee 8 - Structural Stability, TWG 8.4 - Shells, N. 125, 237–258.

199 Greiner, R. and Ofner, R. (2005). A second buckling effect of cylindrical tanks under earthquake, paper presented at the 5th International Conference on Computation of Shell and Spatial Structures (IASS/IACM). Salzburg, Austria.

200 Guerrin, A. (1968). *Traité de béton armé*. Vol. VI: *Réservoirs, châteaux d'eau, piscines*, Paris: Dunod (in French).

201 Gulkan, P. and Sozen, M.A. (1974). Inelastic responses of reinforced concrete structures to earthquake motions, *ACI Journal*, **71**(12), 604–610.

202 Gupta, R.K. (1993). Free vibrations of cylindrical containers, *Journal of Sound and Vibration*, **180**(3), 387–395.

203 Gupta, R.K. (1995). Free vibrations of partially filled cylindrical tanks, *Engineering Structures*, **17**(3), 221–230.

204 Gupta, V.K. and Trifunac, M.D. (1989). Investigation of building response to translational and rotational earthquake excitations, Report No. 89–02, Department of Civil Engineering, University of Southern California.

205 Gürgöze, M. (1996). On the eigenfrequencies of cantilevered beams carrying a tip mass and spring-mass in-span, *International Journal of Mechanical Sciences*, **38**(12), 1295–1306.

206 Habenberger, J. (2001). *Beitrag zur Berechnung von nachgiebig gelagerten Behältertragwerken unter seismischen Einwirkungen*, [Contribution to the analysis of flexibly supported liquid storage tanks under earthquake excitation], PhD thesis, Fakultät Bauingenieurwesen der Bauhaus-Universität Weimar.

207 Habenberger, J. (2008). Berechnung von Flüssigkeitsbehältern unter Erd- bebeneinwirkung, *Stahlbau*, **77**, 42–45.

208 Habenberger, J. and Schwarz, J. (2002). Seismic response of flexibly supported anchored liquid storage tanks, paper presented at the 12th European Conference on Earthquake Engineering, London, Paper Number 328.

209 Habenberger, J. and Schwarz, J. (2003). Simplified models for flexibly supported liquid storage tanks and its application to Eurocode 8, Part 4, *Technical Council on Lifeline Earthquake Engineering Monograph*, **25**, 339–348.

210 Hadjian, A.H. (1982). A re-evaluation of equivalent linear models for simple yielding systems, *Earthquake Engineering and Structural Dynamics*, **10**(6), 759–767.

211 Hamdam, F.H. (2000). Seismic behaviour of cylindrical steel liquid storage tanks, *Journal of Constructional Steel Research*, **53**(3), 307–333.

212 Han, W. and Reddy, B.D. (1999). *Plasticity: Mathematical Theory and Numerical Analysis*, Berlin: Springer.

213 Hanson, R.D. (1973). Behavior of liquid-storage tanks, in: *The Great Alaska Earthquake of 1964*, Ed. Committee on the Alaska Earthquake of the Division of Earth Sciences, Washington, DC: National Research Council, National Academy of Sciences, 331–339.

214 Haroun, M.A. (1980). Dynamic analyses of liquid storage tanks, Research Report EERL 80–04, California Institute of Technology, Pasadena, California.

215 Haroun, M.A. (1983a). Vibration studies and tests of liquid storage tanks, *Earthquake Engineering and Structural Dynamics*, **11**(2), 179–206.

216 Haroun, M.A. (1983b). Behavior of unanchored oil storage tanks: Imperial Valley earthquake, *Journal of Technical Topics in Civil Engineering*, **109**(1), 23–40.

217 Haroun, M.A. (1984). Stress analysis of rectangular walls under seismically induced hydrodynamic loads, *Bulletin of the Seismological Society of America*, **74**(3), 1031–1041.

218 Haroun, M.A. and Abou-Izzeddine, W. (1992a). Parametric study of seismic soil-tank interaction. I: horizontal excitation, *ASCE Journal of Structural Engineering*, **118**(3), 783–797.

219 Haroun, M.A. and Abou-Izzeddine, W. (1992b). Parametric study of seismic soil-tank interaction. II: vertical excitation, *ASCE Journal of Structural Engineering*, **118**(3), 798–812.

220 Haroun, M.A. and Badawi, H.S. (1988). Seismic behaviour of unanchored ground-based cylindrical tanks, in: *Proceedings of the 9th World Conference on Earthquake Engineering*, Paper Number 10–4–7, Vol. VI, 643–648.

221 Haroun, M.A. and Ellaithy, H.M. (1985a). Model for flexible tanks undergoing rocking, *Journal of Engineering Mechanics*, **111**(2), 143–157.

222 Haroun, M.A. and Ellaithy, H.M. (1985b). Seismically induced fluid forces on elevated tanks, *Journal of Technical Topics in Civil Engineering*, **111**(1), 1–15.

223 Haroun, M.A. and Housner, G.W. (1981a). Seismic design of liquid storage tanks, *Journal of the Technical Councils of ASCE*, **107**(1), 191–207.

224 Haroun, M.A. and Housner, G.W. (1981b). Earthquake response of deformable liquid storage tanks, *Journal of Applied Mechanics*, **48**(2), 411–418.

225 Haroun, M.A. and Housner, G.W. (1982). Complications in free vibration analysis of tanks, *ASCE Journal of the Engineering Mechanics Division*, **108**(5), 801–818.

226 Haroun, M.A. and Tayel, M.A. (1985a). Axisymmetrical vibrations of tanks: numerical, *Journal of Engineering Mechanics*, **111**(3), 329–345.

227 Haroun, M.A. and Tayel, M.A. (1985b). Axisymmetrical vibrations of tanks: analytical, *Journal of Engineering Mechanics*, **111**(3), 346–358.

228 Hashemi, S. and Aghashiri, M.H. (2017). Seismic responses of base- isolated flexible rectangular fluid containers under horizontal ground motion, *Soil Dynamics and Earthquake Engineering*, **100**, 159–168.

229 Hashemi, S., Saadatpour, M.M., and Kianoush, M.R. (2013). Dynamic behavior of flexible rectangular fluid containers, *Thin-Walled Structures*, **66**, 23–38.

230 Hasheminejad, S.M., Mohammadi, M.M., and Jarrahi, M. (2014). Liquid sloshing in partly-filled laterally-excited circular tanks equipped with baffles, *Journal of Fluids and Structures*, **44**, 97–114.

231 Heyman, J. (1977). *Equilibrium of Shell Structures*, Oxford: Clarendon Press.

232 Hilburger, M.W., Britt, V.O., and Nemeth, M.P. (2001). Buckling behavior of compression-loaded quasi-isotropic curved panels with a circular cutout, *International Journal of Solids and Structures*, **38**, 1495–1522.

233 Hirde, S., Bajare, A., and Hedaoo, M. (2011). Seismic performance of elevated water tanks, *International Journal of Advanced Engineering Research and Studies*, **1**(I), 78–87.

234 Hirt, C.W., Amsden, A.A., and Cook, J.L. (1997). An arbitrary Lagrangian- Eulerian computing method for all flow speeds, *Journal of Computational Physics*, **135**, 203–216.

235 Hirt, C.W. and Nichols, B.D. (1981). Volume of fluid (VOF) method for the dynamics of free boundaries, *Journal of Computational Physics*, **39**, 201–225.

236 Holl, H. (1987). *Parameteruntersuchung zur Abgrenzung der Anwendbarkeit eines Berechnungskonzeptes für erdbebenbeanspruchte Tankbauwerke*, Vienna: TU Wien.

237 Holmberg, M.W. (2009). Reinforced concrete tank design, *Structure Magazine*, **July**, 22–23.

238 Hoskins, L.M. and Jacobsen, L.S. (1934). Water pressure in a tank caused by a simulated earthquake, *Bulletin of the Seismological Society of America*, **24**(1), 1–32.

239 Hosseinzadeh, N., Sangsari, M.K., and Ferdosiyeh, H.T. (2014). Shake table study of annular baffles in steel storage tanks as sloshing dependent variable dampers, *Journal of Loss Prevention in the Process Industries*, **32**, 299–310.

240 Housner, G.W. (1954). Earthquake pressure on fluid containers, Technical Report NR-081-095, California Institute of Technology, Pasadena, California.

241 Housner, G.W. (1957). Dynamic pressures on accelerated fluid containers, *Bulletin of the Seismological Society of America*, **47**(1), 15–35.

242 Housner, G.W. (1963). The dynamic behaviour of water tanks, *Bulletin of the Seismological Society of America*, **53**(2), 381–387.

243 Housner, G.W. (1963). The behavior of inverted pendulum structures during earthquakes, *Bulletin of the Seismological Society of America*, **53**(2), 403–417.

244 Housner, G.W. (1982). Earthquake Engineering Research - 1982, Report No. CETS-CEER-001B, Washington, DC: National Academy Press.

245 Housner, G.W. and Haroun, M.A. (1979). Vibration tests of full–scale liquid storage tanks, in: *Proceedings of 2nd U.S. National Conference on Earthquake Engineering*, 137–145.

246 Hübner, A., Albiez, M., Kohler, D., and Saal, H. (2007). Buckling of long steel cylindrical shells subjected to external pressure, *Thin-Walled Structures*, **45**(1), 1–7.

247 Hunt, B. and Priestley, N. (1978). Seismic water waves in a storage tank, *Bulletin of the Seismological Society of America*, **68**(2), 487–499.

248 Hwan, S-C., Lee, B-H., and Park, J-C. (2010). Numerical simulation of violent sloshing motion in rectangular tank using improved MPS method, paper presented at the 9th Pacific/Asia Offshore Mechanics Symposium, PA- COMS 2010, Busan, South Korea.

249 Iai, S. and Kameoka, T. (1993). Finite element analysis of earthquake induced damage to anchored sheet pile quay walls, *Soils and foundations*, **33**(1), 71–97.

250 Ibrahim, R.A. (2005). *Liquid Sloshing Dynamics. Theory and Applications*, Cambridge: Cambridge University Press.

251 Ibrahim, R.A. and Pilipchuk, V.N. (2001). Recent advances in liquid sloshing dynamics, *Applied Mechanics Reviews*, **54**(2), 133–199.

252 Iervolino, I. (2003). Analisi quantitativa di rischio sismico nell'industria di processo, PhD thesis, Università degli Studi di Napoli Federico II.

253 Ifrim, M. and Bratu, C. (1969). The effect of seismic action on the dynamic behaviour of elevated water tanks, in: *Proceedings of the 4th World Conference on Earthquake Engineering*, Vol. II, B4–127 to B4–142.

254 Isaacson, M. and Premasiri, S. (2001). Hydrodynamic damping due to baffles in a rectangular tank, *Canadian Journal of Civil Engineering*, **28**(4), 608–616.

255 Ishida, K. and Kobayashi, N. (1984). Nonlinear rocking analysis of unanchored cylindrical tanks (2nd report, dynamic response with consideration of uplift of bottom plates), *Transactions of the Society of Mechanical Engineers, JSME*, **50**(453), 1042–1048 (in Japanese).

256 Ishida, K. and Kobayashi, N. (1988). An effective method of analyzing rocking motion for unanchored cylindrical tanks including uplift, *American Society of Mechanical Engineers (ASME) Journal of Pressure Vessel Technology*, **110**(1), 76–87.

257 Ishiyama, Y. (1982). Motions of rigid bodies and criteria for overturning by earthquake excitations, *Earthquake Engineering and Structural Dynamics*, **10**, 635–650.

258 Ishiyama, Y. (1984). Motions of rigid bodies and criteria for overturning by earthquake excitations, *Bulletin of the New Zealand Society for Earthquake Engineering*, **17**(1), 24–37.

259 Iwan, W.D. (1980). Estimating inelastic response spectra from elastic spectra, *Earthquake Engineering and Structural Dynamics*, **8**(4), 375–388.

260 Iwan, W.D. and Gates, N.C. (1979a). The effective period and damping of a class of hysteretic structures, *Earthquake Engineering and Structural Dynamics*, **7**(3), 199–211.

261 Iwan, W.D. and Gates, N.C. (1979b). Estimating earthquake response of simple hysteretic structures, *ASCE Journal of the Engineering Mechanics Division*, **105**(3), 391–405.

262 Jacobsen, L.S. (1930). Steady forced vibration as influenced by damping, *Transactions, ASME (APM-52-15)*, **52**, 169–181.

263 Jacobsen, L.S. (1949). Impulsive hydrodynamics of fluid inside a cylindrical tank and of fluid surrounding a cylindrical pier, *Bulletin of the Seismological Society of America*, **39**(3), 189–204.

264 Jacobsen, L.S. (1960). Damping in composite structures, in: *Proceedings of the 2nd World Conference on Earthquake Engineering*, Vol. 2, Session 2, 1029–1044.

265 Jacobsen, L.S. and Ayre, R.S. (1951). Hydrodynamic experiments with rigid cylindrical tanks subjected to transient motions, *Bulletin of the Seismological Society of America*, **41**(4), 313–346.

266 Jadhav, M.B. and Jangid, R.S. (2004). Response of base-isolated liquid storage tanks, *Shock and Vibration*, **11**, Article ID 276030.

267 Jain, S.K. and Jaiswal, O.R. (2005a). Modified proposed provisions for aseismic design of liquid storage tanks: Part I - codal provisions, *ASCE Journal of Structural Engineering*, **32**(3), 195–206.

268 Jain, S.K. and Jaiswal, O.R. (2005b). Modified proposed provisions for aseismic design of liquid storage tanks: Part II - commentary and examples, *ASCE Journal of Structural Engineering*, **32**(4), 297–310.

269 Jain, S.K. and Sameer, S.U. (1993). A review of requirements in Indian codes for aseismic design of elevated water tanks, *Bridge & Structural Engineering IABSE*, **23**(1), 1–16.

270 Jaiswal, O.R., Rai, D.C., and Jain, S.K. (2007). Review of seismic codes on liquid-containing tanks, *Earthquake Spectra*, **23**(1), 239–260.

271 Japanese Society of Civil Engineers (2000). Analysis of dynamic interaction between ground and foundation of a structure, *Dynamic Analysis and Earthquake-Resistant Design*, **2**(IV), 258–294.

272 Japanese Society of Civil Engineers (2001). Aboveground storage tank and piping, *Dynamic Analysis and Earthquake-Resistant Design*, **3**(VI), 195–229.

273 Jennings, P.C. (1971). Engineering features of the San Fernando earthquake of February 9, 1971, Technical Report, EERL 71-02, California Institute of Technology, Pasadena, California.

274 Jennings, P.C. and Bielak, J. (1973). Dynamics of building-soil interaction, *Bulletin of the Seismological Society of America*, **63**(1), 9–48.

275 Johansson, B., Maquoi, R., Sedlacek, G., Müller, C., and Beg, D. (2007). Commentary and worked examples to EN 1993-1-5 "Plated Structural Elements", JRC Scientific and Technical Reports.

276 Johns, D.J. (1983). Wind-induced static instability of cylindrical shells, *Journal of Wind Engineering and Industrial Aerodynamics*, **13**(1–3), 261–270.

277 Jolie, M., Hassan, M.M., and El Damatty, A.A. (2013). Assessment of current design procedures for conical tanks under seismic loading, *Canadian Journal of Civil Engineering*, **40**, 1151–1163.

278 Joshi, S.P. (2000). Equivalent mechanical model for horizontal vibration of rigid Intze tanks, *ISET Journal of Earthquake Technology, Technical Note*, **37**(1–3), 39–47.

279 Jullien, J.F. and Limam, A. (1998). Effects of openings on the buckling of cylindrical shells subjected to axial compression, *Thin-Walled Structures*, **31**, 187–202.

280 Jumikis, P.T. and Rotter, J.M. (1983). Buckling of simple ringbeams for bins and tanks, in: *Proceedings of the International Conference on Bulk Materials Storage, Handling and Transportation*, 323–328.

281 Justo, J.L. (1974). Discussion: A note on the interpretation of Coulomb's analysis of the thrust on a rough retaining wall in terms of the limit theorems of plasticity theory, *Géotechnique*, **24**(1), 106–108.

282 Kalantari, A., Nikoomanesh, M.R., and Goudarzi, M.A. (2019). Applicability of mass-spring models for seismically isolated liquid storage tanks, *Journal of Earthquake and Tsunami*, **13**(1), 1950002.

283 Kalnins, A. (1964). Analysis of shells of revolution subjected to symmetrical and nonsymmetrical loads, *Journal of Applied Mechanics*, **31**, 467–476.

284 Kalnins, A. and Godfrey, D.A. (1974). Seismic analysis of thin shell structures, *Nuclear Engineering and Design*, **27**(1), 68–76.

285 Kana, D.D. (1977). Seismic response of flexible cylindrical liquid storage tanks, paper presented at the 4th International Conference on Structural Mechanics in Reactor Technology, SMiRT-4, San Francisco, Paper K5/2.

286 Kana, D.D. (1979). Seismic response of flexible cylindrical liquid storage tanks, *Nuclear Engineering and Design*, **52**(1), 185–199.

287 Kana, D.D. and Dodge, F.T. (1975). Design support modeling of liquid slosh in storage tanks subject to seismic excitation, in: *Proceedings of 2nd ASCE Specialty Conference on Structural Design of Nuclear Plant Facilities*, Vol. I–A, 307–337.

288 Kantorovich, L.V. and Krylov, V.I. (1958). *Approximate Methods of Higher Analysis*, Groningen, the Netherlands: P. Noordhoff Ltd.

289 Kapila, I.P. (1962). Earthquake resistance design of retaining walls, paper presented at 2nd Earthquake Symposium, University of Roorkee, Roorkee, India.

290 Karamanos, S.A., Papaprokopiou, D., and Platyrrachos, M.A. (2009). Finite element analysis of externally-induced sloshing in horizontal-cylindrical and axisymmetric liquid vessels, *American Society of Mechanical Engineers (ASME) Journal of Pressure Vessel Technology*, **131**, 051301.

291 Karamanos, S.A., Patkas, L.A., and Platyrrachos, M.A. (2006). Sloshing effects on the seismic design of horizontal-cylindrical and spherical industrial vessels, *Transactions of the American Society of Mechanical Engineers*, **128**, 328–340.

292 von Karman, T., Sechler, E.E., and Donnell, L.H. (1932). The strength of thin plates in compression, *Transactions, ASME*, **54**, 53–58.

293 von Karman, T. and Tsien, H. (1939). The buckling of spherical shells by external pressure, *Journal of the Aeronautical Sciences*, **7**(2), 43–50.

294 von Karman, T. and Tsien, H. (1941). The buckling of thin cylindrical shells under axial compression, *Journal of the Aeronautical Sciences*, **8**, 302–312; reprinted in *Journal of Spacecraft and Rockets*, **40**(6) (2003), 898–907.

295 Kawano, K., Oda, T., Yoshida, K., Yamamoto, S., Shibuya, T., and Yamada, S. (1980). Damage to oil storage tanks for off Miyagi prefecture in earthquake of June 12, 1978, in: *Proceedings of the 7th World Conference on Earthquake Engineering*, Vol. VIII, 507–510.

296 Kelly, J.M. (1990). Base isolation: linear theory and design, *Earthquake Spectra*, **6**(2), 223–244.

297 Kettler, M. (2008). *Earthquake Design of Large Liquid-Filled Steel Storage Tanks: Comparison of Present Design Regulations, Load-Carrying Behaviour of Storage Tanks*, Berlin: VDM, Verlag Dr. Müller.

298 Keulegan, G.H. and Carpenter, L.H. (1958). Forces on cylinders and plates in an oscillating fluid, *Journal of Research of the National Bureau of Standards*, **60**(5), 423–440, Research Paper 2857.

299 Khansefid, A., Maghsoudi-Barmi, A., and Khaloo, A. (2019). Seismic protection of LNG tanks with reliability based optimally designed combined rubber isolator and friction damper, *Earthquakes and Structures*, **16**(5), 523–532.

300 Khoei, A.R., Gharehbaghi, S.A., Azami, A.R., and Tabarraie, A.R. (2006). SUT-DAM: An integrated software environment for multi-disciplinary geotechnical engineering, *Advances in Engineering Software*, **37**, 728–753.

301 Kianoush, M.R. and Chen, J.Z. (2006). Effect of vertical acceleration on response of concrete rectangular liquid storage tanks, *Engineering Structures*, **28**, 704–715.

302 Kim, J.K., Koh, H.M., and Kwahk, I.J. (1996). Dynamic response of rectangular flexible fluid containers, *Journal of Engineering Mechanics*, **122**(9), 807–817.

303 Kim, M.K., Lim, Y.M., Cho, S.Y., Cho, K.H., and Lee, K.W. (2002). Seismic analysis of base-isolated liquid storage tanks using the BE-FE-BE coupling technique, *Soil Dynamics and Earthquake Engineering*, **22**(9–12), 1151–1158.

304 Kim, N-S. and Lee, D-G. (1995). Pseudodynamic test for evaluation of seismic performance of base-isolated liquid storage tanks, *Engineering Structures*, **17**(3), 198–208.

305 Kim, Y-W. and Lee, Y-S. (2005). Coupled vibration analysis of liquid-filled rigid cylindrical storage tank with an annular plate cover, *Journal of Sound and Vibration*, **279**(1–2), 217–235.

306 Kirchhoff, G. (1891). *Vorlesungen über mathematische Physik*, Leipzig: Druck und Verlag von B.G. Teubner.

307 Klein, F., Hoffman, E.S., and Rice, F. (1981). Application of strength design methods to sanitary structures, *Concrete International*, **3**(4), 35–40.

308 Klein, G.J. (2008). Curved-bar nodes, *Concrete International*, **30**(9), 42–47.

309 Koiter, W.T. (1956). Buckling and postbuckling behaviour of a cylindrical panel under axial compression, Report S476, National Aeronautical Research Institute, **20**, 71–84.

310 Koiter, W.T. (1960). A consistent first approximation in the general theory of thin elastic shells, in: *Proceedings of the IUTAM Symposium on the Theory of Thin Elastic Shells*, 12–33.

311 Knödel, P. and Schulz, U. (1988). Stabilité de cheminées d'acier à ouvertures dans les tuyaux, *Stahlbau*, **57**, 13–21 (in French).

312 Kobayashi, N. and Ishida, K. (1984). Non-linear rocking analysis of unanchored cylindrical tanks (1st report, uplift deformation of bottom plate subjected to overturning moment). *Transactions of the Society of Mechanical Engineers, JSME*, **50**(451), 514–519 (in Japanese).

313 Kock, E. and Olson, L. (1991). Fluid-structure interaction analysis by the finite element method: a variational approach, *International Journal for Numerical Methods in Engineering*, **31**, 463–491.

314 Kollár, L. (1982). Buckling of complete spherical shells and spherical caps, in: *Buckling of Shells: Proceedings of a State-of-the-Art Colloquium, Universität Stuttgart, Germany, May 6-7*, Ed. E. Ramm, Berlin: Springer Verlag, 401–425.

315 Koller, M.G. and Malhotra, P.K. (2004). Seismic evaluation of unanchored cylindrical tanks, paper presented at the 13th World Conference on Earthquake Engineering, Vancouver, Canada, Paper Number 2534.

316 Koshizuka, S. and Oka, Y. (1996). Moving-particle semi-implicit method for fragmentation of incompressible fluid, *Nuclear Science and Engineering*, **123**, 421–434.

317 Kotrasová, K., Leoveanu, I-S., and Kormaníková, E. (2013). A comparative study of the seismic analysis of rectangular tanks according to EC8 and IS 1893, *Buletinul AGIR*, **3**, 120–125.

318 Kraus, H. (1967). *Thin Elastic Shells: An Introduction to the Theoretical Foundations and the Analysis of Their Static and Dynamic Behaviour*, Chichester: John Wiley & Sons.

319 Krenzke, M.A. and Reynolds, T.E. (1967). Structurals research on submarine pressure hulls at the David Taylor Model Basin, *Journal of Hydronautics*, **1**(1), 27–35.

320 Kumar, A. (1981). Studies of dynamic and static response of cylindrical liquid-storage tanks, PhD thesis, Rice University, Houston, TX.

321 Kundurpi, P.S., Samavedam, G., and Johns, D.J. (1975). Stability of cantilever shells under wind loads, *ASCE Journal of the Engineering Mechanics Division*, **101**(EM 5), 517–530.

322 Lamb, H. (1945). *Hydrodynamics*, London: Dover Publications, 370–372.

323 Lancellotta, R. (2004). *Geotecnica*, Bologna: Ed. Zanichelli (in Italian).

324 Lancellotta, R. (2009). *Geotechnical Engineering*, 2nd Edition, Bologna: Ed. Zanichelli.

325 Langhaar, H.L. and Boresi, A.P. (1957). Snap-through and post-buckling behavior of cylindrical shells under the action of external pressure, University of Illinois Engineering Experiment Station Bulletin No. 443, University of Illinois Archives, Record Series number 11/2/801.

326 Langner, C.G. (1963). A preliminary analysis for optimum design of ring and partition antislosh baffles, Technical Report No. 7, NASA CR-52069, Southwest Research Institute.

327 Larkin, T. (2008). Seismic response of liquid storage tanks incorporating soil structure interaction, *Journal of Geotechnical and Geoenvironmental Engineering*, **134**(12), 1804–1814.

328 Lau, D.T. and Clough, R.W. (1989). Static tilt behavior of unanchored cylindrical tanks, Report EERC 89–11, Earthquake Engineering Research Center, College of Engineering, University of California, Berkeley, California.

329 Lau, D.T., Marquez, D., and Qu, F. (1996). Earthquake-resistant design of liquid storage tanks, paper presented at the 11th World Conference on Earthquake Engineering, Acapulco, Mexico, Paper Number 1293.

330 Lee, V.W. and Trifunac, M.D. (1987). Rocking strong earthquake accelerations, *Soil Dynamics and Earthquake Engineering*, **6**, 75–89.

331 Leeds, D.J. (1980). Imperial County, California, Earthquake, October 15, 1979, G.E. Brandow Coordinator, Reconnaissance Report, Earthquake Engineering Research Institute.

332 Lengsfeld, M., Bardia, K., Taagepera, J., Hathaitham, K., La Bounty, D., and Lengsfeld, M. (2007). Analysis of loads for nozzles in API 650 tanks, *American Society of Mechanical Engineers (ASME) Journal of Pressure Vessel Technology*, **129**, 474–481.

333 Leon, G.S. and Kausel, A.M. (1986). Seismic analysis of fluid storage tanks, *ASCE Journal of Structural Engineering*, **112**(1), 1–18.

334 Leonhardt, F. and Mönnig, E. (1973). *Le basi del dimensionamento nelle costruzioni in cemento armato*, Vol. I, Milan: Ed. Tecniche ET (in Italian).

335 Leonhardt, F. and Mönnig, E. (1977). *L'armatura nelle costruzioni in cemento armato: statica, tecnologia e tipologia*, Vol. III, Milan: Ed. Tecniche ET (in Italian).

336 Lev, O.E. and Jain, B.P. (1977). Seismic response of flexible liquid containers, paper presented at the 4th International Conference on Structural Mechanics in Reactor Technology, SMiRT-4, San Francisco, Paper K5/3.

337 Li, H-N., Sun, L-Y., and Wang, S-Y. (2004). Improved approach for obtaining rotational components of seismic motion, *Nuclear Engineering and Design*, **232**(2), 131–137.

338 Liu, D. and Lin, P. (2008). A numerical study of three-dimensional liquid sloshing in tanks, *Journal of Computational Physics*, **227**, 3921–3939.

339 Liu, D. and Lin, P. (2009). Three-dimensional liquid sloshing in a tank with baffles, *Ocean Engineering*, **36**(2), 202–212.

340 Liu, M.B. and Liu, G.R. (2010). Smoothed particle hydrodynamics (SPH): an overview and recent developments, *Archives of Computational Methods in Engineering*, **17**, 25–76.

341 Livaoğlu, R. (2008). Investigation of seismic behaviour of fluid-rectangular tank-soil/foundation systems in frequency domain, *Soil Dynamics and Earthquake Engineering*, **28**(2), 132–146.

342 Livaoğlu, R. and Doğangün, A. (2005). Seismic evaluation of fluid-elevated tank-foundation/soil systems in frequency domain, *Structural Engineering and Mechanics*, **21**(1), 101–119.

343 Livaoğlu, R. and Doğangün, A. (2006). Simplified seismic analysis procedures for elevated tanks considering fluid-structure-soil interaction, *Journal of Fluids and Structures*, **22**(3), 421–439.

344 Long, B. and Garner, B. (2004). *Guide to Storage Tanks and Equipment*, London: Professional Engineering Publishing.

345 Lorenz, R. (1908). Achsensymmetrische Verzerrungen in dünnwandingen Hohlzylindern, *Zeitschrift des Vereines Deutscher Ingenieure (VDI)*, **52**, 1706–1713.

346 Lorenz, R. (1911). Die nicht achsensymmetrische Knickung dünnwandiger Hohlzylindern, *Physikalische Zeitschrift*, **12**, 241–260.

347 Love, A.E.H. (1888). On the small free vibrations and deformations of thin elastic shells, *Philosophical Transactions of the Royal Society A*, **179**, 491–546.

348 Love, A.E.H. (1944 [1892]). *Mathematical Theory of Elasticity*, London: Dover Edition.

349 Low, K.H. (1998). On the eigenfrequencies for mass loaded beams under classical boundary conditions, *Journal of Sound and Vibration*, **215**(2), 381–389.

350 Low, K.H. (2000). A modified Dunkerley formula for eigenfrequencies of beams carrying concentrated masses, *International Journal of Mechanical Sciences*, **42**(7), 1287–1305.

351 LS-Dyna, (2007). *Keyword user's manual*, LS-Dyna Version 971, Livermore Software Technology Corporation.

352 Luco, J.E. and Westmann, R.A. (1971). Dynamic response of circular footings, *ASCE Journal of the Engineering Mechanics Division*, **97**(EM 5), 1381–1395.

353 Luft, R.W. (1984). Vertical accelerations in prestressed concrete tanks, *ASCE Journal of Structural Engineering*, **110**(4), 706–714.

354 Ma, D.C. and Chang, Y.W. (1985). Analysis of seismic sloshing of reactor tanks considering submerged components and seismic isolation, in: *Proceedings of the American Society of Mechanical Engineers (ASME) Pressure Vessels and Piping Conference*, Technical Report CONF-850670-14, **98–7**, 139–147.

355 Ma, D.C., Gvildys, J., and Chang, Y.W. (1984). Sloshing response of a reactor tank with internals, paper presented at American Society of Civil Engineers/Engineering Mechanics Specialty Conference, Laramie, Wyoming, Report CONF-840833-4.

356 Madhukar, G.V. and Madhuri, M.N. (2013). Seismic performance of circular elevated water tank with framed staging system, *International Journal of Advanced Research in Engineering and Technology*, **4**(4), 159–167.

357 Magnucki, K. and Máckiewicz, M. (2006). Elastic buckling of an axially compressed cylindrical panel with three edges simply supported and one edge free, *Thin-Walled Structures*, **44**(4), 387–392.

358 Maity, D., Narayana, T.S., and Saha, U.K. (2009). Dynamic response of liquid storage elastic tanks with baffle, *Journal of Structural Engineering (Madras)*. **36**(3), 172–181.

359 Makris, N. and Konstantinidis, D. (2001). The rocking spectrum and the shortcomings of design guidelines, PEER Report 2001/07, Pacific Earthquake Engineering Research Center, College of Engineering.

360 Makris, N. and Roussos, Y.S. (2000). Rocking response of rigid blocks under near-source ground motions, *Géotechnique*, **50**(3), 243–262.

361 Maleki, A. and Ziyaeifar, M. (2007). Damping enhancement of seismic isolated cylindrical liquid storage tanks using baffles, *Engineering Structures*, **29**(12), 3227–3240.

362 Maleki, A. and Ziyaeifar, M. (2008). Sloshing damping in cylindrical liquid storage tanks with baffles, *Journal of Sound and Vibration*, **311**(1–2), 372–385.

363 Malhotra, P.K. (1991). Seismic response of uplifting liquid storage tanks, PhD thesis, Rice University, Houston, Texas

364 Malhotra, P.K. (1995). Base uplifting analysis of flexibly supported liquid- storage tanks, *Earthquake Engineering and Structural Dynamics*, **24**(12), 1591–1607.

365 Malhotra, P.K. (1996). Seismic uplifting of flexibly supported liquid- storage tanks, paper presented at the 11th World Conference on Earthquake Engineering, Acapulco, Mexico, Paper Number 873.

366 Malhotra, P.K. (1997a). New method for seismic isolation of liquid-storage tanks, *Earthquake Engineering and Structural Dynamics*, **26**(8), 839–847.

367 Malhotra, P.K. (1997b). Method for seismic base isolation of liquid-storage tanks, *ASCE Journal of Structural Engineering*, **123**(1), 113–116.

368 Malhotra, P.K. (1997c). Seismic response of soil-supported unanchored liquid-storage tanks, *ASCE Journal of Structural Engineering*, **123**(4), 440–450.

369 Malhotra, P.K. (1998). Seismic strengthening of liquid-storage tanks with energy-dissipating anchors, *ASCE Journal of Structural Engineering*, **124**(4), 405–414.

370 Malhotra, P.K. (2000). Practical nonlinear seismic analysis of tanks, *Earthquake Spectra*, **16**(2), 473–492.

371 Malhotra, P.K. (2005). Sloshing loads in liquid-storage tanks with insufficient freeboard, *Earthquake Spectra*, **21**(4), 1185–1192.

372 Malhotra, P.K. (2006). Earthquake-induced sloshing in tanks with insufficient freeboard, *Structural Engineering International*, **3**, 222–225.

373 Malhotra, P.K. and Veletsos, A.S. (1994a). Beam model for base-uplifting analysis of cylindrical tanks, *ASCE Journal of Structural Engineering*, **120**(12), 3471–3488.

374 Malhotra, P.K. and Veletsos, A.S. (1994b). Uplifting analysis of base plates in cylindrical tanks, *ASCE Journal of Structural Engineering*, **120**(12), 3489–3505.

375 Malhotra, P.K. and Veletsos, A.S. (1994c). Uplifting response of unanchored liquid-storage tanks, *ASCE Journal of Structural Engineering*, **120**(12), 3525–3547.

376 Malhotra, P.K. and Veletsos, A.S. (1996). Seismic response of unanchored and partially anchored liquid-storage tanks, Research Project 2907-02, Final Report TR-105809, Electric Power Research Institute (EPRI). Palo Alto, California, prepared by Department of Civil Engineering, Rice University, Houston, Texas.

377 Malhotra, P.K., Wenk, T., and Wieland, M. (2000). Simple procedure for seismic analysis of liquid-storage tanks, *Structural Engineering International*, **3**, 197–201.

378 Malushte, S.R. and Whittaker, A.S. (2005). Survey of past base isolation applications in nuclear power plants and challenges to industry/regulatory acceptance, in: *Proceedings of the 18th International Conference on Structural Mechanics in Reactor Technology, SMiRT-18*, Ed. K. Kussmaul, Paper K10-7, 3404–3410.

379 Manevich, A.I. (2001). Coupled instability of cylindrical shells stiffened with thin ribs, in: *Proceedings of the 3rd International Conference on Thin-Walled Structures*, Eds. J. Zaraś, K. Kowal-Michalska, and J. Rhodes, Oxford: Elsevier, 683–691.

380 Manos, G.C. (1986). Earthquake tank-wall stability of unanchored tanks, *ASCE Journal of Structural Engineering*, **112**(8), 1863–1880.

381 Manos, G.C. and Clough, R.W. (1982). Further study of the earthquake response of a broad cylindrical liquid-storage tank model, Report, EERC 82-07, Earthquake Engineering Research Center, College of Engineering, University of California, Berkeley, California.

382 Manos, G.C. and Clough, R.W. (1985). Tank damage during the May 1983 Coalinga earthquake, *Earthquake Engineering and Structural Dynamics*, **13**(4), 449–466.

383 Manos, G.C. and Talaslidis, D. (1988). Response of cylindrical tanks subjected to lateral loads: correlation between analytical and experimental results, in: *Proceedings of the 9th World Conference on Earthquake Engineering*, Paper Number 10-4-11, Vol. VI, 667–672.

384 Marchaj, T.J. (1979). Importance of vertical acceleration in the design of liquid containing tanks, in: *Proceedings of the 2nd U.S. National Conference on Earthquake Engineering*, 146–155.

385 Marguerre, K. (1938). Zur Theorie der gekrümmten Platte grosser Formänderung, in: *Proceedings of the 5th International Congress for Applied Mechanics*, 93–101.

386 Markus, S. (1988). *The Mechanics of Vibration of Cylindrical Shells*, Oxford: Elsevier.

387 Mason, P.A., Amies, H.J., Sangarapillai, G., and Rose, G. (1997). Construction of bunds for oil storage tanks, Report 163, CIRIA Research Project 156, Construction Industry Research and Information Association. Reprinted 2003.

388 Matsuda, I., Mochizuki, T., and Miyano, M. (1982). Consideration on seismic intensity distribution of the Great Kanto Earthquake of 1923, *Geographical Reports of Tokyo Metropolitan University*, **17**, 77–85.

389 Matuo, H. and Ohara, S. (1960). Lateral earth pressure and stability of quay walls during earthquakes, in: *Proceedings of the 2nd World Conference on Earthquake Engineering*, Vol. 1, Session 1, 165–181.

390 Mayer, R. (1912). Über Elastizität und Stabilität des geschlossenen und offenen Kreisbogens, *Zeitschrift für Mathematik und Physik*, 246–321.

391 McLean, R.S. and Moore, W.W. (1936). Computation of the vibration period of steel tank towers, *Bulletin of the Seismological Society of America*, **26**(1), 63–67.

392 Meek, J.W. and Veletsos, A.S. (1972). Dynamic analysis and behaviour of structure–foundation systems, Structural Research at Rice, Report No. 13, Department of Civil Engineering, Rice University, Houston, TX.

393 Meier, S.W. (2001). Today's composite elevated storage tanks, 2002, paper presented at AWWA Conference & Exposition, New Orleans, Louisiana.

394 Mieda, T., Ishida, K., Jitu, K., and Chiba, T. (1993). An experimental study of viscous damping in sloshing mode of cylindrical tanks, in: *Proceedings of the 12th International Conference on Structural Mechanics in Reactor Technology, SMiRT-12*, ed. K. Kussmaul, Paper SD104/1, 231–236.

395 Migliacci, A. (1979). *Progetti di strutture. Parte I*, Milan: Masson Italia Editori (in Italian).

396 Mikishev, G.N. and Dorozhkin, N.Y. (1961). *An experimental investigation of free oscillations of a liquid in containers*, Izvestiya Akademii Nauk SSSR, Mekhanika, Mashinostroenie, Otdeleine Tekhnicheskikh Nauk, No. 4, 48–83 (in Russian). (English translation by D.D. Kana, 1963, Southwest Research Institute).

397 Miles, J.W. (1956). On the sloshing of liquid in a cylindrical tank, Report No. AM6-5, The Ramo-Wooldridge Corporation, Guided Missile Research Division, GM-TR-18.

398 Miles, J.W. (1958). Ring damping of free surface oscillations in a circular tank, *Journal of Applied Mechanics*, **25**(2), 274–276.

399 Miller, C.D. (1977). Buckling of axially compressed cylinders, *ASCE Journal of the Structural Division*, **103**(7), 695–721.

400 Miranda, E. (1999). Approximate seismic lateral deformation demands in multistory buildings, *ASCE Journal of Structural Engineering*, **125**(4), 417–425.

401 Miranda, E. and Reyes, C.J. (2002). Approximate lateral drift demands in multistory buildings with nonuniform stiffness, *ASCE Journal of Structural Engineering*, **128**(7), 840–8495.

402 Miranda, E. and Ruiz-García, J. (2002). Evaluation of approximate methods to estimate maximum inelastic displacement demands, *Earthquake Engineering and Structural Dynamics*, **31**(3), 539–560.

403 Mitra, S. and Sinhamahapatra, K.P. (2007). Slosh dynamics of liquid-filled containers with submerged components using pressure-based finite element method, *Journal of Sound and Vibration*, **304**(1–2), 361–381.

404 Mitra, S., Upadhyay, P.P., and Sinhamahapatra, K.P. (2008). Slosh dynamics of inviscid fluids in two-dimensional tanks of various geometry using finite element method, *International Journal for Numerical Methods in Fluids*, **41**(2), 185–208.

405 Mononobe, N. and Matsuo, H. (1929). On the determination of earth pressures during earthquake, in: *Proceedings of the World Engineering Congress*, Paper Number 388, Vol. 9, 179–187.

406 Moore, T.A. and Wong, E.K. (1984). The response of cylindrical liquid storage tanks to earthquakes, in: *Proceedings of the 8th World Conference on Earthquake Engineering*, Vol. V, 239–246.

407 Morley, L.S.D. (1959). An improvement on Donnell's approximation for thin-walled circular cylinder, *Quarterly Journal of Mechanics and Applied Mathematics*, **12**(1), 88–99.

408 Moslemi, M. (2011). Seismic response of ground cylindrical and elevated conical reinforced concrete tanks, PhD thesis, Ryerson University, Toronto, Ontario, Canada.

409 Moslemi, M., Kianoush, M.R., and Pogorzelski, W. (2011). Seismic response of liquid-filled elevated tanks, *Engineering Structures*, **33**(6), 2074–2084.

410 Mouzakis, T., Nash, W.A., Colonell, J.M., and Wu, C.I. (1975). TANK- FREQ: natural frequencies of cylindrical liquid storage containers, Department of Civil Engineering, University of Massachusetts, Amherst.

411 Munshi, J.A. (1998). *Rectangular Concrete Tanks*, Washington, DC: Portland Cement Association.

412 Munshi, J.A. (2002). *Design of Liquid-Containing Concrete Structures for Earthquake Forces*, Washington, DC: Portland Cement Association.

413 Mushtari, K.M. (1938). Certain generalizations in the theory of thin shells with applications to the problem of elastic stability, *Bulletin*, University of Kazan, **11**(8), 71–150.

414 Myers, P.E. (1997). *Aboveground storage tanks*, New York: McGraw-Hill.

415 Mylonakis, G., Kloukinas, P., and Papantonopoulos, C. (2007). An alternative to the Mononobe-Okabe equations for seismic earth pressures, *Soil Dynamics and Earthquake Engineering*, **27**(10), 957–969.

416 Nachtigall, I., Gebbeken, N., and Urrutia-Galicia, J.L. (2003). On the analysis of vertical circular cylindrical tanks under earthquake excitation at its base, *Engineering Structures*, **25**(2), 201–213.

417 Nadim, F. and Whitman, R.V. (1983). Seismically induced movement of retaining walls, *Journal of Geotechnical Engineering*, **109**(7), 915–931.

418 Narayanan, R.S. and Beeby, A. (2005). *Designers' guide to EN 1992-1-1 and EN 1992-1-2*, Ed. H. Gulvanessian, London: Thomas Telford.

419 Natsiavas, S. (1988). Response and failure of fluid-filled tanks under base axcitation, PhD thesis, California Institute of Technology, Pasadena, California.

420 Natsiavas, S. and Babcock, C.D. (1987). Buckling at the top of a fluid-filled tank during base excitation, *American Society of Mechanical Engineers (ASME) Journal of Pressure Vessel Technology*, **109**, 374–380.

421 Nazarov, A.A. (1949). *On the theory of thin shallow shells*, Report, NACA TM 1426.

422 Neut, A. van der (1932). De elastische stabiliteit van de dunwandigen bol, PhD thesis, Delft University, Faculty of Mechanical, Maritime and Materials Engineering (in Dutch).

423 Newmark, N.M. and Hall, W.J. (1969). Seismic design criteria for nuclear reactor facilities, in: *Proceedings of the 4th World Conference on Earthquake Engineering*, Vol. II, pp. B4-37 to B4-50 (Reprinted in "Selected Papers by Nathan M. Newmark: Civil Engineering Classic", ASCE Committee Report Paper (1975), 757–770).

424 Nielsen, R. and Kiremidjian, Y. (1986). Damage to oil refineries from major earthquakes, *ASCE Journal of Structural Engineering*, **112**(6), 1481–1491.

425 Nikoomanesh, M.R., Moeini, M., and Goudarzi, M.A. (2019). An innovative isolation system for improving the seismic behaviour of liquid storage tanks, *International Journal of Pressure Vessels and Piping*, **173**, 1–10.

426 Nilsson, I.H.E. and Losberg, A. (1976). Reinforced concrete corners and joints subjected to bending moment, *ASCE Journal of the Structural Division*, **102**(ST6), 1229–1254.

427 Niwa, A. (1978). Seismic behaviour of tall liquid storage tanks, Report, EERC 78-04, Earthquake Engineering Research Center, College of Engineering, University of California, Berkeley, California.

428 Niwa, A. and Clough, R.W. (1982). Buckling of cylindrical liquid-storage tanks under earthquake loading, *Earthquake Engineering and Structural Dynamics*, **10**(1), 107–122.

429 Oden, J.T. and Ripperger, E.A. (1981). *Mechanics of Elastic Structures*, 2nd Edition, New York: McGraw Hill.

430 Okabe, S. (1926). General theory of earth pressure and seismic stability of retaining wall and dam, *Journal of Japan Society of Civil Engineers*, **12**(1), 123–134.

431 Ormeño, M., Larkin, T., and Chouw, N. (2012). Comparison between standards for seismic design of liquid storage tanks with respect to soil- foundation-structure interaction and uplift, *Bulletin of the New Zealand Society for Earthquake Engineering*, **45**(1), 40–46.

432 Ormeño, M., Larkin, T., and Chouw, N. (2012). Influence of uplift on liquid storage tanks during earthquakes, *Coupled Systems Mechanics*, **1**(4), 311–324.

433 Ostadan, F. and White, W.H. (1998). Lateral seismic soil pressure. An updated approach, paper presented at 1st US-Japan Soil Structure Interaction (SSI) Workshop, United States Geological Survey, Menlo Park, California.

434 Ozdemir, Z., Souli, M., and Fahjan, Y. (2009). ALE formulation for the evaluation of seismic behaviour of anchored and unanchored tanks, paper presented at 7th European LS-DYNA User's Conference, Salzburg, Austria.

435 Ozdemir, Z., Souli, M., and Fahjan, Y.M. (2010). Application of nonlinear fluid-structure interaction methods to seismic analysis of anchored and unanchored tanks, *Engineering Structures*, **32**, 409–423.

436 Panchal, V.R. and Jangid, R.S. (2008). Variable friction pendulum system for seismic isolation of liquid storage tanks, *Nuclear Engineering and Design*, **238**(6), 1304–1315.

437 Panchal, V.R. and Soni, D.P. (2014). Seismic behaviour of isolated fluid storage tanks: a state-of-the-art review, *KSCE Journal of Civil Engineering*, **18**(4), 1097–1104.

438 Panigrahy, P.K., Saha, U.K., and Maity, D. (2009). Experimental studies on sloshing behavior due to horizontal movement of liquids in baffled tanks, *Ocean Engineering*, **36**(3–4), 213–222.

439 Papaspyrou, S., Karamanos, S.A., and Valougeorgis, D. (2004a). Response of half-full horizontal cylinders under transverse excitation, *Journal of Fluids and Structures*, **19**, 985–1003.

440 Papaspyrou, S., Valougeorgis, D., and Karamanos, S.A. (2004b). Sloshing effects in half-full horizontal cylindrical vessels under longitudinal excitation, *Journal of Applied Mechanics*, **71**, 255–265.

441 Park, J-H., Koh, H-M., and Kim, J.K. (2004). Seismic isolation of pool-type tanks for the storage of nuclear spent fuel assemblies, *Nuclear Engineering and Design*, **199**(1–2), 143–154.

442 Park, K-S., Koh, H-M., and Song, J. (2004). Cost-effectiveness analysis of seismically isolated pool structures for the storage of nuclear spent-fuel assemblies, *Nuclear Engineering and Design*, **231**(3), 259–270.

443 Park, Y.J., Ang, A.H-S., and Wen, Y.K. (1984). Seismic damage analysis and damage-limiting design of R.C. buildings, PhD thesis, Civil Engineering Studies, Structural Research Series N. 516, University of Illinois, Urbana-Champaign.

444 Patel, C.N. and Patel, H.S. (2012). Supporting systems for reinforced concrete elevated water tanks: a state-of-the-art literature review, *International Journal of Advanced Engineering Research and Studies*, **II**(I), 68–71.

445 Patel, C.N., Vaghela, S.N., and Patel, H.S. (2012). Sloshing response of elevated water tank over alternate column proportionality, *International Journal of Advanced Engineering Technology*, **III**(IV), 60–63.

446 Pathmanathan, R. (2006). Numerical modelling of seismic behaviour of earth-retaining walls, MSc dissertation, European School for Advanced Studies in Reduction of Seismic Risk, Rose School IUSS, University of Pavia, Pavia.

447 Patkas, L.A. and Karamanos, S.A. (2007). Variational solutions for externally induced sloshing in horizontal-cylindrical and spherical vessels, *Journal of Engineering Mechanics*, **133**(6), 641–655.

448 Peek, R. (1986). Analysis of unanchored liquid storage tanks under seismic loads, PhD thesis, Earthquake Engineering Research Laboratory, California Institute of Technology, Pasadena, California.

449 Peek, R. (1988). Analysis of unanchored liquid storage tanks under lateral loads, *Earthquake Engineering and Structural Dynamics*, **16**(7), 1087–1100.

450 Peek, R. and El-Bkaily, M. (1991). Postbuckling behavior of unanchored steel tanks under lateral loads, *American Society of Mechanical Engineers (ASME) Journal of Pressure Vessel Technology*, **113**(3), 423–428.

451 Peek, R. and El-Bkaily, M. (1997). Mechanical analogies for liquid storage tanks and action of eccentric gravity loads, *Journal of Engineering Mechanics*, **123**(6), 561–567.

452 Peek, R. and Jenning, P.C. (1988). Simplified analysis of unanchored tanks, *Earthquake Engineering and Structural Dynamics*, **16**(7), 1073–1085.

453 Petrini, L., Pinho, R., and Calvi, G.M. (2004). *Criteri di progettazione antisismica degli edifici*, Pavia: IUSS Press (in Italian).

454 Picone, M. (1960). Sull'equazione integrale non lineare di seconda specie di Fredholm, *Mathematische Zeitschrift*, **74**(1), 119–128.

455 Poole, A.B. (1994). Simplified design and evaluation of liquid storage tanks relative to earthquake loading, paper presented at American Society of Mechanical Engineers (ASME) Pressure Vessels and Piping Conference, Minneapolis, Technical Report CONF-940613-24, 1–6.

456 Popov, E.P. and Medwadowski, S.J. (1981). *Concrete Shell Buckling*, Publication SP-67, Washington, DC: American Concrete Institute.

457 Priestley, M.J.N. (1985). Analysis and design of circular prestressed storage tanks, *PCI Journal*, **30**(4), 64–85.

458 Priestley, M.J.N., Evison, R.J., and Carr, A.J. (1978). Seismic response of structures free to rock on their foundations, *Bulletin of the New Zealand National Society for Earthquake Engineering*, **11**(3), 141–150.

459 Priestley, M.J.N., Wood, J.H., and Davidson, B.J. (1986). Seismic design of storage tanks, *Bulletin of the New Zealand National Society for Earthquake Engineering*, **19**(4), 272–284.

460 Psarropoulos, P.N., Klonaris, G., and Gazetas, G. (2005). Seismic earth pressures on rigid and flexible retaining walls, *Soil Dynamics and Earthquake Engineering*, **25**(7–10), 795–809.

461 Rai, D.C. (2002). Seismic retrofitting of R/C shaft support of elevated tanks, *Earthquake Spectra*, **18**(4), 745–760.

462 Rai, D.C. (2002). Review of code design forces for shaft supports of elevated water tanks, in: *Proceedings of the 12th Symposium on Earthquake Engineering*, 1407–1418.

463 Rai, D.C. (2003). Performance of elevated tanks in Mw 7.7 Bhuj earthquake of January 26th, 2001, *Proceedings of the Indian Academy of Sciences - Earth and Planetary Sciences*, **112**(3), 421–429.

464 Rai, D.C. and Singh, B. (2004). Seismic design of concrete pedestal supported tanks, paper presented at the 13th World Conference on Earthquake Engineering, Vancouver, Canada, Paper Number 230.

465 Ramiah, B.K. and Gupta, D.S.R.M. (1966). Factors affecting seismic design of water towers, *ASCE Journal of the Structural Division*, **92**(4), 13–30.

466 Ramm, E. (Ed.) (1982). *Buckling of Shells: State-of-the-Art Colloquium, Universität Stuttgart, Germany, May 6–7*, Berlin: Springer Verlag.

467 Rammerstorfer, F.G., Auli, W., and Fischer, F.D. (1985). Uplifting and stability of wind-loaded vertical cylindrical shells, *Engineering Computations*, **2**(3), 170–180.

468 Rammerstorfer, F.G., Fischer, F.D., and Scharf, K. (1988). A proposal for the earthquake-resistant design of tanks: results from the Austrian research project, in: *Proceedings of the 9th World Conference on Earthquake Engineering*, Paper Number 10–4–19, Vol. VI, 715–720.

469 Rammerstorfer, F.G., Scharf, K., and Fischer, F.D. (1990). Storage tanks under earthquake loading, *Applied Mechanics Reviews*, **43**(11), 261–282.

470 Rammerstorfer, F.G., Scharf, K., Fischer, F.D., and Seeber, R. (1988). Collapse of earthquake excited tanks, *Res Mechanica*, **25**, 129–143.

471 Rashed, A.A., Rogowsky, D.M., and Elwi, A.E. (1997). Rational design of prestressed and reinforced concrete tanks, Structural Engineering Report No. 220, Department of Civil & Environmental Engineering, University of Alberta, Canada.

472 Rawat, A., Matsagar, V.A., and Nagpal, A.K. (2019). Numerical study of base-isolated cylindrical liquid storage tanks using coupled acoustic- structural approach, *Soil Dynamics and Earthquake Engineering*, **119**, 196–219.

473 Reimbert, M. and Reimbert, A. (1987). *Silos: Theory and Practice*, 2nd Edition, Secaucus, NJ: Lavoisier Publishing Inc.

474 Resinger, F. and Greiner, R. (1982). Buckling of wind loaded cylindrical shells: application to unstiffened and ring-stiffened tanks, in: *Buckling of shells, Proceedings of a State-of-the-Art Colloquium, Universität Stuttgart, Germany, May 6–7*, Ed. E. Ramm, Berlin: Springer Verlag, 305–331.

475 Reynolds, C.E. and Steedman, J.C. (1988). *Reinforced Concrete Designer's Handbook*, 10th Edition, Boca Raton, Florida: CRC Press.

476 Reynolds, C.E., Steedman, J.C., and Threlfall, A.J. (2008). *Reynolds's Reinforced Concrete Designer's Handbook*, 11th Edition, Boca Raton, Florida: CRC Press.

477 Richards, R. and Elms, D.G. (1979). Seismic behavior of gravity-retaining walls, *ASCE Journal of the Geotechnical Engineering Division*, **105**(4), 449–464.

478 Rinne, J.E. (1967). Oil storage tanks, in: *The Prince William Sound, Alaska, Earthquake of 1964 and Aftershocks*, Vol. II, Part A, Ed. J.W. Fergus, Washington, DC: U.S. Department of Commerce, Environmental Science Services Administration, Coast and Geodetic Survey, 245–252.

479 Rosenblueth, E. and Herrera, E. (1964). On a kind of hysteretic damping, *ASCE Journal of the Engineering Mechanics Division*, **90**(EM 4), 37–48.

480 Rotter, J.M. (1986). The analysis of steel bins subject to eccentric discharge, in: *Proceedings of the 2nd International Conference on Bulk Materials Storage, Handling and Transportation*, 264–271.

481 Rotter, J.M. (1990). Local collapse of axially compressed pressurized thin steel cylinders, *ASCE Journal of Structural Engineering*, **116**(7), 1955–1970.

482 Rotter, J.M. (1998a). Cylindrical shells: symmetrical solids loadings, in: *Silos: Fundamentals of Theory, Behaviour and Design*, Eds C.J. Brown and J. Nielsen, Boca Raton, Florida: CRC Press, 346–366.

483 Rotter, J.M. (1998b). Cylindrical shells: unsymmetrical solids loadings and supports, in *Silos: Fundamentals of Theory, Behaviour and Design*, Eds C.J. Brown and J. Nielsen, Boca Raton, Florida: CRC Press, 367–377.

484 Rotter, J.M. (2001). Pressures, stresses and buckling in metal silos containing eccentrically discharging solids, in: *Festschrift Richard Greiner*, Celebration Volume for the 60th birthday of Prof. Richard Greiner, Institute for Steel, Timber and Shell Structures, Technical University Graz, Austria, 85–103.

485 Rotter, J.M. (2002). Shell buckling and collapse analysis for structural design, in: *New Approaches to Structural Mechanics, Shells and Biological Structures*, Eds H.R. Drew and S. Pellegrino, Berlin: Springer, 355–378.

486 Rotter, J.M. (2004). Buckling of cylindrical shells under axial compression, in: *Buckling of Thin Metal Shells*, Eds J.G Teng and J.M. Rotter, London: Spon Press, 71–86.

487 Rotter, J.M. (2005). The practical design of shell structures exploiting different methods of analysis, in *Shell Structures: Theory and Applications*, Eds W. Pietraszkiewicz and C. Szymczak, London: Spon Press, 71–86.

488 Rotter, J.M. (2006). Elephant's foot buckling in pressurised cylindrical shells, *Stahlbau*, **75**(9), 742–747.

489 Rotter, J.M. (2007). Recent advances in the philosophy of the practical design of shell structures, implemented in Eurocodes provisions, paper presented at SEMC 2007, 3rd International Conference on Structural Engineering Mechanics and Computation, Cape Town, South Africa.

490 Rotter, J.M. (2008). Rules for the plastic limit state and plastic reference load assessment, in *Buckling of Steel Shells - European Design Recommendations*, 5th Edition, ECCS Technical Committee 8 - Structural Stability, TWG 8.4 - Shells, N. 125, 83–94.

491 Rotter, J.M. (2011). The new framework for shell buckling design and the European Shell Buckling Recommendations, Fifth Edition, *American Society of Mechanical Engineers (ASME) Journal of Pressure Vessel Technology*, **133**(1), 011203-1/9.

492 Rotter, J.M., Mackenzie, G., and Lee, M. (2016). Spherical dome buckling with edge ring support, *Structures*, **8**, 264–274.

493 Rotter, J.M. and Schmidt, H. (2008). *Buckling of Steel Shells – European Design Recommendations*, 5th Edition, ECCS Technical Committee 8 - Structural Stability, TWG 8.4 - Shells, N. 125.

494 Rungrojsaratis, V. and Ueda, T. (1987). A study of ultimate strength of reinforced concrete corner connection joints subjected to negative moment, *Proceedings of the Annual Conference of Japan Concrete Institute (JCI)*, **9**(2), 687–692.

495 Sadowski, A.J. and Rotter, J.M. (2009). Eccentric discharge buckling of a very slender silo, in: *Proceedings of the International Association for Shell and Spatial Structures (IASS) Symposium*, "Evolution and Trends in Design, Analysis and Construction of Shell and Spatial Structures", Eds A. Domingo and C. Lazaro, 2557–2568.

496 Saha, S.K., Matsagar, V.A., and Jain, A.K. (2013). Comparison of base- isolated liquid storage tank models under bi-directional earthquakes, *Natural Science*, **5**(8A1), 27–37.

497 Sakai, F., Isoe, A., Hirakawa, H., and Mentani, Y. (1988). Experimental study on uplifting behaviour of flat-based liquid storage tanks without anchors, in: *Proceedings of the 9th World Conference on Earthquake Engineering*, Paper Number 10–4–8, Vol. VI, 649–654.

498 Sakai, F., Ogawa, H., and Isoe, A. (1984). Horizontal, vertical and rocking fluid-elastic response and design of cylindrical liquid storage tanks, in: *Proceedings of the 8th World Conference on Earthquake Engineering*, Vol. V, 263–270.

499 Salvadori, M. and Heller, R. (1964). *Le strutture in architettura*, Milan: ETAS KOMPASS (in Italian).

500 Sameer, S.U. and Jain, S.K. (1992). Approximate methods for determination of time period of water tank stagings, *The Indian Concrete Journal*, **66**(12), 691–698.

501 Sameer, S.U. and Jain, S.K. (1994). Lateral-load analysis of frame stagings for elevated water tanks, *ASCE Journal of Structural Engineering*, **120**(5), 1375–1394.

502 Samuelson, L.A. and Eggwertz, S. (2005). *Shell Stability Handbook*, Boca Raton, Florida: CRC Press.

503 Samuelson, L.A., Vandepitte, D., and Paridaens, R. (1987). The background to the ECCS Recommendations for buckling of stringer-stiffened cylinders, in: *Proceedings of the International Colloquium on Buckling of Plate and Shell Structures*, 513–522.

504 Sanders, J.L. (1959). An improved first approximation theory for thin shells, Report, NACA TR R-24.

505 Sarkar, P.K. (1996). Approximate determination of the fundamental frequency of a cantilevered beam with point masses and restraining springs, *Journal of Sound and Vibration*, **195**(2), 229–240.

506 Scharf, K. (1989). Beiträge zur Erfassung des Verhaltens von erdbebenerregten, oberirdischen Tankbauwerken, PhD thesis, and in Contribution to the behaviour of earthquake excited above-ground liquid storage tanks, Institut für Leichtbau und Struktur-Biomechanik, Tecnische Universität Wien, VDI Verlag Edition, 1990, Düsseldorf: Reihe 4: Bauingenieurwesen N. 97.

507 Scharf, K., Rammerstorfer, F.G., and Fischer, F.D. (1989). The scientific background of the Austrian recommendations for earthquake-resistant liquid storage tank design, in *Structural Dynamics and Soil-Structure Interaction*, Eds A.S. Cakmak and I. Herrera, Southampton: Computational Mechanics Publications, 281–313.

508 Schiff, A.J. (1991). *Guide to Post-Earthquake Investigation of Lifelines*, Technical Council on Lifeline Earthquake Engineering, Monograph No. 3.

509 Schmidt, H. and Greiner, R. (2008). Cylindrical shells with ring stiffeners under external pressure, in: *Buckling of Steel Shells: European Design Recommendations*, 5th Edition, ECCS Technical Committee 8 - Structural Stability, TWG 8.4 - Shells, N. 125, 331–351.

510 Schmidt, H. and Rotter, J.M. (2008). Rules for the buckling limit state assessment using global numerical analysis, in: *Buckling of Steel Shells- European Design Recommendations*, 5th Edition, ECCS Technical Committee 8 - Structural Stability, TWG 8.4 - Shells, N. 125, 95–146.

511 Schmidt, H. and Rotter, J.M. (2008). Cylindrical shells of constant wall thickness under general loading, in: *Buckling of Steel Shells - European Design Recommendations*, 5th Edition, ECCS Technical Committee 8 - Structural Stability, TWG 8.4 - Shells, N. 125, 167–216.

512 Schmidt, H. and Samuelson, L. (2008). Cylindrical shells with longitudinal stiffeners under meridional compression, *Buckling of Steel Shells - European Design Recommendations*, 5th Edition, ECCS Technical Committee 8 - Structural Stability, TWG 8.4 - Shells, N. 125, 353–363.

513 Schneider, W., Bohm, S., and Thiele, R. (1999). Failure modes of slender wind-loaded cylindrical shells, in: *Light-Weight Steel and Aluminium Structures*, Eds P. Mäkeläinen and P. Hassinen, Oxford: Elsevier, 407–414.

514 Schneider, W. and Thiele, R. (1995). The stress and strain state in the base area of wind-loaded steel chimneys, in: *Proceedings of the 5th International Offshore and Polar Engineering Conference*, Vol. IV, 104–108.

515 Schneider, W., Timmel, I., and Höhn, K. (2005). The conception of quasicollapse-affine imperfections: a new approach to unfavourable imperfections of thin-walled shell structures, *Thin-Walled Structures*, **43**(8), 1202–1224.

516 Schneider, W. and Zahlten, W. (2004). Load-bearing behaviour and structural analysis of slender ring-stiffened cylindrical shells under quasi-static wind load, *Journal of Constructional Steel Research*, **60**(1), 125–146.

517 Schwerin, E. (1922). Zur Stabilität der dünnwandingen Hohlkugel unter gleichmässigem Aussendruck, *Zeitschrift für Angewandte Matematik und Mechanik (ZAMM)*. **2**, 81–91.

518 Schwind, R.G., Scotti, R.S., and Skogh, J. (1964). The effect of baffles on tank sloshing. Part 1, NASA CR-67426, Lockheed Missiles & Space Company, LMSC-A642961.

519 Schwind, R.G., Scotti, R.S., and Skogh, J. (1967). Analysis of flexible baffles for damping tank sloshing, *Journal of Spacecraft and Rockets*, **4**(1), 47–53.

520 Scott, R.F. (1974). Earthquake-induced earth pressures on retaining walls, in: *Proceedings of the 5th World Conference on Earthquake Engineering*, Session 4, Vol. II, 1611–1620.

521 Seeber, R., Fischer, F.D., and Rammerstorfer, F.G. (1990). Analysis of a three-dimensional tank-liquid-soil interaction problem, *American Society of Mechanical Engineers (ASME) Journal of Pressure Vessel Technology*, **112**(1), 28–33.

522 Seed, H.B. and Whitman, R.V. (1970). Design of earth retaining structures for dynamic loads, in: *Proceedings of ASCE Specialty Conference on Lateral Stresses in the Ground and Design of Earth-Retaining Structures*, 103–147.

523 Seneviratna, G.D.P.K. and Krawinkler, H. (1979). Evaluation of inelastic MDOF effects for seismic design, Report No. 120, The John A. Blume Earthquake Engineering Center, Stanford University, Stanford, California.

524 Shaaban, S.H. and Nash, W.A. (1976). Finite element analysis of a seismically excited cylindrical storage tank, ground supported, and partially filled with liquid, Department of Civil Engineering, University of Massachusetts, Amherst.

525 Shaaban, S.H. and Nash, W.A. (1984). Effect of baffles on seismically excited tanks, in: *Proceedings of the 5th Engineering Mechanics Division Specialty Conference*, 1062–1064.

526 Shao, J., Li, S., Li, Z., and Liu, M. (2015). A comparative study of different baffles on mitigating liquid sloshing in a rectangular tank due to a horizontal excitation, *Engineering Computations*, **32**(4), 1172–1190.

527 Shao, J.R., Li, H.Q., Liu, G.R., and Liu, M.B. (2012). An improved SPH method for modeling liquid sloshing dynamics, *Computers & Structures*, **100–101**(5), 18–26.

528 Sharma, R., Semercigil, S.E., and Turan, Ö.F. (1992). Floating and immersed plates to control sloshing in a cylindrical container at the fundamental mode, Letters to the Editor, *Journal of Sound and Vibration*, **155**(2), 365–370.

529 Shekari, M.R. (2014). Seismic response of liquid-filled tank with baffles, *Journal of Marine Science and Application*, **13**, 299–304.

530 Shekari, M.R., Hekmatzadeh, A.A., and Amiri, S.M. (2019). On the nonlinear dynamic analysis of base-isolated three-dimensional rectangular thin-walled steel tanks equipped with vertical baffle, *Thin-Walled Structures*, **138**, 79–94.

531 Shenton, III, H.W. and Hampton, F.P. (1999). Seismic response of isolated elevated water tanks, *ASCE Journal of Structural Engineering*, **125**(9), 965–976.

532 Shepherd, R. (1972). The two mass representation of a water tower structure, *Journal of Sound and Vibration*, **23**(3), 391–392, IN1, 393–396.

533 Shepherd, R. (1974). The seismic response of elevated water tanks supported on cross braced towers, in: *Proceedings of the 5th World Conference on Earthquake Engineering*, Session 2B, Vol. I, 640–649.

534 Shibata, A. and Sozen, M.A. (1976). Substitute-structure method for seismic design in R/C, *ASCE Journal of the Structural Division*, **102**(1), 1–18.

535 Shibata, A. and Sozen, M.A. (1977). Substitute-structure method to determine design forces in earthquake-resistant reinforced concrete frames, in: *Proceedings of the 6th World Conference on Earthquake Engineering*, Topic 5, Vol. II, 1905–1910.

536 Shih, C-F. (1981). Failure of liquid storage tanks due to earthquake excitation, Technical Report EERL 81–04, California Institute of Technology, Pasadena, California.

537 Shih, C-F. and Babcock, C.D. (1987). Buckling of oil storage tanks in SPPL tank farm during the 1979 Imperial Valley earthquake, *American Society of Mechanical Engineers (ASME) Journal of Pressure Vessel Technology*, **109**(2), 249–255.

538 Shivakumar, P. and Veletsos, A.S. (1995). Dynamic response of rigid tanks with inhomogeneous liquids, *Earthquake Engineering and Structural Dynamics*, **24**, 991–1015.

539 Shrimali, M.K. and Jangid, R.S. (2002a). A comparative study of performance of various isolation systems for liquid storage tanks, *International Journal of Structural Stability and Dynamics*, **2**(4), 573–591.

540 Shrimali, M.K. and Jangid, R.S. (2002b). Earthquake response of liquid storage tanks with sliding systems, *Journal of Seismology and Earthquake Engineering*, **4**(2, 3), 51–61.

541 Shrimali, M.K. and Jangid, R.S. (2002c). Non-linear seismic response of base-isolated liquid storage tanks to bi-directional excitation, *Nuclear Engineering and Design*, **217**(1–2), 1–20.

542 Shrimali, M.K. and Jangid, R.S. (2002d). Seismic response of liquid storage tanks isolated by sliding bearings, *Engineering Structures*, **24**(7), 909–921.

543 Shrimali, M.K. and Jangid, R.S. (2003a). Dynamic analysis of liquid storage tanks with sliding systems, *Advances in Structural Engineering*, **6**(2), 145–158.

544 Shrimali, M.K. and Jangid, R.S. (2003b). Earthquake response of isolated elevated liquid storage steel tanks, *Journal of Constructional Steel Research*, **59**(10), 1267–1288.

545 Shrimali, M.K. and Jangid, R.S. (2003c). Seismic response of base-isolated liquid storage tanks, *Journal of Vibration and Control*, **9**(10), 1201–1218.

546 Shrimali, M.K. and Jangid, R.S. (2004). Seismic analysis of base-isolated liquid storage tanks, *Journal of Sound and Vibration*, **275**(1–2), 59–75.

547 Siekman, J. and Chang, S-C. (1971). On the change of natural frequencies of a sloshing liquid by movable devices, *Acta Mechanica*, **11**(1–2), 73–86.

548 Siller, T.J. and Dolly, M.O. (1992). Design of tied-back walls for seismic loading, *Journal of Geotechnical Engineering*, **118**(11), 1804–1820.

549 Silveira, M.A., Stephens, D.G., and Leonard, H.W. (1961). An experimental investigation of the damping of liquid oscillations in cylindrical tanks with various baffles, Report, NASA TN D-715.

550 Silverman, S. and Abramson, H.N. (1966). Damping of liquid motions and lateral sloshing, in: *The Dynamic Behaviour of Liquids in Moving Containers*, ed. H.N., Abramson, NASA SP-106, 105–143, Chapter 4.

551 Singer, J. (1979). Recent studies on the correlation between vibration and buckling of stiffened shells, *Zeitschrift für Flugwissenschaften und Weltraumforschung*, **3**, 333–343.

552 Singer, J. (1982). Buckling experiments on shells: a review of recent developments, *Solid Mechanics Archives*, **7**, 213–313.

553 Singer, J. (1998). Experimental studies in shell buckling, in: *Stability Analysis of Plates and Shells: A Collection of Papers in Honour of Dr. Manuel Stein*, NASA CP-1998-206280, 19–29.

554 Singer, J. (2004). Stiffened cylindrical shells, in: *Buckling of Thin Metal Shells*, Eds. J.G Teng and J.M. Rotter, London: Spon Press, 286–343.

555 Singer, J., Arbocz, J., and Weller, T. (1998). *Buckling Experiments: Experimental Methods in Buckling of Thin-Walled Structures*. Vol. 1: *Basic, Concepts, Columns, Beams, and Plates*, Chichester: John Wiley & Sons.

556 Singer, J., Arbocz, J., and Weller, T. (2002). *Buckling Experiments: Experimental Methods in Buckling of Thin-Walled Structures*. Vol. 2: *Shells, Built-up Structures, Composites, and Additional Topics*, Chichester: John Wiley & Sons.

557 Soedel, W. (1993). *Vibrations of Shells and Plates*, New York: Marcel Dekker.

558 Song, C-Y. (2002). Buckling of unstiffened cylindrical shell under non-uniform axial compressive stress, *Journal of Zhejiang University SCIENCE*, **3**(5), 520–531.

559 Song, C-Y. and Teng, J.G. (2003). Buckling of circular steel silos subject to code-specified eccentric discharge pressures, *Engineering Structures*, **25**(11), 1397–1417.

560 Southwell, R.V. (1914). On the general theory of elastic stability, *Philosophical Transactions of the Royal Society of London, Series A, Containing Papers of a Mathematical or Physical Character*, **213**, 187–244.

561 Spiegel, M.R. (1968). *Manuale di matematica*, Collana SCHAUM, Ed.

562 Starnes, J.H. (1974). The effects of cutouts on the buckling of thin shells, in: *Thin Shell Structures: Theory, Experiment and Design*, Eds Y.C. Fung and E.E. Sechler, Englewood Cliffs, NJ: Prentice Hall, 289–304.

563 Stein, M., Sanders, J.L., and Crate, H. (1949). Critical stress of ring- stiffened cylinders in torsion, National Advisory Committee for Aeronautics, NACA Technical Note No. 1981.

564 Steinbrugge, K.V. and Flores, R.A. (1963). The Chilean earthquakes of May, 1960: a structural engineering viewpoint, *Bulletin of the Seismological Society of America*, **53**(2), 225–307.

565 Stephen, N.G. (1995). On Southwell's and a novel Dunkerley's method, *Journal of Sound and Vibration*, **181**(1), 179–184.

566 Stephens, D.G. (1966). Flexible baffles for slosh damping, *Journal of Spacecraft and Rockets*, **3**(5), 765–766.

567 Stephens, D.G., Leonard, H.W., and Perry, T.W. Jr. (1962). Investigation of the damping of liquids in right-circular cylindrical tanks, including the effects of a time-variant liquid depth, Report, NASA TN D-1367.

568 Stephens, D.G. and Scholl, H.F. (1967). Effectiveness of flexible and rigid ring baffles for damping liquid oscillations in large-scale cylindrical tanks, Report, NASA TN D-3878.

569 Stewart, J.P., Fenves, G.L., and Seed, R.B. (1997). Seismic soil–structure interaction in buildings. I: Analytical methods, *Journal of Geotechnical and Geoenvironmental Engineering*, **125**(1), 26–37.

570 Stewart, J.P., Kim, S., Bielak, J., Dobry, R., and Power, M.S. (2003). Revisions to soil–structure interaction procedures in NEHRP design provisions, *Earthquake Spectra*, **19**(3), 677–696.

571 Stratta, J.L. (1980). Reconnaissance Report Greenville (Diablo/Livermore). California earthquake sequence, January 1980, *Special Issue of Earthquake Engineering Research Institute Newsletter*, **14**(2), 20–89.

572 Stricklin, G.P. and Baird, J.A. (1966). A survey of ring baffle damping in cylindrical tanks, Report, NASA CR-78985, NASA TN R-185.

573 Suzuki, K. (2002). Report on damage to industrial facilities in the 1999 Kocaeli earthquake, Turkey, *Journal of Earthquake Engineering*, **6**(2), 275–296.

574 Sweedan, A.M.I. (2009). Equivalent mechanical model for seismic forces in combined tanks subjected to vertical earthquake excitation, *Thin-Walled Structures*, **47**, 942–952.

575 Sweedan, A.M.I. and El Damatty, A.A. (2002). Equivalent mechanical model simulating dynamic characteristics of conical steel tanks, paper presented at 4th Structural Specialty Conference of the Canadian Society for Civil Engineering, Montréal, Québec, Canada.

576 Sweedan, A.M.I. and El Damatty, A.A. (2002). Experimental and analytical evaluation of the dynamic characteristics of conical shells, *Thin-Walled Structures*, **40**, 465–486.

577 Sweedan, A.M.I. and El Damatty, A.A. (2003). Experimental identification of the vibration modes of liquid-filled conical tanks and validation of a numerical model, *Earthquake Engineering and Structural Dynamics*, **32**(9), 1407–1430.

578 Sweedan, A.M.I. and El Damatty, A.A. (2005). Equivalent models of pure conical tanks under vertical ground excitation, *ASCE Journal of Structural Engineering*, **131**(5), 725–733.

579 Sweedan, A.M.I. and El Damatty, A.A. (2009). Simplified procedure for design of liquid-storage combined conical tanks, *Thin-Walled Structures*, **47**, 750–759.

580 Tadros, M.K. (1987). Recommended practice for precast prestressed concrete circular storage tanks, *PCI Journal*, July–August, 80–125.

581 Tajirian, F.F. (1993). Seismic isolation of critical components and tanks, iATC-17-1 Seminar on Seismic Isolation, Passive Energy Dissipation and Active Control, **1**, 233–244.

582 Tajirian, F.F. (1998). Base isolation design for civil components and civil structures, paper presented at the Structural Engineers World Congress (SEWC). San Francisco, 19–23 July.

583 Takeda, T., Sozen, M.A., and Nielsen, N.N. (1970). Reinforced concrete response to simulated earthquakes, *ASCE Journal of the Structural Division*, **96**(ST 12). 2557–2573.

584 Tan, C.P. (1969). A study of the design and construction practices of prestressed concrete and reinforced concrete containment vessels, Technical Final Report F-C2121, TID-25176, The Franklin Institute Research Laboratories, Philadelphia, Pennsylvania.

585 Tang, Y. (1992). Hydrodynamic pressure in a tank containing two liquids, Technical Report ANL/CP-75478, American Society of Mechanical Engineers (ASME) Pressure Vessels and Piping Conference, 1–16.

586 Tang, Y. (1993a). Dynamic response of tank containing two liquids, *Journal of Engineering Mechanics*, **119**(3), 531–548.

587 Tang, Y. (1993b). Sloshing displacements in a tank containing two liquids, in: *Proceedings of the American Society of Mechanical Engineers (ASME) Pressure Vessels and Piping Conference*, 179–184.

588 Tang, Y. (1994a). Rocking response of tanks containing two liquids, *Nuclear Engineering and Design*, **152**(1–3), 103–115.

589 Tang, Y. (1994b). SSI effects for a tank containing two liquids, in: Proceedings of the American Society of Mechanical Engineers (ASME) Pressure Vessels and Piping Conference, 67–71.

590 Tang, Y. and Chang, Y.W. (1993). The exact solutions to the dynamic response of tanks containing two liquids, Report ANL/RE–93/2 (Contract W-31109-ENG-38). Argonne National Lab., IL, United States.

591 Tang, Y. and Veletsos, A.S. (1992). Soil-structure interaction effects for laterally excited liquid-tank system, Technical Report ANL/CP-75432, American Society of Mechanical Engineers (ASME) Pressure Vessels and Piping Conference, 1–17.

592 Tang, Z., Zang, Y., and Wan, D. (2015). Numerical study of sloshing in baffled tanks by MPS, paper presented at the 25th International Ocean and Polar Engineering Conference, Kona, Big Island, Hawaii, USA.

593 Tani, S. and Hori, N. (1984). Earthquake response analysis of tanks including hydrodynamic and foundation interaction effects, in: *Proceedings of the 8th World Conference on Earthquake Engineering*, Vol. IV, 817–824.

594 Taniguchi, T. (2004). Rocking behaviour of unanchored flat-bottom cylindrical shell tanks under action of horizontal base excitation, *Engineering Structures*, **26**(4), 415–426.

595 Taniguchi, T. and Ando, Y. (2009a). Fluid pressures on unanchored rigid rectangular tanks under action of uplifting acceleration, *American Society of Mechanical Engineers (ASME) Journal of Pressure Vessel Technology*, **132**(1), 011801.

596 Taniguchi, T. and Ando, Y. (2009b). Fluid pressures on unanchored rigid flat-bottom cylindrical tanks under action of uplifting acceleration, *American Society of Mechanical Engineers (ASME) Journal of Pressure Vessel Technology*, **132**(1), 011802.

597 Taniguchi, T. and Segawa, Y. (2013). Fundamental mechanics of walking of unanchored flat-bottom cylindrical shell model tanks subjected to horizontal harmonic base excitation, *American Society of Mechanical Engineers (ASME) Journal of Pressure Vessel Technology*, **135**(2), 021201.

598 Tayebi, A. and Jin, Y-C. (2015). Development of moving particle explicit (MPE) method for incompressible flows, *Computers & Fluids*, **117**, 1–10.

599 Tedesco, J.W. and Kostem, C.N. (1982). Vibrational characteristics and seismic analysis of cylindrical liquid storage tanks, PhD dissertation, Fritz Laboratory Reports, Paper 505, Report No. 433.5, Civil and Environmental Engineering, Department of Civil Engineering, Lehigh University, Bethlehem, Pennsylvania.

600 Tedesco, J.W., Landis, D.W., and Kostem, C.N. (1989). Seismic analysis of cylindrical liquid storage tanks, *Computers & Structures*, **32**(5), 1165–1174.

601 Teng, J.G. and Chan, F. (1999). New buckling approximations for T- section ringbeams simply-supported at inner edge, *Engineering Structures*, **21**(10), 889–897.

602 Teng, J.G. and Rotter, J.M. (1988). Buckling of restrained monosymmetric rings, *Journal of Engineering Mechanics*, **114**(10), 1651–1671.

603 Teng, J.G. and Rotter, J.M. (Eds) (2004). *Buckling of Thin Metal Shells*, London: Spon Press.

604 Teng, J.G. and Rotter, J.M. (2004). Buckling of thin shells: an overview. In: *Buckling of Thin Metal Shells*, Eds. J.G Teng and J.M. Rotter, London: Spon Press, 1–41.

605 Teng, J.G. and Zhao, Y. (2004). Rings at shell junctions. In: *Buckling of Thin Metal Shells*, Eds. J.G Teng and J.M. Rotter, London: Spon Press, 409–454.

606 Teng, J.G., Zhao, Y., and Lam, L. (2001). Techniques for buckling experiments on steel silo transition junctions, *Thin-Walled Structures*, **39**(8), 685–707.

607 Tennyson, R.C. (1968). The effect of unreinforced circular cutouts on the buckling of circular cylindrical shells under axial compression, *Transactions of the ASME*, **90**, 541–546.

608 Tian, J., Wang, C.M., and Swaddiwudhipong, S. (1999). Elastic buckling analysis of ring-stiffened cylindrical shells under general pressure loading via the Ritz method, *Thin-Walled Structures*, **35**(1), 1–24.

609 Timoshenko, S. (1910). Einige Stabilitätsprobleme der Elastizitätstheorie, *Zeitschrift für Mathematik und Physik*, **58**(4), 337–385.

610 Timoshenko, S. (1930). *Strength of Materials. Part I: Elementary Theory and Problems*, Amsterdam: D. van Nostrand Company.

611 Timoshenko, S. (1930). *Strength of Materials. Part II: Advanced Theory and Problems*, Amsterdam: D. van Nostrand Company.

612 Timoshenko, S. and Gere, J.M. (1936). *Theory of Elastic Stability*, 2nd Edition, New York: McGraw Hill.

613 Timoshenko, S. and Woinowsky-Krieger, S. (1959). *Theory of Plates and Shells*, 2nd Edition. New York: McGraw Hill.

614 Toda, S. (1983). Buckling of cylinders with cutouts under axial compression, *Journal of Experimental Mechanics*, **3**, 414–417.

615 Todhunter, I. and Pearson, K. (1960). *A history of the Theory of Elasticity*. Vol. I and II, London: Dover Publications.

616 Trahair, N.S. and Papangelis, J.P. (1987). Flexural-torsional buckling of monosymmetric arches, *ASCE Journal of Structural Engineering*, **113**(10), 2271–2288.

617 Tse, F.S., Morse, I.E., and Hinkle, R.T. (1978). *Mechanical Vibrations. Theory and Applications*, 2nd Edition, Boston: Allyn and Bacon.

618 Tsipianitis, A. and Tsompanakis, Y. (2019). Impact of damping modeling on the seismic response of base-isolated liquid storage tanks, *Soil Dynamics and Earthquake Engineering*, **121**, 281–292.

619 Uckan, E., Akbas, B., Shen, J., Wen, R., Turandar, K., and Erdik, M. (2014). Seismic performance of elevated steel silos during Van earthquake, October 23, 2011, *Natural Hazards*, DOI: 10.1007/s11069-014-1319-9.

620 Uematsu, Y., Koo, C., and Kondo, K. (2008). Wind loads on open-topped oil storage tanks, paper presented at the 6th International Colloquium on Bluff Bodies Aerodynamics and Applications, Milan, Italy.

621 Uematsu, Y. and Uchiyama, K. (1985). Deflection and buckling behavior of thin, circular cylindrical shells under wind loads, *Journal of Wind Engineering and Industrial Aerodynamics*, **18**(3), 245–261.

622 Varghese, P.C. (2004). *Limit State Design of Reinforced Concrete*, Delhi: Prentice Hall of India.

623 Vathi, M., Pappa, P., and Karamanos, S.A. (2013). Seismic response of unanchored liquid storage tanks, in: *Proceedings of the ASME (American Society of Mechanical Engineers) Pressure Vessels & Piping Division Conference*, PVP2013, PVP2014-97700, **8**: Seismic Engineering

624 Veletsos, A.S. (1973). Seismic effects in flexible liquid storage tanks, in: *Proceedings of the 5th World Conference on Earthquake Engineering*, Session 2, Vol. I, 630–639.

625 Veletsos, A.S. (1977). Dynamics of structure-foundation systems, in: *Structural and Geotechnical Mechanics: A Volume Honoring N.M. Newmark*, Ed. W.J. Hall, Englewood Cliffs, NJ: Prentice Hall, 333–361.

626 Veletsos, A.S. (1984). Seismic response and design of liquid storage tanks, in: *Guidelines for the Seismic Design of Oil and Gas Pipeline Systems*, New York: ASCE Technical Council on Lifeline Earthquake Engineering, 255–370 and Appendix C, 443–461.

627 Veletsos, A.S. and Kumar, A. (1984). Dynamic response of vertically excited liquid storage tanks, in: *Proceedings of the 8th World Conference on Earthquake Engineering*, Vol. VII, 453–460.

628 Veletsos, A.S. and Meek, J.W. (1973). Dynamic behaviour of building–foundation systems, Structural Research at Rice, Report No. 20, Department of Civil Engineering, Rice University, Houston, Texas.

629 Veletsos, A.S. and Meek, J.W. (1974). Dynamic behaviour of building–foundation systems, *Earthquake Engineering and Structural Dynamics*, **3**, 121–138.

630 Veletsos, A.S. and Nair, V.V.D. (1974). Seismic interaction of structures on hysteretic foundations, Structural Research at Rice, Report No. 21, Department of Civil Engineering, Rice University, Houston, Texas.

631 Veletsos, A.S. and Nair, V.V.D. (1975). Seismic interaction of structures on hysteretic foundations, *ASCE Journal of the Structural Division*, **101**(ST1), 109–129.

632 Veletsos, A.S. and Shivakumar, P. (1993). Sloshing response of layered liquids in rigid tanks, *Earthquake Engineering and Structural Dynamics*, **22**, 801–821.

633 Veletsos, A.S. and Shivakumar, P. (1994). Reply by authors to Y.W. Chang, *Earthquake Engineering and Structural Dynamics*, **23**, 801–805.

634 Veletsos, A.S. and Shivakumar, P. (1995). Hydrodynamic effects in rigid tanks containing layered liquids, *Earthquake Engineering and Structural Dynamics*, **24**, 835–860.

635 Veletsos, A.S. and Shivakumar, P. (1997). Tanks containing liquids or solids, in *Computer Analysis and Design of Earthquake Resistant Structures: A Handbook*, eds D.E. Beskos and S.A. Anagnostopoulos, Southampton: Computational Mechanics Publications, Vol. 3, 725–774.

636 Veletsos, A.S. and Tang, Y. (1986a). Interaction effects in vertically excited steel tanks, in: *Proceedings of the Third Conference organized by the Engineering Mechanics Division of the American Society of Civil Engineers*, Eds G.C. Hart and R.B. Nelson, 636–643.

637 Veletsos, A.S. and Tang, Y. (1986b). Dynamics of vertically excited liquid storage tanks, *ASCE Journal of Structural Engineering*, **112**(6), 1228–1246.

638 Veletsos, A.S. and Tang, Y. (1987). Rocking response of liquid storage tanks, *Journal of Engineering Mechanics*, **113**(11), 1774–1792.

639 Veletsos, A.S. and Tang, Y. (1988). Soil-structure interaction effects for vertically excited tanks, in: *Proceedings of the 9th World Conference on Earthquake Engineering*, Paper Number 10–4–5, Vol. VI, 631–636.

640 Veletsos, A.S. and Tang, Y. (1990). Soil-structure interaction effects for laterally excited liquid storage tanks, *Earthquake Engineering and Structural Dynamics*, **19**(4), 473–496.

641 Veletsos, A.S., Tang, Y., and Tang, H.T. (1992). Dynamic response of flexibly supported liquid-storage tanks, *ASCE Journal of Structural Engineering*, **118**(1), 264–283.

642 Veletsos, A.S. and Verbic, B. (1973). Vibration of viscoelastic foundations, *Earthquake Engineering and Structural Dynamics*, **2**, 87–102.

643 Veletsos, A.S. and Verbic, B. (1974). Basic response functions for elastic foundations, *ASCE Journal of the Engineering Mechanics Division*, **100**(EM 2), 189–202.

644 Veletsos, A.S. and Wei, Y.T. (1971). Lateral and rocking vibration of footings, *Journal of the Soil Mechanics and Foundations Division*, **97**(SM9), 1227–1248.

645 Veletsos, A.S. and Yang, J.Y. (1976). Dynamics of fixed–base liquid-storage tanks, in: *Proceedings of U.S.-Japan Seminar on Earthquake Engineering Research with Emphasis on Lifeline Systems*, 317–341.

646 Veletsos, A.S. and Yang, J.Y. (1977). Earthquake response of liquid-storage tanks, ASCE Advanced in Civil Engineering through Engineering Mechanics, in: *Proceedings of the 2nd Annual Engineering Mechanics Division Specialty Conference*, 1–24.

647 Veletsos, A.S. and Younan, A.H. (1992). Dynamic soil pressures on rigid vertical walls, Department of Nuclear Energy, Brookhaven National Laboratory, Upton, New York, Technical Report BNL-52357.

648 Veletsos, A.S. and Younan, A.H. (1994). Dynamic modeling and response of soil-wall systems, *Journal of Geotechnical Engineering*, **120**(12), 2155–2179.

649 Veletsos, A.S. and Younan, A.H. (1997). Dynamic response of cantilever retaining walls, *Journal of Geotechnical and Geoenvironmental Engineering*, **123**(2), 161–172.

650 Vesenjak, M., Müllerschön, H., Hummel, A., and Ren, Z. (2004). Simulation of fuel sloshing - Comparative study, LS-DYNA Anwenderforum, Bamberg, G-I-1/8.

651 Virella, J.C., Godoy, L.A., and Suárez, L.E. (2003). Influence of the roof on the natural periods of empty steel tanks, *Engineering Structures*, **25**(7), 877–887.

652 Virella, J.C., Godoy, L.A., and Suárez, L.E. (2006). Fundamental modes of tank-liquid systems under horizontal motions, *Engineering Structures*, **28**, 1450–1461.

653 Vlasov, V.Z. (1951). Basic differential equations in general theory of elastic shells, NACA Technical Memorandum, TM 1241, translation of "Osnovnye differentsialnye uravnenia obshche teorii uprugikh obolochek" from Prikladnaya Matamatika i Mekhanika (1944), **8**, 109–140.

654 Vlasov, V.Z. (1964). *General theory of shells and its application to engineering*, NASA Technical Translation, NASA–TT-F-99, translation of "Obshchaya teoriya obolochek i yeye prilozheniya v tekhnike", Gosudarstvennoye Izdatel'stvo Tekhniko Teoreticheskey Literatury, Moscow-Leningrad (1949).

655 Vronay, D.F. and Smith, B.L. (1970). Free vibration of circular cylindrical shells of finite length, *AIAA Journal*, **8**(3), 601–603.

656 Walker, A.C. and Sridharan, S. (1980). Analysis of the behaviour of axially compressed stringer-stiffened cylindrical shells, *Proceedings Institute of Civil Engineers (ICE)*, **69**(2), 447–472.

657 Walker, A.C. and Sridharan, S. (1981). Discussion: Analysis of the behaviour of axially compressed stringer-stiffened cylindrical shells, *Proceedings Institute Civil Engineers (ICE)*, **71**(2), 563–569.

658 Wan, F.Y.M. and Weinitschke, H.J. (1988). On shells of revolution with the Love-Kirchhoff hypotheses, *Journal of Engineering Mathematics*, **22**(4), 285–334.

659 Wang, L.Y. (2003). Seismic design standards and guidelines of steel and concrete liquid storage tanks, *Technical Council on Lifeline Earthquake Engineering Monograph*, **25**, 327–338.

660 Wang, Y-P., Teng, M-C., and Chung, K-W. (2001). Seismic isolation of rigid cylindrical tanks using friction pendulum bearings, *Earthquake Engineering and Structural Dynamics*, **30**(7), 1083–1099.

661 Warnitchai, P. and Pinkaew, T. (1998). Modelling of liquid sloshing in rectangular tanks with flow-dampening devices, *Engineering Structures*, **20**(7), 593–600.

662 Wei, Y-T. (1971). Steady state response of certain foundation systems, PhD thesis, Rice University, Houston, Texas.

663 Werner, P.W. and Sundquist, K.J. (1949). On hydrodynamic earthquake effects, *Transactions American Geophysical Union*, **30**(5), 636–657.

664 Westergaard, H.M. (1933). Water pressures on dams during earthquakes, *Transactions of the American Society of Civil Engineers*, **98**(2), 418–433.

665 Whittaker, A.S. and Constantinou, M. (2006). Seismic isolation of bridges and mission-critical infrastructure, University of Buffalo, Workshop, 23rd Annual International Bridge Conference, Pittsburgh.

666 Whittaker, D. and Jury, R.D. (2000). Seismic design loads for storage tanks, paper presented at the 12th World Conference on Earthquake Engineering, New Zealand, Paper Number 2376.

667 Whittaker, D. and Saunders, D. (2008). Revised NZSEE Recommendations for seismic design of storage tanks, paper presented at 2008 NZSEE Conference, Paper Number 04.

668 Wilson, E.L. and Khalvati, M. (1983). Finite elements for the dynamic analysis of fluid-solid systems, *International Journal for Numerical Methods in Engineering*, **19**, 1657–1668.

669 Winterstetter, T.A. and Schmidt, H. (2002). Stability of circular cylindrical steel shells under combined loading, *Thin-Walled Structures*, **40**(10), 893–909.

670 Wood, J.H. (1973). Earthquake-induced soil pressures on structures, Technical Report EERL 73–05, California Institute of Technology, Pasadena, California.

671 Wood, J.H. (1975). Earthquake-induced pressures on a rigid wall structure, *Bulletin of the New Zealand National Society for Earthquake Engineering*, **8**(3), 175–186.

672 Wozniak, R.S. and Mitchell, W.W. (1978). Basis of seismic design provisions for welded steel oil storage tanks, Session on Advanced in Storage Tank Design, API Refining 43rd Midyear Meeting, Toronto, Ontario, Canada.

673 Wu, C.I., Mouzakis, T., Nash, W.A., and Colonell, J.M. (1975). Natural frequencies of cylindrical liquid storage containers, a computer program distributed by NISEE/Computer Applications, 540/W81/1975, Grant GI39644, Department of Civil Engineering, University of Massachusetts, Amherst.

674 Wu, G.X., Ma, Q.W., and Taylor, R.E. (1998). Numerical simulation of sloshing waves in a 3D tank based on a finite element method, *Applied Ocean Research*, **20**, 337–355.

675 Wunderlich, W. (2008). Spherical shells under uniform external pressure, in: *Buckling of steel shells - European Design Recommendations*, 5th Edition, ECCS Technical Committee 8 - Structural Stability, TWG 8.4 - Shells, N. 125, 309–318.

676 Wunderlich, W. and Albertin, U. (2000). Analysis and load-carrying behaviour of imperfection sensitive shells, *International Journal for Numerical Methods in Engineering*, **47**(1–3), 255–273.

677 Wunderlich, W. and Albertin, U. (2002). Buckling behaviour of imperfect spherical shells, *International Journal of Non-Linear Mechanics*, **37**(4–5), 589–604.

678 Yamada, S., Oda, T., Yoshida, K., Yamamoto, S., Kawano, K., and Shibuya, T. (1980). Damage analysis of oil storage tanks for off Miyagi prefecture earthquake of June 12, 1978, in: *Proceedings of the 7th World Conference on Earthquake Engineering*, Vol. VIII, 415–422.

679 Yamamoto, S., Kawano, K., Shimizu, N., Umebayashi, S., and Yamagata, M. (1984). Radiation damping of cylindrical liquid storage tank resting on elastic body, in: *Proceedings of the 8th World Conference on Earthquake Engineering*, Vol. V, 231–238.

680 Yang, J.Y. (1976). Dynamic behavior of fluid-tank systems, PhD thesis, Rice University, Houston, Texas.

681 Yang, Y., Tang, Z., and Wan, D. (2014). Numerical simulations of 3D liquid sloshing flows by MPS method, paper presented at the 11th Pacific/Asia Offshore Mechanics Symposium, PACOMS 2014, Shanghai, China.

682 Yang, Y-Q., Tang, Z-Y., and Wan, D-C. (2015). Numerical study on liquid sloshing in horizontal baffled tank by MPS method, *Shuidonglixue Yanjiu yu Jinzhan/Chinese Journal of Hydrodynamics Ser. A*, **30**(2), 146–153.

683 Zaharia, R. and Taucer, F. (2008). Equivalent period and damping for EC8 spectral response of SDOF ring-spring hysteretic models, JRC Scientific and Technical Reports.

684 Zeiny, A. (2003). Factors affecting the nonlinear seismic response of unanchored tanks, paper presented at16th ASCE Engineering Mechanics Conference, University of Washington, Seattle.

685 Zhang, Y-X. and Wan, D-C. (2012). Numerical simulation of liquid slosh- ing in low-filling tank by MPS, *Shuidonglixue Yanjiu yu Jinzhan/Chinese Journal of Hydrodynamics Ser. A*, **27**(1), 100–107.

686 Zhu, E., Mandal, P. and Calladine, C.R. (2002). Buckling of thin cylindrical shells: an attempt to resolve a paradox, *International Journal of Mechanical Sciences*, **44**(8), 1583–1601.

687 Zienkiewicz, O.C. and Taylor, R.L. (1997). *The Finite Element Method*, Vol. 1: *Basic Formulation and Linear Problems*, 4th Edition, New York: McGraw Hill.

688 Zienkiewicz, O.C. and Taylor, R.L. (2000). *The Finite Element Method*, Vol. 2: *Solid Mechanics*, 4th Edition, New York: McGraw Hill.

689 Zoelly, R. (1915). Über ein Knickungsproblem an der Kugelschale, PhD Dissertation, University of Zürich.

Index